Electrochemical Water Splitting
Materials and Applications

Edited by

Inamuddin[1,2,3], Rajender Boddula[4], Rizwana Mobin[5] and Abdullah M. Asiri[1,2]

[1]Chemistry Department, Faculty of Science, King Abdulaziz University, Jeddah 21589, Saudi Arabia

[2]Centre of Excellence for Advanced Materials Research, King Abdulaziz University, Jeddah 21589, Saudi Arabia

[3]Department of Applied Chemistry, Faculty of Engineering and Technology, Aligarh Muslim University, Aligarh-202 002, India

[4]CAS Key Laboratory of Nanosystem and Hierarchical Fabrication, National Center for Nanoscience and Technology, Beijing 100190, PR China

[5]Department of Industrial Chemistry, Govt. College for Women, Cluster University, Srinagar, Jammu and Kashmir-190006, India

Published by **Materials Research Forum LLC**
Millersville, PA 17551, USA

Published as part of the book series
Materials Research Foundations
Volume 59 (2019)
ISSN 2471-8890 (Print)
ISSN 2471-8904 (Online)

Print ISBN 978-1-64490-044-4
eBook ISBN 978-1-64490-045-1

Distributed worldwide by

Materials Research Forum LLC
105 Springdale Lane
Millersville, PA 17551
USA
http://www.mrforum.com

Manufactured in the United States of America
10 9 8 7 6 5 4 3 2 1

Table of Contents

Preface

The increase in the worldwide energy demand, the quick utilization of fossil fuels and their related climate issues are creating serious research enthusiasm for the development of an alternative renewable resource of energy. Water splitting by electrochemical approach is regarded as a promising technology for generating a renewable resource of energy which can resolve the approaching energy and environmental emergencies. Since, it creates clean hydrogen energy with zero carbon dioxide emission. The sustainable and versatile generation of hydrogen energy through water splitting requires the development of efficient and robust electrocatalysts. Electrochemical water splitting has pulled impressive enthusiasm for as long as a couple of decades, and a variety of dynamic electrocatalysts used for hydrogen generation are being investigated and developed by photochemists, electrochemists, engineers and scientific experts in the related areas of energy and environmental science.

This book describes the rapidly expanding field of electrochemical energy generation by making hydrogen from water splitting. It provides in-depth cutting-edge research on fundamental aspects, key factors, mechanisms, theoretical insights and experimental evidence for electrochemical water splitting involving elementary reactions such as hydrogen evolution reaction (HER), and oxygen evolution reaction (OER). Various types of electrocatalysts, including noble metals, earth-abundant metals, metal-organic frameworks, carbon nanomaterials, polymers with their methods of syntheses, catalytic performances as well as characterization methods, are discussed in detail. This book is the collective efforts of the top researchers working in the field of water electrolysis having various backgrounds and expertise. This book is a unique reference for students, professors, scientists and R&D industrial specialists working in the field of energy and environmental sciences.

Inamuddin[1,2,3], Rajender Boddula[4], Rizwana Mobin[5] and Abdullah M. Asiri[1,2]

[1]Chemistry Department, Faculty of Science, King Abdulaziz University, Jeddah 21589, Saudi Arabia

[2]Centre of Excellence for Advanced Materials Research, King Abdulaziz University, Jeddah 21589, Saudi Arabia

[3]Department of Applied Chemistry, Faculty of Engineering and Technology, Aligarh Muslim University, Aligarh-202 002, India

[4]CAS Key Laboratory of Nanosystem and Hierarchical Fabrication, National Center for Nanoscience and Technology, Beijing 100190, PR China

[5]Department of Industrial Chemistry, Govt. College for Women, Cluster University, Srinagar, Jammu and Kashmir-190006, India

Electrochemical Water Splitting: Materials and Applications   Materials Research Forum LLC
Materials Research Foundations **59** (2019) 1-36   doi: https://doi.org/10.21741/9781644900451-1

Chapter 1

Transition Metal-Based Electrocatalysts for Oxygen-Evolution Reaction beyond Ni, Co, Fe

H.L.S. Santos[1], J.A. Dias[1], M.A.S. Andrade Jr[1*], L.H. Mascaro[1]

[1] Federal University of Sao Carlos, Rodovia Washington Luiz, km 235, São Carlos, SP 13565-905, Brazil

*marcos_asaj@hotmail.com

Abstract

Nickel, Cobalt, and iron-based electrocatalyst are the most widely used materials applied to highly efficient oxygen evolution reaction. This chapter focuses on the progress of transition metal based electrocatalysts other than Ni, Co, and Fe for this application, such as alloys, oxides, perovskite oxides, also the transition metal carbides, phosphides, and nitrides which present very competitive properties to these most exploited metal-based materials.

Keywords

Transition Metal, Alloys, Oxides, Perovskites, Nitrides, Carbides, Phosphides

Contents

1.   Introduction

The electrolysis of water has caused great enthusiasm in recent decades because of the important role in chemical energy storage using renewable sources. This reaction has been considered as a potential solution for human energy problems due to its theoretical energy density. However, it presents a limited overall efficiency as a consequence of sluggish kinetics in the oxygen evolution reaction (OER) [1,2].

It is notable that the more important electrocatalysts for OER are: on one hand, the noble metals Ru, Ir and Pd and their oxides with relative scarcity and high cost and on the other hand, the first-row transition metal oxides (NiO_x, CoO_x, and FeO_x) with their low cost, earth abundance and good intrinsic activity [3-5]. However, there are other transition metal-based electrodes for the oxidation of water with good intrinsic activity and cost effectiveness, which are less exploited, due to some shortcomings such as low conductivity and difficulties of synthesis.

In this chapter, a discussion of general aspects of the transition metal-based materials for the oxygen evolution reaction electrocatalysis such as alloys, simple oxide, perovskite oxides, and also the transition metal carbides, phosphides, and nitrides which present very competitive properties to the most exploited metal-based electrocatalyst.

2.   Towards transition metal alloys beyond Ni, Co and Fe applied for OER

The noble metals and their alloys are among the few materials that have chemical resistance and stability required for OER applications, regarding the extreme reaction conditions to proceed with appreciable kinetics [6]. Within this class, the traditional binary and ternary noble alloys have been extensively investigated and used as high-performance catalysts in oxygen electrode reactions mainly in fuel cells and air-metal batteries [2,7,8]. As of late, growing interest in noble-non-noble metal alloys as electrocatalysts for oxidation of water has been motivated by the scarcity and high cost of these metals [6,9,10]. The synthesis of such materials has gained notoriety because it not only reduces the expensive metals used but can increase the intrinsic activity of these electrocatalysts due to synergistic effects between the elements [11]. In general, these alloys are synthesized in the form of bimetallic nanostructures that have three-dimensional geometry, which results in greater molecular accessibility, and better catalytic activity in relation to the bulk materials [10,12].

According to theoretical studies, the synergistic effects on the intrinsic activity of nanostructured heterometallic electrocatalysts are derived from the changes of surface electronic states, more precisely they are related to the stress and the effective number of atomic coordination on the surface [13,14]. Ir-Cu nanostructured bimetallic alloys exemplify well the synergistic effects of these electrocatalysts for OER. Chao Wang and co-authors [15], synthesized Cu-Ir polyhedral nanoparticles (NPs) from Cu as a template by the modified galvanic substitution mechanism. The Cu atoms in the nanostructures were replaced spontaneously by the Ir atoms forming Ir-Cu nanocages. In terms of electrocatalytic activity, among the alloys synthesized by Wang, the $Cu_{1.11}Ir$ ratio was the one that presented the best activity for OER. A low overpotential of 286 mV at 10 mA cm^{-2} using 0.05 mol L^{-1} H_2SO_4 aqueous solution was obtained for this material. The stability test showed that all Ir-Cu ratios were more stable than the Ir nanoparticles at 1 mA cm^{-2} during 5 h to OER.

Although the noble-non-noble metal alloys have a great activity, the pronounced leaching of the less noble component in the strongly acidic media in OER conditions, have limited their applications as electrocatalyst [16,17]. On the other hand, irregular and porous structured-noble alloys with the high surface area are interesting materials to improve the catalytic activity. In this way, the chemical de-alloying strategy has been applied to obtain highly porous nanostructures with high stability and activity for OER. As a practical example, Yecan Pi et al. [16] produced highly porous Ir-Cu three-dimensional nanoparticles (P-IrCu) by de-alloying Cu from pristine solid $IrCu_x$ nanoparticles (S-IrCu) using nitric acid, asillustrated in Fig.1(a) [16]. These three-dimensional structures showed a high OER activity in H_2SO_4 solution. The most active porous $IrCu_{1.4}$ nanoparticles are showing 12.8 mA cm^{-2} at 1.55 V vs. reversible hydrogen electrode (RHE). The P-CuIr nanoparticles exhibited a huge improvement in the catalytic activity in comparison to S-CuIr, as shown in Fig. 1(b) [16]. The higher catalytic currents of these materials are credited to the larger surface area of the porous 3D structures [16].

A brief comparison of OER kinetic and catalytic activity parameters for Ir binary alloys (M-Ir) that are considered the best catalysts for this reaction can be seen in Table 1 [7,15,16,18-22]. Among the outstanding alloys, $IrCu_{1.11}$ shows a better catalytic activity for the OER in acidic medium when compared to noble metal alloys such as PtIr and IrPd, and, even to, Ir with first-row transition metal as IrNi and IrCo. Another point that stands out is the excellent catalytic activity of the IrW alloy supported on carbon that presented only 301 mV of overpotential.

Table 1: *Kinetic and activity parameters for OER on MT-Ir alloys in acid medium.*

Alloys	electrolytes	Overpotential [mV] at 10 mA cm^{-1}	Tafel slope [mV/dec]	Ref.
PtIr	0.5 M H$_2$SO$_4$	340	81	[7]
IrPd	0.5 M H2SO$_4$	300	58	[18]
IrNi	0.1 M HClO$_4$	314	49	[19]
IrCo	0.1 M HClO$_4$	304	77	[20]
IrCu$_{1.11}$	0.05 M H$_2$SO$_4$	286	44	[15]
IrCu$_{1.4}$	0.05 M H$_2$SO$_4$	311	54	[16]
IrFe$_{0.41}$/C	0.5 M HClO$_4$	278	56	[21]
IrW/C	0.1 M HClO$_4$	301	57	[22]

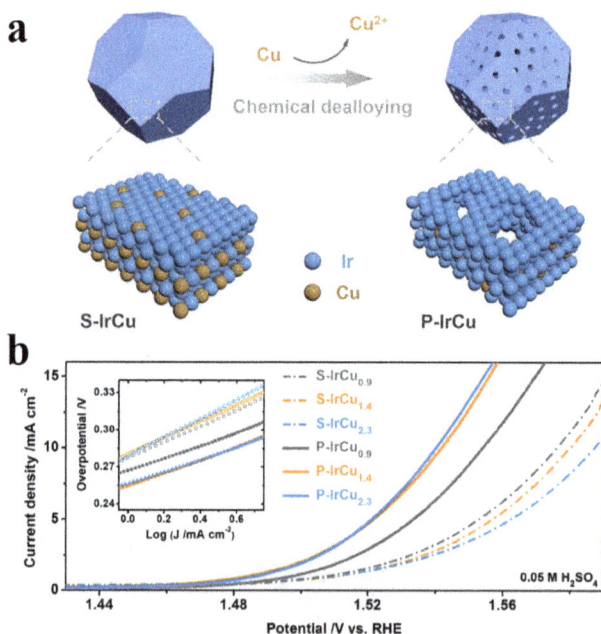

Fig. 1: *(a) Illustration of the forming porous IrCu NCs via chemical de-alloying of solid IrCu in acid medium; (b) Polarization curves of porous IrCux NCs and solid IrCux NCs electrodes in 0.05 mol L-1 H$_2$SO$_4$. In inset the Tafel plots [16]. Adapted with permission.*

Unlike the noble-non-noble metal alloys, lead alloys do not present a good intrinsic activity for OER. However, they have been constantly studied as electrocatalysts for this reaction because of their importance in the well-known lead-acid batteries, since the success of the application of such devices is intimately connected to water electrolysis [23]. In the past decades, silver has been used as an additive to lead to increase its electrochemical responses for OER. Monahov et al. [24] has intensively explored the impacts of silver on the Pb-Ag alloys for Pb/PbO_2 electrodes. It was verified that low silver concentration apparently could not affect the electrochemical performance of Pb-Ag alloys. However, the highest content of silver (0.28%) leads to a better performance in acidic conditions at room and even at higher temperatures. In fact, alloys with a high content of Ag usually contain eutectic Ag-Pb phases besides to an α-Pb solid solution, which may influence on OER activity of the produced electrodes. Moreover, it was observed for these cases that silver was also oxidized to Ag^+ and incorporated toPbO_2. The good performance of silver-dopedelectrodes is assigned to the changing of centers for OER, leading to a reduction of overvoltage and depletion of activation energy compared to pure Pb/PbO_2.

Another element that has been studied in order to increase the electrocatalytic activity of lead acid battery anode is the bismuth. For example, Li et al. [25] produced lead-bismuth alloys with a bismuth content of 0-8 wt% and noticed that only alloys with bismuth content greater than 0.1% affected the OER. This fact was attributed to the dissolution and the adsorption of bismuth on the surface of the lead dioxide formed in high overpotential, which, consequently, increases the kinetic parameters in comparison to non-adsorbed electrodes.

As above mentioned, few metallic alloys with chemical resistance and stability can withstand the OER drastic conditions, such as, the noble metal and noble-non-noble metal alloys, and lead alloys which in contrast have high overpotentials. The challenge of finding non-noble metal electrocatalysts is enormous but necessary for further development of renewable energy storage devices.

3.   Metal oxides for OER beyond Ni, Co, and Fe

3.1   Transition metal binary oxide-based electrocatalyst

Metal oxides are undoubtedly the most common electrocatalysts for OER in basic media because of their highest stability compared to other materials [26]. Metal oxide such as RuO_2 had been taken as reference for OER electrodes because of high performance and great steadiness at extremes pH values. For example, an overpotential of only 300 mV is required at 10 mA cm^{-2} in basic media for a thin coating of RuO_2 [27,28]. Ni, Co, and Fe-based binary and ternary oxides are currently among the most studied catalysts due to their inexpensive costs and good activity [29-33]. Albeit numerous other transition metal oxides have been studied as electrocatalysts for OER, they present some shortcomings such as poor electrical conductivity and less available catalytic active sites. Nonetheless, there are some investigations towards copper, titanium, and manganese oxides electrodes applied to OER [34-36].

Manganese oxides are considered good catalysts for reactions involving multiple steps of electron exchange because they have diversified redox properties and are inexpensive and abundant. Manganese oxides are among the most studied materials for catalytic, photocatalytic, and electrocatalytic OER [37-39]. Unlike NiO_x, CoO_x, and FeO_x metal oxides, the manganese oxides are chemically stable in both acidic and basic media [38]. However, a good intrinsic activity for OER depends on MnO_x stoichiometry and crystalline phase. Studies have shown that Mn_2O_3 is the most active phase for this reaction. For example, Ramírez et al. [40] evaluated the structure-function relationship between three different manganese oxides for the OER catalytic activity. They observed that the electrocatalytic activity of these oxides in basic solution follows the order $Mn_2O_3 > MnO_x > Mn_3O_4$; where for the Mn_2O_3 films presented an onset overpotential of only 170 mV, and for MnO_x and Mn_3O_4, the onsets of 230 mV and 290 mV, respectively as shown in the Fig. 2(a) [40]. The best activity of Mn_2O_3 also was confirmed by Differential Electrochemical Mass Spectroscopy (DEMS) that showed the highest values of O_2 production in an overpotential more positive on Mn_2O_3 compared with MnO_x and Mn_3O_4 (Fig. 2(b) [40]). DEMS is a reliable technique as it excludes the current values due to oxidative processes of the materials [40].

The proposed mechanism for OER on MnO_x and Mn_2O_3 can be seen as scheme Fig. 2(c). The oxygen vacancies in the manganese oxide films react with hydroxide ionsat the electrode interface. When the onset overpotential is reached, the surface of the electrode reacts with the holes electrochemically created, producing oxygen and water molecules and recovering the oxygen vacancies [40]. In addition, the catalytic activity of these materials can vary according to the methods of film deposition, annealing temperature, size, and shape of the particles [39-41].

The best catalytic activity for the Mn_2O_3 in comparison to the other phases can be explained considering their crystalline structure. The orthorhombic structure of the Mn_2O_3 has a high number of coordination units, where the MnO_6 octahedron shares 12 edges and 6 corners. The Mn_3O_4 exhibits a semi-covalent nature in the tetrahedral sites, and a short O-O distance (2.55 Å) in their octahedral sites, apical bonds. This crystalline structure has a smaller assortment of bond lengths and, consequently, a lower reach of dissimilar Mn-O lengths in the OER multi-steps [40].

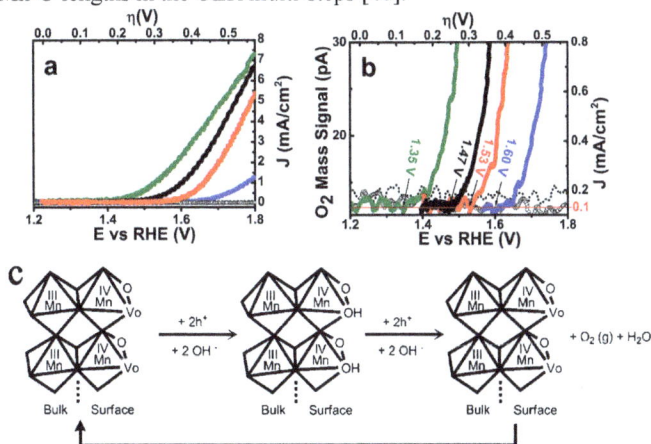

Fig. 2: (a) Polarizations curves and (b) oxygen mass signal from DEMS as a function of E vs RHE RuO₂ (green), MnOₓ (red), α-Mn₂O₃ (black), and Mn₃O₄ (blue) films. (c) Illustration of manganese oxide octahedral (MnO₆) and the surface interaction with the electrolyte on the OER conditions. Adapted with permissions of ref. [40]. Accessed in https://pubs.acs.org./doi/abs/10.1021%2Fjp500939d. Further permission related to the material excerpted should be directed to the ACS.

Although the manganese oxides present a good OER electrocatalytic activity, they are still far from high enough reactive kinetics to be commercially applied because of their poor electrical conductivity ($\sim 10^{-6}$ S cm^{-1} for MnO_2) [40,42,43]. On the other hand, modifications such as doping or self-supporting are done to increase the conductivity of these materials [35,44]. In this direction, Masa and co-workers [44] supported nitrogen doped carbon manganese oxides (Mn_xO_y /NC) using a monocyclic complex as the source of Mn. To obtain Mn_xO_y/NC, the Mn complex was impregnated with nitrogen-doped carbon and subjected to two heat treatments, one under an inert atmosphere of He and another in an oxidizing atmosphere of O_2. This material presented very good activity for OER in alkaline medium (450 mV of overpotential at 10 mA cm^{-2}), having only 4 and 2 mV higher compared to overpotentials to RuO_2 and IrO, respectively [44]. This data shows us that manganese oxides are excellent electrocatalysts for OER because of their diversified redox properties, easy preparation, and affordable price.

Electrochemical Water Splitting: Materials and Applications   Materials Research Forum LLC
Materials Research Foundations **59** (2019) 1-36   doi: https://doi.org/10.21741/9781644900451-1

Copper oxide is other emerging candidate for OER in large scale due to its low cost compared to the Ni, Co, and noble metals. Although the copper oxides exhibit poor electrical conductivity, this can easily be overcome by using highly conductive substrate. The affordability and easy-preparation of copper oxide nanostructured electrodes, with surface area higher than that for the bulk material, allow their use as an efficient electrocatalyst for OER [45]. In this perspective, $Cu@(Cu(OH)_2$-$CuO)$ core-shell nanostructures grown over copper flakes showed high efficiency for OER in alkaline media presenting only 417 mV of overpotential at 10 mA cm^{-2} [46]. The catalysts were simply produced, following some steps of galvanostatic anodization in alkaline media, followed by chemical reduction and new galvanostatic anodization. The electrode was quite stable in alkaline media (KOH 0.1 M) since the electrocatalytic polarization profile is the same, even after 800 cycles [47].

The utilization of carbon-based materials as conductive support as a way for improving the activity of low conductive materials has been reported in several works [45,48]. Particularly, carbon nanotube has been used to support the active metal oxides to enhance the area for mass exchange. Carbon nanotube/CuO composites have been successfully produced with good electrochemical results. In this case, 420 mV of overpotential at 10 mA.cm^{-2} were acquired in 1mol L^{-1} KOH aqueous solution [45]. The good electrochemical responses were attributed to the high mass exchange of electrolyte throughout the porous composite.

Among the copper compounds, cuprous oxide (Cu_2O) has been demonstrating interesting properties for OER application. Although Cu_2O has been widely applied to photoelectrochemistry [49], new researches have confirmed its potential for OER. $Cu_2O@C$ core-shell structures have been prepared to avoid particle aggregation and to enhance the surface area [50]. It had an onset of 1.48 V in 1 mol L^{-1} KOH solution. This value is considerably lower than the commercial Cu_2O and carbon black at the same experimental conditions (onset potential of 1.58 and 1.90 V, respectively) [50]. $Cu_2O@C$ core/shell structures had good OER stability; the catalytic current decreased only 7% during 100 h at 1.57 V vs. RHE. Otherwise, films of a physical mixture of Cu_2O/C had low stability and maintained only 30% of the initial current after 2h of constant electrolysis. The good performance of the $Cu_2O@C$ was assigned to the strong adhesive interaction between the substrate and the catalyst. Instead, the Cu_2O/C physical mixture presented a visual peeling off indicating its poor adhesion [50].

TiO_2 is one of the most widely studied oxides for environmental concerns. This semiconductor has good optical absorption, high chemical stability, and it is inexpensive [51]. Although, it has been generally used in photodegradation of several organic molecules, pollutants and as a photocatalyst for water splitting [51,52], TiO_2 presents a poor activity for OER due to its sluggish kinetics and high overpotential [53]. However, recent works have reported the inclusion of other transition metal ions into the TiO_2 lattice as a way for increasing the electro and photoelectrochemical activity for water oxidation [54-56]. Aliovalent doping is expected to create structural modification such as oxygen vacancies, which reduces the threshold for light absorption and may enhance its performance as an OER catalyst [53,57]. A study developed by Kim and co-workers [57] shows the optimal dopant element onto TiO_2 two-dimensional nanostructures for both photo and electrochemical water oxidation. In this study, it was assumed that the mechanism for OER occurs by four proton-coupled electron transfers (PCET) intermediate steps showed in Fig. 3(b) [57]. Based on the spin-polarized DFT, a Vulcan-like graph of activity was proposed by associating the free energy of the first PCET (step $A \rightarrow B$: $\Delta G_{H_2O/OH^*}$) with the overpotential for the OER, as shown in Fig. 3a[57]. In terms of catalytic activity, those elements close to the peak of the volcano profile promote the reduction of the activation energies to the rate determining step and lead the system to near the ideality. In Fig. 3b [57], the influence of Rh, Nb and Pd dopants can be observed in each reaction step. It was noted that the influence of dopants in the free energy is mainly in PCET: $A \rightarrow B$ and $C \rightarrow D$ act smoothing these energy barriers providing a higher electrocatalytic activity for TiO_2. An explanation was suggested for the tendency of the differences between the energy barriers observed for the different dopants, considering the variations of geometries and the total of electrons that surround each atomic species. Therefore, for dopants with low oxidation states such as Ru (most common oxidation state Ru^{3+}), the formation of -OOH bonds, with species O_2^{-2}, reduces the energy barrier of the reaction limiting step. On the other hand, for dopants with high oxidation state like Nb (most common oxidation state Nb^{5+}) promote the formation of weakly adsorbed $-OH^-$ species that also reduce the energy barrier and lead to a higher electrocatalytic activity compared to pure TiO_2 [57].

The easy preparation and doping of high surface area TiO_2 nanostructures using simple and inexpensive techniques such as anodization have attracted great attention [55,56,58,59]. However, although transition-metal-doped TiO_2 nanostructures present high surface areas, this material present higher OER overpotential (650-1000 mV) when compared to other inexpensive transition metal oxides with good activity, such as NiO_x and CoO_x[59-61]. The kinetic and activity parameters of the binary oxides used as the catalyst for the OER are listed in Table 2 [40,50,62-67]. Among these materials, manganese and copper oxides exhibit similar activities to Ni, Co, and Fe oxides, which are recognized as good electrocatalysts for the electrochemical water oxidation.

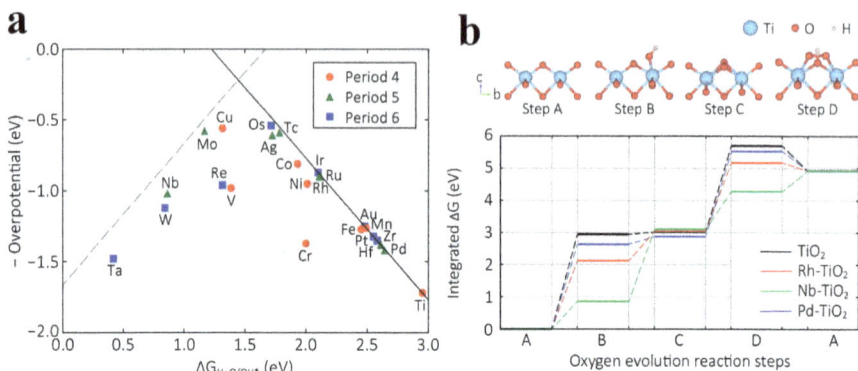

Fig. 3: (a) Volcano plot for transition metal doped TiO_2 nanostructures; (b) Gibbs free energy changes according to the OER steps on TiO_2, $Rh-TiO_2$, $Pd-TiO_2$, and $Nb-TiO_2$. Adapted with permissions of Ref.[57]. Copyright 2017 American Chemical Society.

Table 2: Kinetic and activity parameters of the binary oxides used as the catalyst for the OER.

Oxides	electrolytes	Overpotential [mV] at a particular current density	Tafel slope [mV/dec]	Ref.
Fe_3O_4	1 M KOH	360@ 10 mA cm^{-2}	87	[62]
Co_3O_4	1 M KOH	400@10 mA cm^{-2}	49	[63]
NiO	1 M NaOH	540@10 mA cm^{-2}	210.4	[64]
Mn_2O_3	1 M KOH	440@10 mA cm^{-2}	-	[40]
$\alpha-MnO_2$	0.1 M KOH	490@10 mA cm^{-2}	77.5	[65]
Mn_3O_4	1 M KOH	520@ 5 mA cm^{-2}	-	[40]
CuO	1 M KOH	450@ 2 mA cm^{-2}	60	[66]
Cu_2O	1 M KOH	720@ 8 mA cm^{-2}	166	[50]
TiO_2	0.1 M KOH	740@ 0.3 mA cm^{-2}	-	[67]

3.2 Perovskites oxides electrocatalysts

Among the transition metal oxides, compounds with perovskite structure have been showing a great interest for OER, with reactions mainly conducted in alkaline media [68]. Perovskite is a class of compounds with the same structure of $CaTiO_3$ [69].Fig. 4presents a typical scheme for a perovskite structure with ABX_3 minimum formulae, in which the coordination of both B (Fig.4a) and A (Fig.4b) cations are emphasized [69]. These materials present high activity and chemical stability, and because of that are promising electrocatalysts.

Nowadays, a large variety of compounds are recognized to assume this structure [70]. More than one cation with different sizes and charges is present in a perovskite. The smallest and most positively charged cation, B^{4+}, occupies the interstitial octahedral sites of a close-packed structure [71], forming BO_6 subunits. On the other hand, the largestcation, A^{2+}, occupies the twelve-coordinated sites [72]. Oxygen is the most common Xanion that promotes charge neutrality. However, other anions are also allowed [69,70,73].

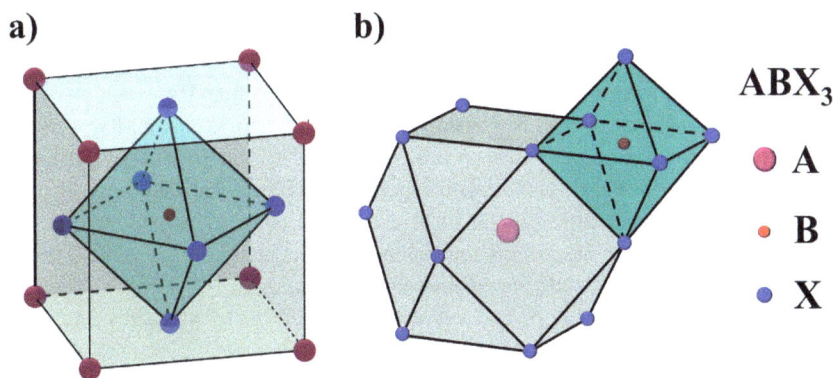

Fig. 4: Schemes for a typical ABX₃ perovskite structure: (a) B cation showing octahedral coordination; (b) A cation surrounded by a cuboctahedron of X anion.

The ABO_3 minimum formula is widely used to describe general oxide perovskites. Perovskite compounds with rare earth or alkaline earth element as the A cation and a transition metal ion as B are particularly applied to electrocatalytic OER [74,75]. Many variations are permitted for this structure, for example, $LaBO_3$ and YBO_3, with B and At rivalents. In addition, $CaBO_3$ and $SrBO_3$, and their derived compounds are very common perovskites for this application [75].

The high activity provided by perovskites for OER is mainly determined by the great capability for OH^- adsorption. Moreover, the rate of oxygen desorption on the catalyst surface can affect the kinetics [76,77]. These characteristics allow enhancement of mass and charge transfer rates, improving the catalyst activity, and making these compounds attractive for this application. Perovskites usually present considerable stability for high corrosive alkaline media; however, after several OER cycles, the catalyst degradation can occur mainly by amorphization [78].

The use of these compounds is also desirable because it is possible to modify the structure and composition in order to modify their properties [74,79,80]. In fact, several theoretical and experimental studies have been performed to understand the parameters which affect their efficiency. Till now, the impacts of these parameters are still not established in the literature [75,81]. However, some researchers have emphasized that the physicochemical characteristics of the central cation are deeply linked to the catalytic performance of the perovskite [75].

Some studies have reported the influence of, e.g. filling status from the transition metal at the octahedral position for the electrocatalytic activity of perovskites [82-85]. Suntivich et al. [83] proposed a model based on orbitals occupancy and OER activity, showing relationships between Fe, Co, Ni-based compounds and other transition-metal elements. A volcano behaviorwas attributed for OER activity of perovskites by varying the e_g descriptor filling surface from the B site cation. Experimentally, it was verified that the OER efficiency approaches the maximum value of e_g at about 1.2 electrons. Therefore, further studies for developing perovskites with e_g close to unity were fully encouraged [83,84,86]. Other works have applied the same theory for several pure perovskites and solid solutionswith results presented in Fig.5 [75]. Even though showing the dispersion of the data, the volcano tendency (represented in dashed line) was visualized when plotting overpotential η versus electron number e_g (Fig. 5a).

The relation between OER catalytic activity and orbital occupancy is justified by the influence of intermediate formation on the catalyst surface, which can be impacted by the e_g filling status. The first part of the volcano plot is mainly limited by deprotonation of oxyhydroxide species, to form peroxide. On the other side, the limiting rate is mainly determined by the formation of O–O bonds of OOH intermediate in its second part. The e_g descriptor value close to unity seems to favor both limiting reactions [83]. Moreover, a high covalency of oxygen and B cation bonds also positively contribute to OER, since the transference of charge between central cation and adsorbates is facilitated [83,85,86].

Therefore, based on these previous results, other studies have demonstrated that the same trend observed in alkaline media is also applied to the neutral pH solution [87]. Some contributions were introduced to comprehend the stability of perovskites as electrocatalysts. The oxygen p band center close to the Fermilevel plays animportant role on the catalyst stability. When they are too close, the catalysts seem to be more efficient for OER. However, they present a tendency to leach the A cation and are more susceptible to become amorphouseven at neutral pH medium. Otherwise, the perovskites have presented considerable stability at low potentials, but segregation of the B cation at high potential seems to be favorable [87].

Although there is wide acceptance for the relationship between e_g electron number and OER activity, this theory has been demonstrated not to be a general rule. Solid solutions containing fractional values of e_g have demonstrated an inversed volcano-like trend, as observed in red squares of Fig.5b [75]. Therefore, this model may contain some limitations, requiring attention before its use.

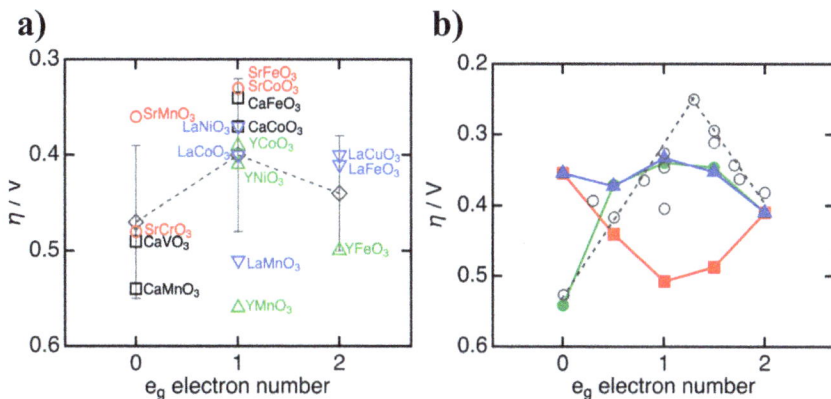

Fig. 5: Relation between overpotential η and e_g orbital occupancy for several perovskites: (a) ABO_3 simple perovskites – $CaBO_3$ (black), $SrBO_3$ (red), $LaBO_3$ (blue) and YBO_3 (green). (b) Solid solutions – $CaMn_{1-x}Fe_xO_3/Ca_{1-x}La_xFeO_3$ (green), $SrMn_{1-x}Fe_xO_3/Sr_{1-x}La_xFeO_3$ (blue) and $Sr_{1-x}La_xMnO_3/LaMn_{1-x}Fe_xO_3$ (red). Gray circles refer to the data from the previous work performed by Suntivich et al. [83] for comparison. Adapted with permission from Ref.[75]. Copyright 2018 American Chemical Society.

A systematic study of several perovskites with elements from the third period as the central cation was made and few tendencies were proposed. It was demonstrated that the overpotential decreased from the lightest up to the heaviest elements, in general. The increasing in the transition metal valency can also affect the overpotential in almost all cases. Based on ABO_3 general formulae, trivalent B^{3+} transition metals with A = Y^{3+}, La^{3+} lead to a higher overpotential than perovskites with tetravalent central atom B^{4+} (in this last case, A = Ca^{2+}, Sr^{2+})[75].

Another descriptor proposed to influence the catalytic activity is the charge-transfer energy between electrocatalyst and adsorbate. The tendency between this parameter and the OER activity of perovskites with B cation from the third period is shown in Fig. 6a [75].A trend of lowering overpotential when reducing the charge-transfer energy is observed to those perovskites. In bulk, the charge-transfer energy (named as Δ) is the difference between the unoccupied 3-d ($\varepsilon_{3d\text{-}un}$) and O-$2p$ (ε_{2p}) band centers. During electrocatalysis, the adsorbate interacts with the catalyst, requiring a new value of energy for charge transfer, Δ' (Fig.6-b). This parameter is related to the charge transfer barrier. As previously observed for the orbital occupancy theory, the authors emphasize that a single descriptor is not able to approach all the cases. Other parameters such as the unoccupied $3d$ and O-$2p$ levels, for instance, were observed to individually perform their influence for OER. However, even containing some restrictions, these efforts in proposing relations between the composition of perovskites and OER activity are very important, since they provide a guide to develop highly efficient electrocatalysts.

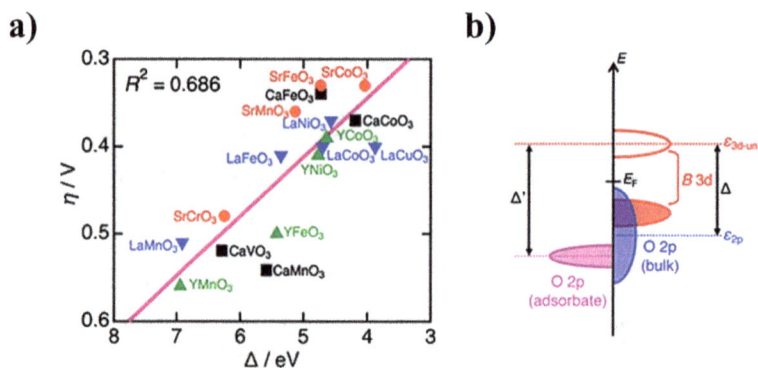

Fig. 6: Systematic study of OER activity of perovskites with transition metal from the third period: (a) variation of overpotentials η versus charge transfer energy Δ – CaBO₃ (black), SrBO₃ (red), LaBO₃ (blue) and YBO₃ (green). (b) Schemes of band centers in bulk electrocatalyst and the same with adsorbates. The charge transfer energy for each case, Δ and Δ' respectively, is also presented. Adapted with permissions from Ref.[75]. Copyright 2018 American Chemical Society.

Besides those relations established mainly for stoichiometric perovskites, other phenomena can also influence the OER activity. The presence of charged defects is one of the most important. Oxygen vacancies, $V_O^{\cdot\cdot}$ (represented in Kröger-Vink notation [88]), are recognized to be deeply related to the electrocatalyst activity. These defects act as donor or acceptor centers, facilitating the charge flux between the adsorbed species and electrocatalyst, improving its activity [74,82].

Several strategies have been performed to create these charged defects in perovskites. The processes include annealing under a reductive atmosphere [82] and quenching [89]. Besides, partial replacement or doping with aliovalent ions is one of the most reliable methods for this purpose to control defects creation [71,72].

Kim et al. [82] has reported the production of an oxygen-deficient $Ca_2Mn_2O_5$ perovskite – equivalent to $CaMnO_{2.5}$ – by annealing the $CaMnO_3$ in a reductive atmosphere. By this process, Mn^{4+} from the stoichiometric perovskite was converted to Mn^{3+}, while oxygen was expelled from the perovskite lattice. The annealed and pristine compounds were then deposited on carbon electrodes and the catalytic performance was evaluated in O_2-saturated alkaline media (Fig. 7a). In comparison to $CaMnO_3$, it has revealed a prominent enhancement of electrocatalytic activity than the oxygen-deficient compound. The Tafel slopes were 149 and 197mV.dec^{-1} for $Ca_2Mn_2O_5$/C and $CaMnO_3$/C, respectively, confirming the positive impact of oxygen-deficiency for OER electrocatalysts.

The explanation for the better performance observed for the $Ca_2Mn_2O_5$ perovskite is based on the oxygen-vacant structure and electronic configuration of manganese ions. Mn^{3+} ($3d^4$) possess a $t^3_{2g}\,e^1_g$ configuration, while Mn^{4+} ($3d^3$) is t^3_{2g}. Therefore, the high spin state orbital with one electron e^1_g for Mn^{3+} can bond with OH^- easier than Mn^{4+}, favoring the OER. Moreover, as observed in Fig. 7b, subunits of MnO_5 are appropriated for OH^- interaction, forming OH^-–MnO_5 intermediates. This correspondence between adsorbates and central ions seems to be more propitious for the oxygen-vacant MnO_5 than to the MnO_6 subunits, which possess complete polyhedra [82].

Double perovskites are other possible variation to be applied as electrocatalysts (Fig. 8). These compounds present minimum formulae $A'_xA_{1-x}BO_3$ when more than one cation occupies the highly coordinated sites, A and A' [90]. The double perovskite structure can be defect-free or defective, depending on the A and A' valences. In the first case, the natural sites of the highly coordinated cation are intercalated between A and A' ions. The modeling is more complex to defected-double perovskites due to lattice imperfections in the structure. Considering $V_O^{\cdot\cdot}$ as structural defects, the A' cation can occupy distinct positions: i) at oxygen deficient environment and ii) at normal A perovskite lattice. Experimentally, it was demonstrated that A' may also occupy other specific environments characteristics of these ions [74].

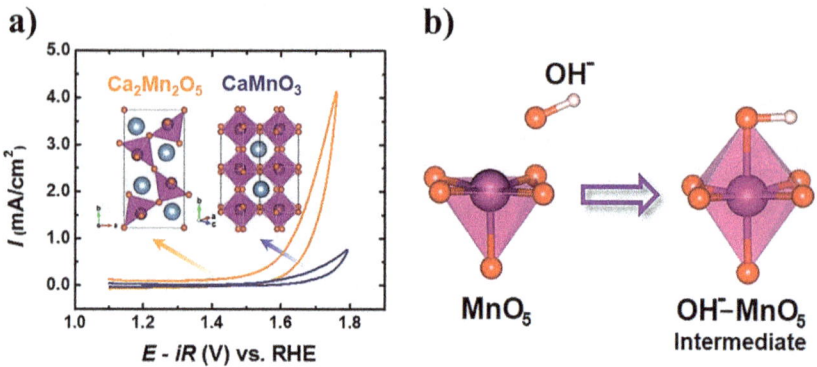

Fig. 7: *Comparison between OER activity of oxygen-deficient and stoichiometric perovskites: (a) Voltammetry for $Ca_2Mn_2O_5$/C and $CaMnO_3$/C electrocatalysts. (b) Possible impacts of oxygen-vacant MnO_5 subunits to the OER rates (alkaline media). Adapted with permissions of Ref.[82]. Copyright 2014 American Chemical Society.*

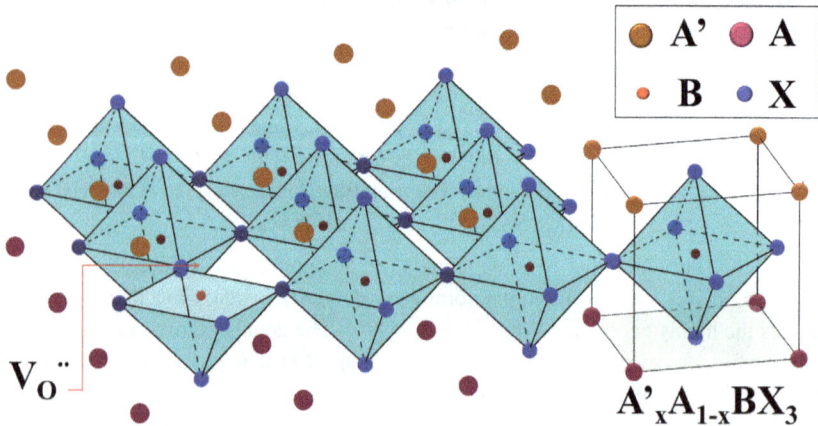

Fig. 8: *Scheme of a double perovskite with $A'_xA_{1-x}BX_3$ minimum formulae, being oxygen the X anion. Oxygen vacancy ($V_o^{\cdot\cdot}$) was emphasized as structural imperfection, allowing different environments for the A' cation.*

$CaCu_3Ti_4O_{12}$ is a double-perovskites explored for OER. In this material, Ca^{2+} and Cu^{2+} occupy the highly coordinated positions, where Ti^{4+} is octahedrally coordinated [91]. It presented an overpotential of 140mV at 0.5 mA.cm^{-2}, lower than that for several other perovskites from literature. It includes the iron-based $SrFeO_3$, $CaFeO_3$, and $CaCu_3Fe_4O_{12}$ compounds from previous work [92]. This compound naturally presents a high density of oxygen vacancies and elevated covalency for metal-oxygen bonds, which justifies its high OER electrocatalytic activity [74].

Besides composition and structure, particle and grain size, and morphology also deeply affect the electrocatalyst performance. The production of porous and nanostructured perovskites have been displayed as an interesting way to enhance the OER rates [76,82]. For example, perovskites over carbonaceous materials are an applicable way to improve the catalyst surface area, which can favor the mass transport, and the charge exchange [93,94].

New catalysts based on perovskites have been a hot topic for electrocatalytic OER. Perovskite hydroxides have been recently applied to the oxygen evolution reaction. These perovskites typically display a distorted ReO_3 structure, which can be simply produced by hydrothermal or co-precipitation [95]. Another point of emerging interest is the study of ferroelectricity to complement the knowledge on the electrochemical performance of perovskites.

Ferroelectricity is the phenomenon of spontaneous polarization which arises from a non-centrosymmetric structure. The electric dipoles can be lined up by applying an external field, analogous to ferromagnetism [71]. The remaining polarization was already proved to affect the catalytic performance of the perovskites [96-98].

An interesting example is the case of $Bi_{0.5}Na_{0.5}TiO_3$ [96]. This perovskite has been produced by a solid-state reaction, pelleted, and submitted to different intensities of the electric field, varying from 10 to 60 kV.cm^{-1}, to polarize the material. The OER results obtained using the poled perovskite were significantly superior to that to unpoled material. In fact, a gradual enhancement of OER activity was observed according to the intensity of the applied field. Tafel slope decreased from 85mVdec^{-1} to 39mVdec^{-1} when comparing the unpoled and poled perovskite at 60 kVcm^{-1}, respectively. The improved electrocatalytic performance by poling the ferroelectric catalyst was attributed to a most efficient charge transfer and better facility for species adsorption with opposite charges than the surface. These results open the possibility of ferroelectric application as a new variety of high performance electrocatalysts.

A summary of the most remarkable perovskites with transition-metal elements beyond Fe, Co, and Ni for catalytic OER are presented in Table 3 [74,75,82,96,99,100]. Kinetic parameters for simple ABO_3, oxygen-deficient, and ferroelectric perovskites are shown for comparison.

Table 3: Kinetic and activity parameters of perovskites with B beyond Fe, Co, Ni applied as catalysts for OER.

Perovskite	Electrolytes	Overpotential [mV] at 1 mA cm^{-1}	Tafel slope [mV/dec]	Ref.
$Ca_2Mn_2O_5/C$	O_2-saturated 0.1 M KOH	460	149	[82]
$CaMnO_3/C$	O_2-saturated 0.1 M KOH	590	197	[82]
Poled-$Bi_{0.5}Na_{0.5}TiO_3$	1 M KOH	635	39	[96]
$La_{0.6}Sr_{0.4}MnO_3$	1 M KOH	–	110	[99]
$La_{0.6}Sr_{0.4}MnO_3$ films	Ar-saturated 0.1M KOH	514	61	[100]
$CaCu_3Ti_4O_{12}$	0.1M KOH	320[1]	49[1]	[74]
$SrMnO_3$	0.1M KOH	360	113	[75]
$LaCuO_3$	0.1M KOH	610	127	[75]

[1]Measured at 1600 rpm.

Therefore, perovskites are very versatile and efficient electrocatalysts for OER. Even though the most common perovskites for this purpose based on Fe, Co, and Ni elements, several other materials have shown comparable results. The production of oxygen-vacant and double perovskites, with high covalency for B–O bonds, are the main strategies to improve their catalytic activity. The control of morphology and grain size has also impacted their electrocatalytic performance. On the other side, amorphization for a long time of electrocatalysis and considerable charge resistivity are disadvantages to be overcome when applying these materials. Perovskite hydroxides and polarized ferroelectrics are new tendencies for this class of materials, which have already been demonstrated propitious for electrocatalytic application.

4.   Transition-metals carbides, nitrides, and phosphides applied for OER

4.1   Carbides

Transition metal carbides present good chemical stability for electrocatalytic application, mainly in acidicconditions [101,102]. The class of compounds also present elevated conductivity, as well as good mechanical stability [103-105]. These compounds can present several stoichiometries and structures with M_xC_y minimum formulae [105]. The most common stoichiometry assumed bytransitionmetal carbides are summarized in Fig. 9 [102].

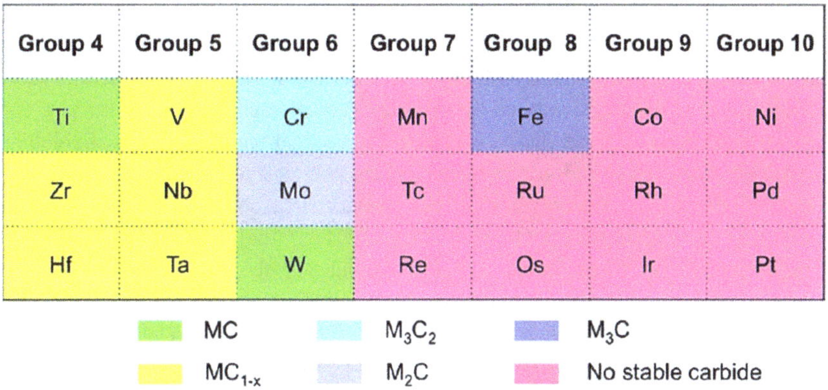

Group 4	Group 5	Group 6	Group 7	Group 8	Group 9	Group 10
Ti	V	Cr	Mn	Fe	Co	Ni
Zr	Nb	Mo	Tc	Ru	Rh	Pd
Hf	Ta	W	Re	Os	Ir	Pt

▇ MC	▇ M_3C_2	▇ M_3C	
▇ MC_{1-x}	▇ M_2C	▇ No stable carbide	

Fig. 9: *Summary of the most common transition metal carbides and their stoichiometries based on data from Ref. [105]. Reproduced with permission from Ref.[102]. Copyright 2013 American Chemical Society.*

Among those, tantalum [106,107], niobium [106,108], titanium [104,106], molybdenum [109,110] and tungsten [106, 111] carbides are often used as supporting materials for electrochemical application in general. Although they are widely used as support, recently works have reported their application as active electrocatalysts [112,113]. For example, boron- and nitrogen-doped molybdenum carbide have presented high electrocatalytic activity for OER [113].

Anjum et al. [113] have demonstrated that undoped Mo_2C presents a Tafel slope of 84mVdec^{-1}. Although this result is very similar to that 82mVdec^{-1} obtained from IrO_2, used as reference under alkaline media, Mo_2C is an earth abundant element-based compound in comparison to IrO_2. The doped carbides show superior activity than that for the afore mentioned materials. Decreased Tafel slopes were obtained for N-Mo_2C (66mVdec^{-1}) and N,B-Mo_2C (61mVdec^{-1}) catalysts. The current density in the system comprised by N,B-Mo_2C was quite stable during 20h of electrocatalysis. It is mainly attributed to partial surface oxidation, in which the carbide is converted into molybdenum oxides (MoO_2 and MoO_3), which present a considerable activity for OER. It is meaningful that transition metal carbides can be converted to their correspondent oxides at positive potentials and in alkaline conditions [101]. The as-formed oxide on the catalyst surface may be good for the electrocatalytic performance [113]. On the other side, the metal oxide can dissolve after several cycles of the test in alkaline conditions, consequently leading to a performance depletion [101].

An interesting material which combines the properties of transition-metal carbides and nitrides has been recently produced by Ma et al. [114]. Graphitic carbon nitride, g-C_3N_4, coupled to titanium carbide, Ti_3C_2, have shown excellent results for OER. Exhibiting a conductive framework and hydrophilic surface, the charge, and transport mass are facilitated on this electrocatalyst. Elevated surface area obtained by exfoliation also contributed by facilitating OH⁻ adsorption. In fact, the polarization curves have demonstrated high current density at high potential than those obtained to IrO_2/C. The synergistic combination also highly increased the activity compared to the single g-C_3N_4 and Ti_3C_2 compounds. Another important characteristic was the elevated stability in alkaline media: the current density was almost the same even after five thousand cycles during electrocatalysis.

In summary, carbides are propitious to be applied as support in acid media. In neutral and alkaline electrolytes, they tend to lose stability because of their conversion into oxides, followed by dissolution. These compounds can be applied as active electrocatalysts, such in case of B, N-doped Mo_2C, and g-C_3N_4/Ti_3C_2ones. Particularly for the second case, its development breaks down barriers for using carbides and nitrides- hybrid materials as highly active and stableelectrocatalysts for OER.

4.2   Nitrides

Transition metal nitrides are a class of compounds with a M_xN_y typical formula with useful electrochemical applications, for example niobium [115], titanium [1] and manganese [116] nitrides. However, the most efficient for OER are nickel and cobalt based nitrides [117,118]. Tafel slopes varying from 45mVdec⁻¹ [118] to 64 mVdec⁻¹ [119] have been reported to Ni_3N. For cobalt-based nitrides, the activity depends on the cobalt valency and the catalyst morphology: CoN (70 mVdec⁻¹ [120]), Co_2N (80 mVdec⁻¹ [121]), Co_3N (72 mVdec⁻¹[121]) and Co_4N (58 mVdec⁻¹ [121]) have been reported. Mixed system with iron such as Ni_3FeN and Fe_2Ni_2N/Carbon nanotubes have also displayed notorious results (46 [119] and 38 mVdec⁻¹ [122], respectively).

Beyond these elements, nitrides are still little explored for OER. Manganese nitride has been recently produced with satisfactory results. However, a single phase could not be obtained for electrochemical evaluation [116]. The use of titanium nitride as support for iridium species has demonstrated good results mainly attributed to the elevated conductivity achieved for the IrO_2@Ir/TiN catalysts [123]. Titanium nitride and its derived material, titanium oxynitrides, TiO_xN_y, have also been evaluated as active electrocatalysts for OER [1]. Nevertheless, no considerable quantity of oxygen was detected by mass spectrometry during the experiments, suggesting a poor catalytic activity.

Although the use of nitrides as electrocatalysts is still mainly focused on Ni, Co, and Fe elements, new insights have been demonstrating the potentiality for using other transition metal nitrides mainly as support. Further studies exploring other transition metal nitrides as active catalysts are still necessary.

4.3   Phosphides

Transition metal phosphides are comprised of a group of compounds with importance for OER in alkaline media. T_xP (where T is a transition metal) is the most common stoichiometry for phosphides applied as electrocatalysts. The most prominent ones for this interesting application are based on Fe, Co, and Ni elements, similarly to nitrides [124]. Cobalt phosphide has presented asingular importance, with electrocatalytic activity changing according to the morphology, presence of support and heteroatoms, type and concentration of electrolyteand cobalt valency [124]. The Tafel slope reported for CoP nanowires varies from 78 [125] to 64 mVdec^{-1} [126], higher than Co_2P with similar morphology (52mVdec^{-1} [126]). When the shape is changed to the needle, Co_2P presents a value of 50 mVdec^{-1} [127], the same obtained to CoP on carbon nanotubes [128]. Partial substitution with iron, $Co_{0.7}Fe_{0.3}P$, has also demonstrated its impact on Tafel slope: a value as low as 27 mVdec^{-1} was already obtained [129]. Similar studies are also found for Ni and Fe-based phosphides [130-134]. It is meaningful that these comparisons are shown only for illustration since the OER activity is very sensitive to the electrolyte utilized, its concentration and other peculiarities fromthetests.

Copper phosphide is a promising candidate for OER catalysis beyond Ni, Fe and Co compounds [135,136]. The semimetallic Cu_3P nanoarrays were produced by Hou et al. [135] on copper foil, showing good electrocatalytic results. An overpotential of 412mV was reached at a current density of 50 mA.cm^{-2} for the surface oxidized catalyst (Cu-P@ Cu_3P/Copper Foil – CF), which is comparable to the results obtained for cobalt-based phosphides. Tafel slope of 63 mVdec^{-1} was estimated for this catalyst, lower than those presented by CuO_x/CF (128 mVdec^{-1}) and RuO_2/CF (75 mVdcc^{-1}). In fact, the authors emphasized that the surface-oxidized copper phosphide nanoarrays are the most active copper-based catalyst for OER up to date of their study. The high activity demonstrated by this material is justified by the elevated area provided by the catalyst morphology, favoring mass transfer. The semimetallic Cu_3P-core also contributed, providing fast charge transference between catalyst and adsorbate. Further studies have demonstrated great stability of Cu_3P on nickel foil up to 20 h of test [136].

Therefore, transition metal phosphides are great candidates for application as active electrocatalysts. Even though the iron, nickel and cobalt-based phosphides are the most notorious ones for this application, new researches have introduced copper phosphide as a propitious alternative to those.

Conclusions

The best catalyst for oxygen evolution reaction is still noble-metal-based catalyst due to their stability and activity in acidic medium, but their application is limited by their high cost and scarcity. On the other side, transition metal-based compounds have played a key role in the development of an efficient and cost-effective water splitting technology, exclusively to OER under alkaline conditions. Although over the past decades, Ni, Co, and Fe-based electrocatalysts have attracted extensive interest for this application, other metal transition compounds beyond the afore mentioned elements have also been shown competitively. Based on the brief discussion of the selection of the described electrocatalysts in this chapter, it is important to point out that: among the metal alloys applied for OER, Cu- and W-Ir alloys have shown most promising electrocatalytic activity than those alloys based on Ni and Co, and even than the noble metals PtIr and IrPd; Mn and Cu oxides exhibit similar activities in comparison to Ni, Co, and Fe oxides; perovskite oxides are very versatile and efficient electrocatalyst for OER; and that other inorganic materials, such as, metal transition carbides, phosphides and nitrides have been introduced to OER. It is important to go beyond Ni, Co and Fe -based electrocatalyst to look for OER electrocatalyst which can efficiently work at all pH values.

References

[1]   C. Gebauer, P. Fischer, M. Wassner, T. Diemant, Z. Jusys, N. Hüsing, R.J. Behm, Performance of titanium oxynitrides in the electrocatalytic oxygen evolution reaction, Nano Energy. 29 (2016) 136–148. https://doi.org/10.1016/j.nanoen.2016.05.034

[2]   A.T.N. Nguyen, J.H. Shim, Facile one-step synthesis of Ir-Pd bimetallic alloy networks as efficient bifunctional catalysts for oxygen reduction and oxygen evolution reactions, J. Electroanal. Chem. 827 (2018) 120–127. https://doi.org/10.1016/j.jelechem.2018.09.012

[3]   T. Reier, M. Oezaslan, P. Strasser, Electrocatalytic oxygen evolution reaction (OER) on Ru, Ir, and Pt catalysts: A comparative study of nanoparticles and bulk materials, ACS Catal. 2 (2012) 1765–1772. https://doi.org/10.1021/cs3003098

[4]   L. Trotochaud, J.K. Ranney, K.N. Williams, S.W. Boettcher, Solution-cast metal oxide thin film electrocatalysts for oxygen evolution, J. Am. Chem. Soc. 134 (2012) 17253–17261. https://doi.org/10.1021/ja307507a

[5]   S. Trasatti, Electrocatalysis by oxides - Attempt at a unifying approach, Electroanalysis. 111 (1980) 125–131. https://doi.org/10.1016/S0022-0728(80)80084-2

[6]   N.M. Alyami, A.P. Lagrow, K.S. Joya, J. Hwang, K. Katsiev, D.H. Anjum, Y. Losovyj, L. Sinatra, J.Y. Kim, O.M. Bakr, Tailoring ruthenium exposure to enhance the performance of fccplatinum@ruthenium core-shell electrocatalysts in the oxygen evolution reaction, Phys. Chem. Chem. Phys. 18 (2016) 16169–16178. https://doi.org/10.1039/C6CP01401A

[7]   Y.W. Zhang, T. Zhang, S.C. Li, W. Zhu, Z.P. Zhang, J. Gu, Shape-tunable Pt-Ir alloy nanocatalysts with high performance in oxygen electrode reactions, Nanoscale. 9 (2017) 1154–1165. https://doi.org/10.1039/C6NR08359E

[8]   K.C. Neyerlin, G. Bugosh, R. Forgie, Z. Liu, P. Strasser, Combinatorial study of high-surface-area binary and ternary electrocatalysts for the oxygen evolution reaction, J. Electrochem. Soc. 156 (2009) B363. https://doi.org/10.1149/1.3049820

[9]F. Wang, K. Kusada, D. Wu, T. Yamamoto, T. Toriyama, S. Matsumura, Y. Nanba, M. Koyama, H. Kitagawa, Solid-solution alloy nanoparticles of the immiscible iridium-copper system with a wide composition range for enhanced electrocatalytic applications, Angew. Chem. Int. Ed. 57 (2018) 4505–4509. https://doi.org/10.1002/anie.201800650

[10]   J. Pei, J. Mao, X. Liang, C. Chen, Q. Peng, D. Wang, Y. Li, Ir-Cu nanoframes: One-pot synthesis and efficient electrocatalysts for oxygen evolution reaction, Chem. Commun. 52 (2016) 3793–3796. https://doi.org/10.1039/C6CC00552G

[11]   H. Lv, D. Li, D. Strmcnik, A.P. Paulikas, N.M. Markovic, V.R. Stamenkovic, Recent advances in the design of tailored nanomaterials for efficient oxygen reduction reaction, Nano Energy. 29 (2016) 149–165. https://doi.org/10.1016/j.nanoen.2016.04.008

[12]   J. Ding, X. Zhu, L. Bu, J. Yao, J. Guo, S. Guo, X. Huang, Highly open rhombic dodecahedral PtCu nanoframes, Chem. Commun. 51 (2015) 9722–9725. https://doi.org/10.1039/C5CC03190G

[13]   H.L. Jiang, Q. Xu, Recent progress in synergistic catalysis over heterometallic nanoparticles, J. Mater. Chem. 21 (2011) 13705–13725. https://doi.org/10.1039/c1jm12020d

[14]   A. Groß, Reactivity of bimetallic systems studied from first principles, Top. Catal. 37 (2006) 29–39. https://doi.org/10.1007/s11244-006-0005-x

[15]   C. Wang, Y. Sui, G. Xiao, X. Yang, Y. Wei, G. Zou, B. Zou, Synthesis of Cu-Ir nanocages with enhanced electrocatalytic activity for the oxygen evolution reaction, J. Mater. Chem. A. 3 (2015) 19669–19673. https://doi.org/10.1039/C5TA05384F

[16]   Y. Pi, J. Guo, Q. Shao, X. Huang, Highly efficient acidic oxygen evolution electrocatalysis enabled by porous Ir–Cu nanocrystals with three-dimensional electrocatalytic surfaces, Chem. Mater. 30 (2018) 8571–8578. https://doi.org/10.1021/acs.chemmater.8b03620

[17]   C. Spöri, J.T.H. Kwan, A. Bonakdarpour, D.P. Wilkinson, P. Strasser, The stability challenges of oxygen evolving catalysts: towards a common fundamental understanding and mitigation of catalyst degradation, Angew. Chemie - Int. Ed. 56 (2017) 5994–6021. https://doi.org/10.1002/anie.201608601

[18]   T. Zhang, S. Liao, L. Dai, J. Yu, W. Zhu, Y. Zhang, Ir-pd nanoalloys with enhanced surface-microstructure-sensitive catalytic activity for oxygen evolution reaction in acidic and alkaline media, Sci. China-Materials. 61 (2018) 926–938. https://doi.org/10.1007/s40843-017-9187-1

[19]   H. Jin, Y. Hong, J. Yoon, A. Oh, N.K. Chaudhari, H. Baik, Lanthanide metal-assisted synthesis of rhombic dodecahedral MNi (M = Ir and Pt) nanoframes toward efficient oxygen evolution catalysis, Nano Energy. 42 (2017) 17–25. https://doi.org/10.1016/j.nanoen.2017.10.033

[20]   J. Feng, F. Lv, W. Zhang, P. Li, K. Wang, C. Yang, B. Wang, Y. Yang, J. Zhou, F. Lin, G.C. Wang, S. Guo, Iridium-based multimetallic porous hollow nanocrystals for efficient overall-water-splitting catalysis, Adv. Mater. 29 (2017) 1–8. https://doi.org/10.1002/adma.201703798

[21]   L. Fu, P. Cai, G. Cheng, W. Luo, Colloidal synthesis of iridium-iron nanoparticles for electrocatalytic oxygen evolution, Sustain. Energy Fuels. 1 (2017). https://doi.org/10.1039/C7SE00113D

[22]   F. Lv, J. Feng, K. Wang, Z. Dou, W. Zhang, J. Zhou, C. Yang, M. Luo, Y. Yang, Y. Li, P. Gao, S. Guo, Iridium-tungsten alloy nanodendrites as pH-universal water-splitting electrocatalysts, ACS Cent. Sci. 4 (2018) 1244–1252. https://doi.org/10.1021/acscentsci.8b00426

[23] J.E. Manders, L.T. Lam, R. De Marco, J.D. Douglas, R. Pillig, D.A.J. Rand, Battery performance enhancement with additions of bismuth, J. Power Sources. 48 (1994) 113–128. https://doi.org/10.1016/0378-7753(94)80012-X

[24] B. Monahov, D. Pavlov, D. Petrov, Influence of Ag as alloy additive on the oxygen evolution reaction on Pb/PbO_2 electrode, J. Power Sources. 85 (2000) 59–62. https://doi.org/10.1016/S0378-7753(99)00383-3

[25] W.S. Li, H.Y. Chen, X.M. Long, F.H. Wu, Y.M. Wu, J.H. Yan, C.R. Zhang, Oxygen evolution reaction on lead-bismuth alloys in sulfuric acid solution, J. Power Sources. 158 (2006) 902–907. https://doi.org/10.1016/j.jpowsour.2005.11.048

[26] Y. Matsumoto, E. Sato, Electrocatalytic properties of transition metal for oxygen evolution reaction, Mater. Chem. Phys. 14 (1986) 397–426. https://doi.org/10.1016/0254-0584(86)90045-3

[27] S. Park, Y. Shao, J. Liu, Y. Wang, Oxygen electrocatalysts for water electrolyzers and reversible fuel cells: Status and perspective, Energy Environ. Sci. 5 (2012) 9331–9344. https://doi.org/10.1039/c2ee22554a

[28] F. Song, L. Bai, A. Moysiadou, S. Lee, C. Hu, L. Liardet, X. Hu, Transition metal oxides as electrocatalysts for the oxygen evolution reaction in alkaline solutions: An application-inspired renaissance, J. Am. Chem. Soc. 140 (2018) 7748–7759. https://doi.org/10.1021/jacs.8b04546

[29] Y. Wang, T. Zhou, K. Jiang, P. Da, Z. Peng, J. Tang, B. Kong, W. Bin Cai, Z. Yang, G. Zheng, Reduced mesoporous Co_3O_4 nanowires as efficient water oxidation electrocatalysts and supercapacitor electrodes, Adv. Energy Mater. 4 (2014) 1–7. https://doi.org/10.1142/9789814513289_0001

[30] J. Yang, J.K. Cooper, F.M. Toma, K.A. Walczak, M. Favaro, J.W. Beeman, L.H. Hess, C. Wang, C. Zhu, S. Gul, J. Yano, C. Kisielowski, A. Schwartzberg, I.D. Sharp, A multifunctional biphasic water splitting catalyst tailored for integration with high-performance semiconductor photoanodes, Nat. Mater. 16 (2017) 335–341. https://doi.org/10.1038/nmat4794

[31] K. Fominykh, J.M. Feckl, J. Sicklinger, M. Döblinger, S. Böcklein, J. Ziegler, L. Peter, J. Rathousky, E.W. Scheidt, T. Bein, D. Fattakhova-Rohlfing, Ultrasmall dispersible crystalline nickel oxide nanoparticles as high-performance catalysts for electrochemical water splitting, Adv. Funct. Mater. 24 (2014) 3123–3129. https://doi.org/10.1002/adfm.201303600

[32]   M. Chen, Y. Wu, Y. Han, X. Lin, J. Sun, W. Zhang, R. Cao, An iron-based film for highly efficient electrocatalytic oxygen evolution from neutral aqueous solution, ACS Appl. Mater. Interfaces. 7 (2015) 21852–21859. https://doi.org/10.1021/acsami.5b06195

[33]   C.G. Morales-Guio, L. Liardet, X. Hu, Oxidatively electrodeposited thin-film transition metal (oxy)hydroxides as oxygen evolution catalysts, J. Am. Chem. Soc. 138 (2016) 8946–8957. https://doi.org/10.1021/jacs.6b05196

[34]   J. Du, F. Li, Y. Wang, Y. Zhu, L. Sun, Cu_3P/CuO core-shell nanorod arrays as high-performance electrocatalysts for water oxidation, Chem. Electro. Chem. 5 (2018) 2064–2068. https://doi.org/10.1002/celc.201800323

[35]   Z. Ye, T. Li, G. Ma, Y. Dong, X. Zhou, Metal-ion (Fe, V, Co, and Ni)-doped MnO_2 ultrathin nanosheets supported on carbon fiber paper for the oxygen evolution reaction, Adv. Funct. Mater. 27 (2017) 1–8. https://doi.org/10.1002/adfm.201704083

[36]   C. Hao, H. Lv, C. Mi, Y. Song, J. Ma, Investigation of mesoporous niobium-doped TiO_2 as an oxygen evolution catalyst support in an SPE water electrolyzer, ACS Sustain. Chem. Eng. 4 (2016) 746–756. https://doi.org/10.1021/acssuschemeng.5b00531

[37]   M.M. Najafpour, R. Mostafalu, M. Hołyńska, F. Ebrahimi, B. Kaboudin, Nano-sized Mn_3O_4 and β-MnOOH from the decomposition of β-cyclodextrin-Mn: 2. the water-oxidizing activities, J. Photochem. Photobiol. B Biol. 152 (2015) 112–118. https://doi.org/10.1016/j.jphotobiol.2015.02.011

[38]   M. Huynh, D.K. Bediako, D.G. Nocera, A functionally stable manganese oxide oxygen evolution catalyst in acid, J. Am. Chem. Soc. 136 (2014) 6002–6010. https://doi.org/10.1021/ja413147e

[39]   K.L. Pickrahn, S.W. Park, Y. Gorlin, H.B.R. Lee, T.F. Jaramillo, S.F. Bent, Active MnO_x electrocatalysts prepared by atomic layer deposition for oxygen evolution and oxygen reduction reactions, Adv. Energy Mater. 2 (2012) 1269–1277. https://doi.org/10.1002/aenm.201200230

[40]   A. Ramírez, P. Hillebrand, D. Stellmach, M.M. May, P. Bogdanoff, S. Fiechter, Evaluation of MnO_x, Mn_2O_3, and Mn_3O_4 electrodeposited films for the oxygen evolution reaction of water, J. Phys. Chem. C. 118 (2014) 14073–14081. https://doi.org/10.1021/jp500939d

[41] F. Jiao, H. Frei, Nanostructured manganese oxide clusters supported on mesoporous silica as efficient oxygen-evolving catalysts, Chem. Commun. 46 (2010) 2920–2922. https://doi.org/10.1039/b921820c

[42] X. Liu, C. Chen, Y. Zhao, B. Jia, A review on the synthesis of manganese oxide nanomaterials and their applications on lithium-ion batteries, J. Nanomater. 2013 (2013). https://doi.org/10.1155/2013/736375

[43] D. Bélanger, T. Brousse, J.W. Long, Manganese oxides: Battery materials make the leap to electrochemical capacitors, Electrochem. Soc. Interface. 17 (2008) 49–52.

[44] J. Masa, W. Xia, I. Sinev, A. Zhao, Z. Sun, S. Grützke, P. Weide, M. Muhler, W. Schuhmann, Mn_xO_y/NC and Co_xO_y/NC nanoparticles embedded in a nitrogen-doped carbon matrix for high-performance bifunctional oxygen electrodes, Angew. Chemie - Int. Ed. 53 (2014) 8508–8512. https://doi.org/10.1002/anie.201402710

[45] A. Chinnappan, D. Ji, C. Baskar, X. Qin, S. Ramakrishna, 3-Dimensional MWCNT /CuO nanostructures use as an electrochemical catalyst for oxygen evolution reaction, J. Alloys Compd. 735 (2018) 2311–2317. https://doi.org/10.1016/j.jallcom.2017.11.390

[46] N. Cheng, Y. Xue, Q. Liu, J. Tian, L. Zhang, A.M. Asiri, X. Sun, $Cu/(Cu(OH)_2$-CuO) core/shell nanorods array: in-situ growth and application as an efficient 3D oxygen evolution anode, Electrochim. Acta. 163 (2015) 102–106. https://doi.org/10.1016/j.electacta.2015.02.099

[47] T. Ma, J. Bai, Q. Wang, C. Li, The novel synthesis of a continuous tube with laminated g-C_3N_4 nanosheets for enhancing photocatalytic activity and oxygen evolution reaction performance, Dalton Trans. 4 (2018) 10240–10248. https://doi.org/10.1039/C8DT01898G

[48] M.P. Browne, C. Domínguez, P.E. Colavita, Emerging trends in metal oxide electrocatalysis: bifunctional oxygen catalysis, synergies and new insights from in situ studies, Curr. Opin. Electrochem. 7 (2018) 208–215. https://doi.org/10.1016/j.coelec.2017.09.012

[49] A. Pérez-tomás, A. Mingorance, D. Tanenbaum, Metal oxides in photovoltaics: all-oxide, ferroic, and perovskite solar cells, in: M.L. Cantu (Ed.), The future of semiconductor oxides in next-generation solar cells, Elsevier Inc., Amsterdam, 2018, pp. 267–356. https://doi.org/10.1016/B978-0-12-811165-9.00008-9

[50]   H. Zhang, Z. Zhang, N. Li, W. Yan, Z. Zhu, Cu_2O@C core/shell nanoparticle as an electrocatalyst for oxygen evolution reaction, J. Catal. 352 (2017) 239–245. https://doi.org/10.1016/j.jcat.2017.05.019

[51]   G. Wang, H. Wang, Y. Ling, Y. Tang, X. Yang, R.C. Fitzmorris, C. Wang, J.Z. Zhang, Y. Li, Hydrogen-treated TiO_2 nanowire arrays for photoelectrochemical water splitting, Nano Lett. 11 (2011) 3026–3033. https://doi.org/10.1021/nl201766h

[52]   C. Chen, W. Ma, J. Zhao, Semiconductor-mediated photodegradation of pollutants under visible-light irradiation, Chem. Soc. Rev. 39 (2010) 4206–4219. https://doi.org/10.1039/b921692h

[53]   N. Roy, Y. Sohn, K.T. Leung, D. Pradhan, Engineered electronic states of transition metal doped TiO_2 nanocrystals for low overpotential oxygen evolution reaction, J. Phys. Chem. 118 (2014) 29499-29506. https://doi.org/10.1021/jp508445t

[54]   L. Cai, I.S. Cho, M. Logar, A. Mehta, J. He, C.H. Lee, P.M. Rao, Y. Feng, J. Wilcox, F.B. Prinz, X. Zheng, Sol-flame synthesis of cobalt-doped TiO_2 nanowires with enhanced electrocatalytic activity for oxygen evolution reaction, Phys. Chem. Chem. Phys. 16 (2014) 12299–12306. https://doi.org/10.1039/C4CP01748J

[55]   H. Yoo, Y.W. Choi, J. Choi, Ruthenium oxide-doped TiO_2 nanotubes by single-step anodization for water-oxidation applications, Chem. Cat. Chem. 7 (2015) 643–647. https://doi.org/10.1002/cctc.201402787

[56]   J. Chattopadhyay, R. Srivastava, P.K. Srivastava, Ni-doped TiO_2 hollow spheres as electrocatalysts in water electrolysis for hydrogen and oxygen production, J. Appl. Electrochem. 43 (2013) 279–287. https://doi.org/10.1007/s10800-012-0509-y

[57]   N. Kim, E.M. Turner, Y. Kim, S. Ida, H. Hagiwara, T. Ishihara, E. Ertekin, Two-dimensional TiO_2 nanosheets for photo and electro-chemical oxidation of water: predictions of optimal dopant species from first-principles, J. Phys. Chem. C. 121 (2017) 19201–19208. https://doi.org/10.1021/acs.jpcc.7b04725

[58]   X. Li, M. Zhang, Y. Zhang, C. Yu, W. Qi, J. Cui, Y. Wang, Y. Qin, J. Liu, X. Shu, Y. Chen, T. Xie, Y. Wu, Controlled synthesis of MnO_2@TiO_2 hybrid nanotube arrays with enhanced oxygen evolution reaction performance, Int. J. Hydrogen Energy. 43 (2018) 14369–14378. https://doi.org/10.1016/j.ijhydene.2018.06.027

[59]   H. Yoo, M. Kim, Y.-T. Kim, K. Lee, J. Choi, Catalyst-doped anodic TiO_2 nanotubes: binder-free electrodes for (photo)electrochemical reactions, Catalysts. 8 (2018) 555. https://doi.org/10.3390/catal8110555

[60] F. Lu, M. Zhou, Y. Zhou, X. Zeng, First-row transition metal based catalysts for the oxygen evolution reaction under alkaline conditions: basic principles and recent advances, Small. 13 (2017) 1–18. https://doi.org/10.1002/smll.201701931

[61] G. Wu, W. Chen, X. Zheng, D. He, Y. Luo, X. Wang, J. Yang, Y. Wu, W. Yan, Z. Zhuang, X. Hong, Y. Li, Hierarchical Fe-doped NiO_x nanotubes assembled from ultrathin nanosheets containing trivalent nickel for oxygen evolution reaction, Nano Energy. 38 (2017) 167–174. https://doi.org/10.1016/j.nanoen.2017.05.044

[62] C. He, X. Kong, M. Jiang, X. Lei, Metal Ni-decorated Fe_3O_4 nanoparticles: A new and efficient electrocatalyst for oxygen evolution reaction, Mater. Lett. 222 (2018) 138–141. https://doi.org/10.1016/j.matlet.2018.03.142

[63] J.A. Koza, Z. He, A.S. Miller, J.A. Switzer, Electrodeposition of crystalline Co_3O_4-A catalyst for the oxygen evolution reaction, Chem. Mater. 24 (2012) 3567–3573. https://doi.org/10.1021/cm3012205

[64] F. Basharat, U.A. Rana, M. Shahid, M. Serwar, Heat treatment of electrodeposited NiO films for improved catalytic water oxidation, RSC Adv. 5 (2015) 86713–86722. https://doi.org/10.1039/C5RA17041A

[65] Y. Meng, W. Song, H. Huang, Z. Ren, S.Y. Chen, S.L. Suib, Structure-property relationship of bifunctional MnO_2 nanostructures: Highly efficient, ultra-stable electrochemical water oxidation and oxygen reduction reaction catalysts identified in alkaline media, J. Am. Chem. Soc. 136 (2014) 11452–11464. https://doi.org/10.1021/ja505186m

[66] Y. Deng, A.D. Handoko, Y. Du, S. Xi, B.S. Yeo, In situ raman spectroscopy of copper and copper oxide surfaces during electrochemical oxygen evolution reaction: identification of CuIII oxides as catalytically active species, ACS Catal. 6 (2016) 2473–2481. https://doi.org/10.1021/acscatal.6b00205

[67] D.M. Jang, I.H. Kwak, E.L. Kwon, C.S. Jung, H.S. Im, K. Park, J. Park, Transition-metal doping of oxide nanocrystals for enhanced catalytic oxygen evolution, J. Phys. Chem. C. 119 (2015) 1921–1927. https://doi.org/10.1021/jp511561k

[68] K. Elumeeva, J. Masa, J. Sierau, F. Tietz, M. Muhler, W. Schuhmann, Perovskite-based bifunctional electrocatalysts for oxygen evolution and oxygen reduction in alkaline electrolytes, Electrochim. Acta. 208 (2016) 25–32. https://doi.org/10.1016/j.electacta.2016.05.010

[69]   R.J.D. Tilley, Perovskites: structure-property relationships, first ed., Wiley, West Sussex, 2016. https://doi.org/10.1002/9781118935651

[70]   I. Borriello, G. Cantele, D. Ninno, Ab initio investigation of hybrid organic-inorganic perovskites based on tin halides, Phys. Rev. B - Condens. Matter Mater. Phys. 77 (2008). https://doi.org/10.1103/PhysRevB.77.235214

[71]   W.D. Kingery, H. K. Bowen, D.R. Uhlmann, Introduction to Ceramics, second ed., Wiley, New York, 1960. https://doi.org/10.1016/j.poly.2018.09.031

[72]   E. Omari, M. Omari, D. Barkat, Oxygen evolution reaction over copper and zinc co-doped $LaFeO_3$ perovskite oxides, Polyhedron. 156 (2018) 116–122.

[73]   P. Gao, M.Grätzel, M.K. Nazeeruddin, Organohalide lead perovskites for photovoltaic applications, Energy Environ. Sci. 7 (2014) 2448. https://doi.org/10.1039/C4EE00942H

[74]   H.S. Kushwaha, A. Halder, P. Thomas, R. Vaish, $CaCu_3Ti_4O_{12}$: abifunctional perovskite electrocatalyst for Oxygen Evolution and Reduction reaction in alkaline medium, Electrochim. Acta. 252 (2017) 532–540. https://doi.org/10.1016/j.electacta.2017.09.030

[75]   I. Yamada, A. Takamatsu, K. Asai, T. Shirakawa, H. Ohzuku, A. Seno, T. Uchimura, H. Fujii, S. Kawaguchi, K. Wada, H. Ikeno, S. Yagi, Systematic study of descriptors for Oxygen Evolution Reaction catalysis in perovskite oxides, J. Phys. Chem. C. 122 (2018) 27885–27892. https://doi.org/10.1021/acs.jpcc.8b09287

[76]   M. Wan, H. Zhu, S. Zhang, H. Jin, Y. Wen, Building block nanoparticles engineering induces multi-element perovskite hollow nanofibers structure evolution to trigger enhanced oxygen evolution, Electrochim. Acta. 279 (2018) 301–310. https://doi.org/10.1016/j.electacta.2018.05.077

[77]   H. Liu, J. Yu, J. Sunarso, C. Zhou, B. Liu, Y. Shen, Mixed protonic-electronic conducting perovskite oxide as a robust oxygen evolution reaction catalyst, Electrochim. Acta. 282 (2018) 324–330. https://doi.org/10.1016/j.electacta.2018.06.073

[78]   D.S. Bick, A. Kindsmüller, G. Staikov, F. Gunkel, D. Müller, T. Schneller, R. Waser, I. Valov, Stability and degradation of perovskite electrocatalysts, Electrochim. Acta. (2016). https://doi.org/10.1016/j.electacta.2016.09.116

[79]   Y. Bai, T. Siponkoski, J. Peräntie, H. Jantunen, J. Juuti, Ferroelectric, pyroelectric, and piezoelectric properties of a photovoltaic perovskite oxide, Appl. Phys. Lett. 110 (2017). https://doi.org/10.1063/1.4974735

[80]　B.E. Hayden, F.K. Rogers, Oxygen reduction and oxygen evolution on $SrTi_{1-x}Fe_xO_{3-y}$ (STFO) perovskite electrocatalysts, 819 (2018) 275–282. https://doi.org/10.1016/j.jelechem.2017.10.056

[81]　M.H. Seo, H.W. Park, D.U. Lee, M.G. Park, Z. Chen, Design of highly active perovskite oxides for Oxygen Evolution Reaction by combining experimental and ab Initio Studies, ACS Catal. 5 (2015)4337–4344. https://doi.org/10.1021/acscatal.5b00114

[82]　J. Kim, X. Yin, K. Tsao, S. Fang, H. Yang, $Ca_2Mn_2O_5$ as oxygen-deficientperovskite electrocatalyst for Oxygen Evolution Reaction, J. Am. Chem. Soc. 136 (2014) 14646–14649. https://doi.org/10.1021/ja506254g

[83]　J. Suntivich, K.J. May, H.A. Gasteiger, J.B. Goodenough, Y. Shao-Horn, A perovskite oxide optimized for Oxygen Evolution catalysis from molecular orbital principles, Science. 334 (2011) 1383–1385. https://doi.org/10.1126/science.1212858

[84]　J. Gracia, Spin dependent interactions catalyse the oxygen electrochemistry, Phys. Chem. Chem. Phys. 19 (2017) 20451–20456. https://doi.org/10.1039/C7CP04289B

[85]　W.T. Hong, M. Risch, K.A. Stoerzinger, A. Grimaud, Toward the rational design of non-precious transition metal oxides for oxygen electrocatalysis, Energy Env. Sci. 8 (2015) 1404–1427. https://doi.org/10.1039/C4EE03869J

[86]　Y. Zhou, S. Sun, J. Song, S. Xi, B. Chen, Y. Du, A.C. Fischer, F. Cheng, X. Wang, H. Zhang, Z. Xu, Enlarged Co-O covalency in octahedral sites leading to highly efficient spinel oxides for oxygen evolution reaction, Adv. Mater. 30 (2018) 1802912. https://doi.org/10.1002/adma.201802912

[87]　B. Han, M. Risch, Y. Lee, C. Ling, H. Jia, Y. Shao-horn, Activity and stability trends of perovskite oxides for oxygen evolution catalysis at neutral pH, Phys. Chem. Chem. Phys.17 (2015) 22576–22580. https://doi.org/10.1039/C5CP04248H

[88]　F.A. Kröger, H.J. Vink, Relations between the concentrations of imperfections in crystalline solids, Solid State Phys. - Adv. Res. Appl. 3 (1956) 307–435. https://doi.org/10.1016/S0081-1947(08)60135-6

[89]　A.I. Becerro, C. McCammon, F. Langenhorst, F. Seifert, R. Angel, Oxygen vacancy ordering in $CaTiO_3$-$CaFeO_{2.5}$perovskites: from isolated defects to infinite sheets, Phase Transitions. 69 (1999) 133–146. https://doi.org/10.1080/01411599908208014

[90]　A. Grimaud, K.J. May, C.E. Carlton, Y. Lee, M. Risch, W.T. Hong, J. Zhou, Y. Shao-　　horn, Double perovskites as a family of highly active catalysts for oxygen

evolution in alkaline solution, Nature Commun. 4 (2013) 2439.
https://doi.org/10.1038/ncomms3439

[91] J.H. Clark, M.S. Dyer, R.G. Palgrave, C.P. Ireland, J.R. Darwent, J.B. Claridge,
M.J. Rosseinsky, Visible light photo-oxidation of model pollutants using $CaCu_3Ti_4O_{12}$:
an experimental and theoretical study of optical properties, electronic structure, and
Selectivity, J. Am. Chem. Soc. 133 (2011) 1016–1032.
https://doi.org/10.1021/ja1090832

[92] S. Yagi, I. Yamada, H. Tsukasaki, A. Seno, M. Murakami, H. Fujii, H. Chen, N.
Umezawa, H. Abe, N. Nishiyama, S. Mori, Covalency-reinforced oxygen evolution
reaction catalyst, Nature Commun. 6 (2015) 8249.
https://doi.org/10.1038/ncomms9249

[93] C. Alegre, E. Modica, M. Rodlert-bacilieri, F.C. Mornaghini, T. Avanzate, E.
Nicola, S. Santa, Enhanced durability of a cost-effective perovskite-carbon catalyst for
the oxygen evolution and reduction reactions in alkaline environment, Int. J. Hydrog.
Energy 42 (2017) 28063–28069. https://doi.org/10.1016/j.ijhydene.2017.03.216

[94] J. Hu, Q. Liu, L. Shi, Z. Shi, H. Huang, Silver decorated $LaMnO_3$
nanorod/graphene composite electrocatalysts as reversible metal-air battery electrodes,
Appl. Surf. Sci. 402 (2017) 61–69. https://doi.org/10.1016/j.apsusc.2017.01.060

[95] D. Chen, M. Qiao, Y. Lu, L. Hao, D. Liu, C. Dong, Y. Li, S. Wang, Preferential
cation vacancies in perovskite hydroxide for the Oxygen Evolution Reaction, Angew.
Chemie - Int. Ed. 57 (2018) 8691–8696. https://doi.org/10.1002/anie.201805520

[96] H.S. Kushwaha, A. Halder, R. Vaish, Ferroelectric electrocatalysts: a new class of
materials for oxygen evolution reaction with synergistic effect of ferroelectric
polarization, J. Mat. Sci. 53 (2018) 1414–1423. https://doi.org/10.1007/s10853-017-
1611-7

[97] A. Kakekhani, S. Ismail-beigi, E.I. Altman, Ferroelectrics: A pathway to
switchable surface chemistry and catalysis, Surf. Sci. 650 (2015) 302–306.
https://doi.org/10.1016/j.susc.2015.10.055

[98] S. Park, C.W. Lee, M. Kang, S. Kim, H.J. Kim, J.E. Kwon, S. Y. Park, C. Kang,
K.S. Hong, K.T. Nam, A ferroelectric photocatalyst for enhancing hydrogen evolution:
polarized particulate suspension, Phys. Chem. Chem. Phys. 16 (2014) 10408–10413.
https://doi.org/10.1039/C4CP01267D

[99] N.K. Singh, B. Lal, R.N. Singh, Electrocatalytic properties of perovskite-type
$La_{1-x}Sr_xMnO_3$ obtained by a novel sol–gel route for O_2 evolution in KOH solutions,

Int. J. Hydrogen Energy. 27 (2002) 885–893. https://doi.org/10.1016/S0360-3199(02)00008-3

[100] J. Scholz, M. Risch, G. Wartner, C. Luderer, V. Roddatis, C. Jooss, Tailoring the Oxygen Evolution activity and stability using defect chemistry, Catalysts. 7 (2017) 139. https://doi.org/10.3390/catal7050139

[101] M.C. Weidman, D. V Esposito, Y. Hsu, J.G. Chen, Comparison of electrochemical stability of transition metal carbides (WC, W_2C, Mo_2C) over a wide pH range, J. Power Sources. 202 (2012) 11–17. https://doi.org/10.1016/j.jpowsour.2011.10.093

[102] Y. Liu, T.G. Kelly, J.G. Chen, W.E. Mustain, Metal carbides as alternative electrocatalyst supports, ACS Catal. 3 (2013) 1184–1194. https://doi.org/10.1021/cs4001249

[103] S. Zhou, G. Zhou, S. Jiang, P. Fan, H. Hou, Flexible and refractory tantalum carbide-carbon electrospun nanofibers with high modulus and electric conductivity, Mater. Lett. 200 (2017) 97–100. https://doi.org/10.1016/j.matlet.2017.04.115

[104] Y. Wang, J. Wang, G. Han, C. Du, Q. Deng, Y. Gao, G. Yin, Y. Song, Pt decorated Ti_3C_2MXene for enhanced methanol oxidation reaction, Ceram. Int. 45 (2018) 2411–2417. https://doi.org/10.1016/j.ceramint.2018.10.160

[105]D.J. Ham, J.S. Lee, Transition metal carbides and nitrides as electrode materials for low temperature fuel cells, Energies. 2 (2009) 873–899. https://doi.org/10.3390/en20400873

[106] F. Karimi, B.A. Peppley, Metal carbide and oxide supports for iridium-based Oxygen Evolution Reaction electrocatalysts for polymer-electrolyte-membrane water electrolysis, Electrochim. Acta. 246 (2017) 654–670. https://doi.org/10.1016/j.electacta.2017.06.048

[107] I.M. Petrushina, E. Christensen, K. Bouzek, C.B. Prag, J.E.T. Andersen, J. Polonsky, N.J. Bjerrum, Tantalum carbide as a novel support material for anode electrocatalysts in polymer electrolyte membrane water electrolysers, Int. J. Hydrogen Energy. 37 (2011) 2173–2181. https://doi.org/10.1016/j.ijhydene.2011.11.035

[108] Y. Nabil, S. Cavaliere, I.A. Harkness, D.J. Jones, Novel niobium carbide/carbon porous nanotube electrocatalyst supports for proton exchange membrane fuel cell cathodes, J. Power Sources. 363 (2017) 20–26. https://doi.org/10.1016/j.jpowsour.2017.07.058

[109] T. Li, Z. Tang, K. Wang, W. Wu, S. Chen, C. Wang, Palladium nanoparticles grown on B-Mo_2C nanotubes as dual functional electrocatalysts for both oxygen

reduction reaction and hydrogen evolution reaction, Int. J. Hydrogen Energy. 43
(2018) 4932–4941. https://doi.org/10.1016/j.ijhydene.2018.01.107

[110] S. Saha, J. Andrés, C. Rodas, S. Tan, D. Li, Performance evaluation of platinum-
molybdenum carbide nanocatalysts with ultralow platinum loading on anode and
cathode catalyst layers of proton exchange membrane fuel cells, J. Power Sources. 378
(2018) 742–749. https://doi.org/10.1016/j.jpowsour.2017.12.062

[111] Y. Liu, W.E. Mustain, Evaluation of tungsten carbide as the electrocatalyst support
for platinum hydrogen evolution/oxidation catalysts, Int. J. Hydrogen Energy. 37
(2012) 8929–8938. https://doi.org/10.1016/j.ijhydene.2012.03.044

[112] L. Liao, X. Bian, J. Xiao, B. Liu, M.D. Scanlon, H.H. Girault, Nanoporous
molybdenum carbide wires as an active electrocatalyst towards the oxygen reduction
reaction, Phys. Chem. Chem. Phys. 16 (2014) 10088–10094.
https://doi.org/10.1039/C3CP54754J

[113] M.A.R. Anjum, M.H. Lee, J.S. Lee, Boron- and nitrogen-codopedmolybdenum
carbide nanoparticles imbedded in a BCN network as a bifunctional electrocatalyst for
hydrogen and oxygen evolution reactions, ACS Catal. 8 (2018) 8296.
https://doi.org/10.1021/acscatal.8b01794

[114] T.Y. Ma, J.L. Cao, M. Jaroniec, S.Z. Qiao, Interacting carbon nitride and titanium
carbide nanosheets for high-performance oxygen evolution, Angew. Chemie - Int. Ed.
55 (2016) 1138–1142. https://doi.org/10.1002/anie.201509758

[115] H. Cui, G. Zhu, X. Liu, F. Liu, Y. Xie, C. Yang, T. Lin, H. Gu, F. Huang, Niobium
nitride Nb_4N_5as a new high-performance electrode material for supercapacitors, Adv.
Sci. 2 (2015) 150026. https://doi.org/10.1002/advs.201500126

[116] E. Davari, D.G. Ivey, Synthesis and electrochemical performance of manganese
nitride as an oxygen reduction and oxygen evolution catalyst for zinc–air secondary
batteries, J. Appl. Electrochem. 47 (2017) 815–827. https://doi.org/10.1007/s10800-
017-1084-z

[117] M.S. Balogun, Y. Huang, W. Qiu, H. Yang, H. Ji, Y. Tong, Updates on the
development of nanostructured transition metal nitrides for electrochemical energy
storage and water splitting, Mater. Today. 20 (2017) 425–451.
https://doi.org/10.1016/j.mattod.2017.03.019

[118] K. Xu, P. Chen, X. Li, Y. Tong, H. Ding, X. Wu, W. Chu, Z. Peng, C. Wu, Y. Xie,
Metallic nickel nitride nanosheets realizing enhanced electrochemical water oxidation,
J. Am. Chem. Soc. 137 (2015) 4119–4125. https://doi.org/10.1021/ja5119495

[119] X. Jia, Y. Zhao, G. Chen, L. Shang, R. Shi, X. Kang, G.I.N. Waterhouse, L. Wu, C. Tung, T. Zhang, Ni_3FeN nanoparticles derived from ultrathin NiFe-layered double hydroxide nanosheets :an efficient overall water splitting electrocatalyst, Adv. Energy Mater. 6 (2016) 1502585. https://doi.org/10.1002/aenm.201502585

[120] Y. Zhang, B. Ouyang, J. Xu, G. Jia, S. Chen, R.S. Rawat, H.J. Fan, Rapid synthesis of cobalt nitride nanowires: highly efficient and low-cost catalysts for Oxygen Evolution, Angew. Chemie - Int. Ed. 637616 (2016) 8670–8674. https://doi.org/10.1002/anie.201604372

[121] P. Chen, K. Xu, Y. Tong, X. Li, S. Tao, Z. Fang, W. Chu, X. Wu, C. Wu, Cobalt nitrides as a class of metallic electrocatalysts for the oxygen evolution reaction, Inorg. Chem. Front. 3 (2016) 236–242. https://doi.org/10.1039/C5QI00197H

[122] X. Chen, P. Gao, H. Liu, J. Xu, B. Zhang, Y. Zhang, Y. Tang, C. Xiao, In situ growth of iron-nickel nitrides on carbon nanotubes with enhanced stability and activity for oxygen evolution reaction, Electrochim. Acta. 267 (2018) 8–14. https://doi.org/10.1016/j.electacta.2018.01.192

[123] G. Li, K. Li, L. Yang, J. Chang, R. Ma, Z. Wu, J. Ge, Boosted performance of Ir species by employing TiN as the support toward Oxygen Evolution Reaction, ACS Appl. Mater. Interfaces. 10 (2018) 38117–38124. https://doi.org/10.1021/acsami.8b14172

[124] L. Peng, S. Shoaib, A. Shah, Z. Wei, Recent developments in metal phosphide and sulfide electrocatalysts for oxygen evolution reaction, Chinese J. Catal. 39 (2018) 1575–1593. https://doi.org/10.1016/S1872-2067(18)63130-4

[125] W. Li, X. Gao, D. Xiong, F. Xia, J. Liu, W. Song, J. Xu, S.M. Thalluri, M.F. Cerqueira, X. Fu, L. Liu, Vapor–solid synthesis of monolithic single- crystalline CoP nanowire electrodes for efficient and robust water electrolysis, Chem. Sci. 8 (2017) 2952–2958. https://doi.org/10.1039/C6SC05167G

[126] Z. Jin, P. Li, D. Xiao, Metallic Co_2P ultrathin nanowires distinguished from CoP as robust electrocatalysts for overall, Green Chem. 18 (2016) 1459–1464. https://doi.org/10.1039/C5GC02462E

[127] A. Dutta, A.K. Samantara, S.K. Dutta, B.K. Jena, N. Pradhan, Surface-oxidized dicobaltphosphide nanoneedles as a nonprecious, durable, and efficient OER catalyst, ACS Energy Lett. 1 (2016) 169–174. https://doi.org/10.1021/acsenergylett.6b00144

[128] C. Hou, S. Cao, W. Fu, Y. Chen, Ultrafine CoP nanoparticles supported on carbon nanotubes as highly active electrocatalyst for both Oxygen and Hydrogen Evolution in

basic media, ACS Appl. Mater. Interfaces. 7 (2015) 28412–28419.
https://doi.org/10.1021/acsami.5b09207

[129] M. Jiang, J. Li, X. Cai, Y. Zhao, L. Pan, Q. Cao, D. Wang, Y. Du, Ultrafine bimetallic phosphide nanoparticles embedded in carbon nanosheets: two-dimensional metal–organic framework-derived non-noble electrocatalysts for the highly efficient oxygen evolution reaction, Nanoscale. 10 (2018) 19774–19780.
https://doi.org/10.1039/C8NR05659E

[130] D. Xiong, X. Wang, W. Li, L. Liu, Facile synthesis of iron phosphide nanorods for efficient and durable electrochemical oxygen evolution, Chem. Commun. 52 (2016) 8711–8714. https://doi.org/10.1039/C6CC04151E

[131] P. He, X. Yu, X. Wen, D. Lou, Carbon-incorporated nickel–cobalt mixed metal phosphide nanoboxeswith enhanced electrocatalytic activity for oxygen evolution, Angew. Chemie - Int. Ed. 56 (2017) 3897–3900.
https://doi.org/10.1002/anie.201612635

[132] C. Zhang, Y. Xie, H. Deng, C. Zhang, J. Su, Y. Dong, J. Lin, Ternary nickel iron phosphide supported on nickel foam as a high-efficiency electrocatalyst for overall water splitting, Int. J. Hydrogen Energy. 43 (2018) 7299–7306.
https://doi.org/10.1016/j.ijhydene.2018.02.157

[133] G. Liu, D. He, R. Yao, Y. Zhao, J. Li, Enhancing the water oxidation activity of Ni$_2$P nanocatalysts by iron-doping and electrochemical activation, Electrochim. Acta. 253 (2017) 498–505. https://doi.org/10.1016/j.electacta.2017.09.057

[134] H. Man, C. Tsang, M.M. Li, J. Mo, B. Huang, L. Yoon, S. Lee, Y. Leung, K. Wong, S. Chi, E. Tsang, Tailored transition metal-doped nickel phosphide nanoparticles for the electrochemical oxygen evolution reaction (OER), Chem. Commun. 54 (2018) 8630–8633. https://doi.org/10.1039/C8CC03870H

[135] C. Hou, Q. Chen, C. Wang, F. Liang, Z. Lin, W. Fu, Y. Chen, Self-supported cedar like semimetallic Cu$_3$P nanoarrays as a 3D High-Performance Janus electrode for both Oxygen and Hydrogen Evolution under basic conditions, ACS Appl. Mater. Interfaces. 8 (2016) 23037–23048. https://doi.org/10.1021/acsami.6b06251

[136] S.M. Pawar, B.S. Pawar, P.T. Babar, A. Talha, A. Ahmed, H.S. Chavan, Y. Jo, S. Cho, J. Kim, A.I. Inamdar, J. Hyeok, H. Kim, H. Im, Electrosynthesis of copper phosphide thin films for efficient water oxidation, Mater. Lett. 241 (2019) 243–247.
https://doi.org/10.1016/j.matlet.2019.01.118

Chapter 2

Fe-Based Electrocatalysts for Oxygen-Evolution Reaction

Yanyan Yang[1,2], Shasha Li[1], Xuli Ma[1], Xiaogang Hao[1], Abuliti Abudula[2], Guoqing Guan[2]*

[1] Department of Chemical Engineering, Taiyuan University of Technology, Taiyuan 030024, China

[2] Energy Conversion Engineering Laboratory, Institute of Regional Innovation (IRI), Hirosaki University, 2-1-3, Matsubara, Aomori 030-0813, Japan

* guan@hirosaki-u.ac.jp

Abstract

Iron is the most common element by mass on the earth, and iron-based catalysts have attracted remarkable attention on account of their low cost as well as high activity. In the water electrolysis process, besides conventional precious metal-based catalysts, iron (Fe)-based materials are also becoming most promising catalysts for oxygen evolution reaction (OER). In this chapter, the states of the art of main Fe-based electrocatalysts, including their preparation methods, performances, and the strategies for the activity improvement are reviewed. In addition, the catalytic mechanisms of Fe-based electrocatalysts are also analyzed for giving insight into the intrinsic active catalytic sites. Finally, future research on how to enhance the performance of Fe-based OER electrocatalysts is discussed.

Keywords

Oxygen Evolution Reaction, Fe-Based Catalysts, Electrocatalysis, Water Splitting, Mechanism

Contents

1.   Introduction

Electrochemical water splitting for the generation of hydrogen and oxygen could be a promising strategy to store unstable power such as wind, solar, and tide and surplus power [1]. In the water splitting process, the sluggish oxygen evolution reaction (OER) determines the efficiency of water electrolysis owing to the multi-step four electron-proton transfer property [2, 3]. To date, it is still a crucial issue to explore effective, low-cost OER electrocatalysts which can replace the noble metal based eletrocatalysts (such as IrO_2 and RuO_2) to satisfy the economical production of hydrogen from water splitting [3-8].

Recently, a large amount of the transition metal-based electrocatalysts has been reported for electrochemical water oxidation in the literature. Among them, Fe-based materials attract special attention since iron is the most common element by mass on the earth, and Fe-based electrocatalysts always exhibit high activity [9-11]. Especially, the rich redox property of an iron element, which is beneficial for the oxygen activation, makes Fe-based materials active for water oxidation [12]. However, compared with cobalt (Co) and nickel (Ni)-based electrocatalysts, Fe-based ones are less explored electrocatalysts for water oxidation [13,14]. One possible reason is their worse electrical conductivity [10,11,15,16]. To date, various approaches have been considered to improve the performances of the Fe-based electrocatalysts. For example, the conductivity and charge transport can be enhanced via immobilizing Fe-based electrocatalysts in highly conductive substrates or doping with other elements [17,18]. In this chapter, the oxygen evolution reaction mechanism is introduced at first and then, the recent progress on the development of main Fe-based OER electrocatalysts including Fe oxides, Fe-(oxy)hydroxides, Fe-based layered double hydroxides (LDHs) and other Fe-based composites, are briefly summarized. Finally, future research on how to enhance the performance of Fe-based OER electrocatalysts is discussed.

Electrochemical Water Splitting: Materials and Applications   Materials Research Forum LLC
Materials Research Foundations **59** (2019) 37-58   doi: https://doi.org/10.21741/9781644900451-2

2. Mechanism of oxygen evolution reaction

The mechanisms of OER on various electrocatalysts should be different, and until now, the reaction routes on the surface of catalysts still remain disputed due to the complexity [7]. In general, the OER reaction involves four proton-electron coupled processes, and the possible reactions occur in acidic and alkaline solutions and can be expressed as follows:

In acidic media,

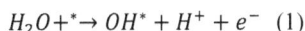

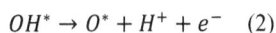

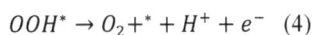

$$H_2O + ^* \rightarrow OH^* + H^+ + e^- \quad (1)$$

$$OH^* \rightarrow O^* + H^+ + e^- \quad (2)$$

$$O^* + H_2O \rightarrow OOH^* + H^+ + e^- \quad (3)$$

$$OOH^* \rightarrow O_2 + ^* + H^+ + e^- \quad (4)$$

In alkaline media,

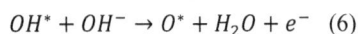

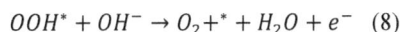

$$OH^- + ^* \rightarrow OH^* + e^- \quad (5)$$

$$OH^* + OH^- \rightarrow O^* + H_2O + e^- \quad (6)$$

$$O^* + OH^- \rightarrow OOH^* + e^- \quad (7)$$

$$OOH^* + OH^- \rightarrow O_2 + ^* + H_2O + e^- \quad (8)$$

where * means the active site of catalyst and adsorbed intermediates of O, OH and OOH.

According to the above basic reaction equations, the OER could occur within four steps. The initial step is the formation of OH^* accompanied with the first electron transfer. The second and third ones are transformations of OH^* to O^*, and then to OOH^*. The last step is the generation of O_2. Herein, each step discharges one electron.

The overpotential of OER is determined by the reaction Gibbs free energies (ΔG_i, i=1, 2, 3, 4.) of all four steps. Theoretically, the ΔG of each elemental reaction is equal to 1.23 eV vs reversible hydrogen electrode (RHE) [7,19]. However, as demonstrated by Exner et al. [20] for RuO_2 (110) in the third step, a large Gibbs energy change of 0.5 eV was found. Therefore, the overpotential of OER can be defined as [19]:

$$\eta = \frac{\max(\Delta G_i)}{e} - 1.23 \quad (9)$$

Tafel slope is another important parameter for evaluating the electrocatalytic activity. It can be obtained from the Tafel equation:

$$\eta = a + b(j) \quad (10)$$

in which, b: Tafel slope; j: current density.

Electrochemical Water Splitting: Materials and Applications   Materials Research Forum LLC
Materials Research Foundations **59** (2019) 37-58   doi: https://doi.org/10.21741/9781644900451-2

For Fe-based electrocatalysts, it is assumed that the valence state change of iron is involved in the OER mechanism. Zhang et al. [21] described the oxygen evolution reaction over the iron oxide in the alkaline electrolyte as follows:

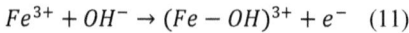

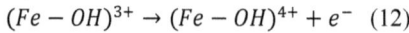

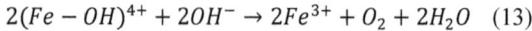

$$Fe^{3+} + OH^- \rightarrow (Fe - OH)^{3+} + e^- \quad (11)$$

$$(Fe - OH)^{3+} \rightarrow (Fe - OH)^{4+} + e^- \quad (12)$$

$$2(Fe - OH)^{4+} + 2OH^- \rightarrow 2Fe^{3+} + O_2 + 2H_2O \quad (13)$$

Enman et al. [22] demonstrated that the partial Fe oxidation and the shortening of Fe-O bond length occurred during the oxygen evolution over $Co(Fe)O_xH_y$ based on operando X-ray absorption spectroscopy (XAS) as well as density functional theory (DFT) calculations. Meanwhile, they proposed that the oxidized Fe plays a key role in the OER over $Co(Fe)O_xH_y$ depended on the density functional theory + U(DFT+U)calculation results. Herein, when the OER is occurring on $Co(Fe)O_xH_y$, Fe^{3+} is further oxidized, but Co^{3+} is not [14]. Many recently theoretical and experimental studies further indicate that Fe may be the active site in Fe-based electrocatalysts [13,22-27]. However, this consideration remains debatable. For example, Chen et al. [28] thought that although the observed Fe^{4+} species had an important role during the water oxidation over NiFe-oxide electrocatalysts, they could not be kinetically qualified to act as the active sites.

3.   Fe-based catalysts for OER

3.1   Fe-based oxides catalysts

Iron oxides (such as Fe_2O_3 and Fe_3O_4) usually show very poor catalytic activity for OER due to their low conductivity, although their redox potentials are much lower than the theoretical potential of OER [29].To solve this problem, hybridizing iron oxides with other conductive materials or building three-dimensional (3D) heterostructures have been considered. Bandal et al. [29] reported that iron(III) oxide (Fe_2O_3) hollow nanorods decorated on carbon nanotubes (Fe_2O_3/CNTs) showed higher catalytic activity than the pure Fe_2O_3, CNTs, and their physical mixture. It is considered that the combination with the CNTs could reduce the electrical resistance of the composite. Yang et al.[17]   also fabricated a composite(Fe_3O_4@Co_9S_8/rGO) made of iron(II) iron(III) oxide (Fe_3O_4), cobalt sulfide (Co_9S_8) and reduced graphene oxide(rGO), and revealed that synergistic effect among Fe, Co-based species and rGO nanosheets with excellent conductivity improved the catalytic activity. Recently, owing to the high surface area, rapid electron and ion transfer, Fe-based electrocatalysts with a rational 3D heterostructure have also attracted numerous attentions. For example, Cheng et al. [30] prepared a 3D ternary Fe_2O_3@Ni_2P/Ni(PO$_3$)$_2$hybrid. As shown in Fig 1, this 3D hybrid catalyst exhibited an excellent activity (with potentials of 1.57 V@500 mA/cm^2 and 1.60 V @1000 mA/cm^2, respectively) due to the synergetic effect between the Fe_2O_3 and $Ni_2P/Ni(PO_3)_2$.

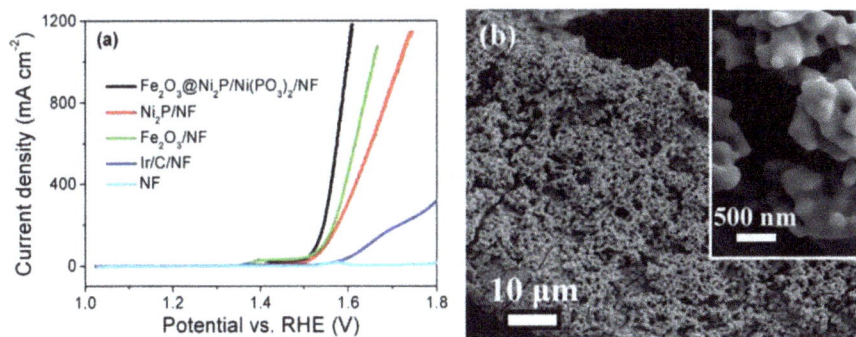

Fig. 1: *(a) Polarization curves of 3D $Fe_2O_3@Ni_2P/Ni(PO_3)_2/NF$, Ni_2P/NF, Fe_2O_3/NF, Ir/C/NF, and NF. (b) FESEM image of 3D $Fe_2O_3@Ni_2P/Ni(PO_3)_2/NF$, inset: the enlarged FESEM image of $Fe_2O_3@Ni_2P/Ni(PO_3)_2/NF$. Reprinted with permission from ref. 30, copyright 2019, The Royal Society of Chemistry.*

Incorporating additional cation (e.g., Ni, Co, and Mn) to form mixed metal oxides is another efficient strategy to improve the catalytic performance of iron oxide based electrocatalysts. As reported by Louie et al. [31] the OER current density of Ni-Fe mixed oxide containing 60% Ni performed about three- and two-orders of the magnitude of sole Fe and Ni films, respectively.

Dong et al. [32] fabricated a Ni-Fe-O mesoporous nanowire composite and found that this composite exhibited excellent OER catalytic property compared to IrO_2 in basic electrolyte. Moreover, they constructed an alkaline water electrolyzer with Ni-Fe-O material in both anode and cathode, which displayed favorable electrolysis performance (1.64 V@10 mA/cm^2) [32]. Also, in the case of Co-Fe mixed oxide, Bandal et al. [33] compared electrochemical property of FeCoO deposited on nickel foam (NF) with CoO-NF and RuO_2 electrocatalysts, and found that for attaining a high current density (i.e.,50 mA/cm^{2}or 100 mA/cm^2), the applied potential of FeCoO-NF was much lower than those of CoO-NF and RuO_2.

3.2   Fe-based (oxy)hydroxides catalysts

Ferric oxyhydroxide (FeOOH) is also considered as a promising electrocatalyst for OER [34]. For instance, Luo et al. [9] prepared a well-crystallized α-FeOOH film via chemical bath deposition method. After rapid heat treatment, the α-FeOOH revealed electrocatalytic activity as the amorphous γ-FeOOH with good stability. However, even after calcination at 300 °C for 5 min, an onset potential of α-FeOOH was about 1.6 V_{RHE}@2.5 mA/cm^2. Zou and coworkers [25] systematically investigated the effects of deposition method, film thickness, and substrate on the OER catalysis over Fe (oxy)hydroxide in the basic electrolyte, and found that Fe played an important role in catalysis of OER process. Furthermore, Fe incorporation increased the catalytic performances of NiOOH and CoOOH. In addition, they indicated that the intrinsic activity of FeOOH itself was also high. In their previous work, they also sequenced the activity of various metal (oxy)hydroxide as $NiFeO_xH_y> CoFeO_xH_y> FeO_xH_y> CoO_xH_y> NiO_xH_y> MnO_xH_y$ [27].

Similar to the iron oxide catalysts, the apparent activity of pure FeO_xH_y is also restricted by its poor electrical conductivity, and combining it with the conductive additives is also an attractive method to improve the electrical conductivity. Recently, many efforts have been contributed to controllably hybridizing the FeO_xH_y with the conductive materials to form diverse structures, e.g., core-shell [11], nanotube arrays[10, 34] and vertically aligned nanosheet arrays [35]. These studies confirmed that the ordered structure could have a critical influence on the property of FeO_xH_y-based electrocatalysts. Fig. 2 shows a typical example, by which the vertically aligned FeOOH/NiFe-LDHs-NF nanostructure exhibited the best catalytic activity (η_{10}=208 mV, η_{100}=247 mV), owing to the enhanced electrochemical surface area and rapid ion-electron transfer brought by this structure [35].

In general, Fe-containing mixed-metal (oxy)hydroxides, such as $CoFeO_xH_y$ and $NiFeO_xH_y$, always have higher intrinsic OER activity than the pure Co-(oxy)hydroxides and Ni-(oxy)hydroxides. It is reported that Fe-modified NiOOH has at least a hundred folds activity for OER as the pure NiOOH. Even in near-neutral borate buffer solutions, $NiFeO_xH_y$ [36] and $CoFeO_xH_y$ [27] also exhibit much higher activity than NiO_xH_y and CoO_xH_y, respectively. Fig. 3 shows the catalytic mechanism of OER over mixed metal (oxy)hydroxides [26,27], in which OER occurs on the Fe sites and meanwhile, the NiOOH host guarantees the electronic transport and the chemical stabilization of Fe [22]. Friebel et al. [23] provided further evidence for this hypothesis. Operando X-ray absorption spectroscopy results of $Ni_{1-x}Fe_xOOH$ indicate that the octahedral sites are occupied by Fe^{3+} with short Fe-O bond in (oxy)hydroxide structure. From the density functional theory (DFT) calculations, they found that this structural motif brings optimal adsorption energy for the OER intermediates. These results are strongly consistent with the hypothesis that Fe serves as the active site in $NiFeO_xH_y$. After that, for $CoFeO_xH_y$, Enman et al. [14] also used operando X-ray absorption spectroscopy and DFT methods to provide a clear evidence for Fe acting as the active site. However, the catalytic mechanism for Fe-containing (oxy)hydroxides is still disputed. More efforts should be still devoted to unraveling the intrinsic mechanisms for developing highly efficient OER electrocatalysts.

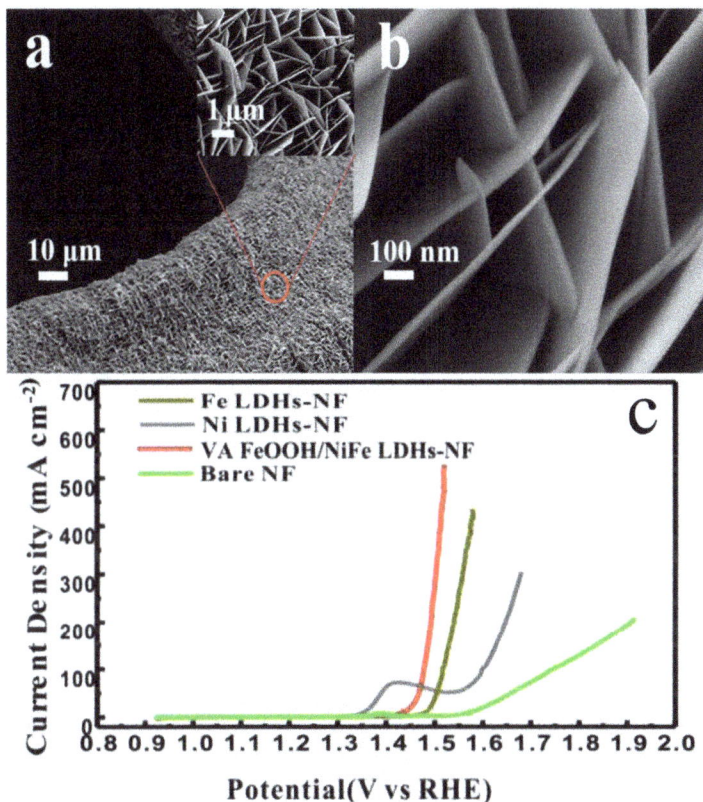

Fig. 2: *(a, b) SEM images of VA FeOOH/NiFe LDHs-NF with different magnifications (inset in a: higher magnification); (c) Polarization curves of VA FeOOH/NiFe LDHs-NF, Ni LDHs-NF, Fe LDHs-NF, and Bare NF for OER with iR-correction. Reprinted with permission from ref. 35, copyright 2016, American Chemical Society.*

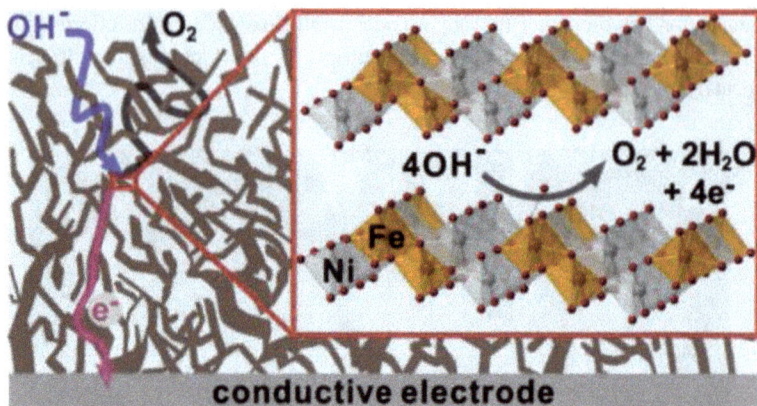

Fig. 3: *Schematic representation of mechanism for the OER on mixed-metal (oxy)hydroxides. Reprinted with permission from ref. 25, copyright 2015, American Chemical Society.*

3.3   Fe-based lamellar layered double hydroxide catalysts

Lamellar layered double hydroxide (LDH) consists of positively charged layers and charge-balancing interlayer anions. Herein, the positively charged layers are always constructed by divalent or monovalent cations (e.g., Li^+, Ni^{2+}, Mg^{2+}, Co^{2+}, Cu^{2+}, Zn^{2+}) and trivalent cations (e.g., Al^{3+}, Co^{3+}, Fe^{3+}, Mn^{3+}, Cr^{3+}).Owing to the chemical versatility, layered open structure, and intrinsically high activity, LDHs are considered as promising candidates for electrochemical water splitting [7,37]. In particular, iron containing LDHs generally exhibited excellent activity for OER [19]. It is found that NiFe LDH has the best activity among various LDHs by comparing OER performances of NiFe LDHs with other LDHs [38,39].

However, the low conductivity and limited specific surface area of LDHs restricted their practical applications as the OER catalysts. To solve these problems, it is proposed to rebuild the 2D nanoarray to the 3D architecture or assist with the conductive materials. Ma et al. [40] prepared NiFe-LDH nanosheets on the $Cu(OH)_2$ nanorod arrays for efficient exposure of more active sites to enhance the catalytic activity. Compared to the pure LDH and $Cu(OH)_2$, this core-shell nanorod array coated electrode showed higher catalytic activity. Furthermore, they demonstrated that the morphologies of $Cu(OH)_2$@NiFe-LDH nanorod array are different by using different preparation methods, which can influence the catalytic activity [40]. Yu et al. [41] designed a Ni-Fe LDH on

3D electrochemically reduced graphene oxide (3D-ErGO) composite with a larger electrochemical active surface area, which exhibited a lower overpotential (259 mV) and a smaller Tafel slope (39 mV/dec) than NiFe-LDH and 3D-ErGO.

Intercalation or exfoliation of NiFe-LDHs could further enhance the intrinsic activity on account of the increase of active edge sites and enlargement of surface area. As shown in Fig. 4, our research group developed an *in-situ* intercalation method for enlarging the interlayer spacing of electrodeposited NiFe-LDH. It can be obviously found that after immersing the NiFe-LDH in formamide at 80 ℃ for 3 h, the interlayer distance of NiFe-LDH increased from 0.78 to 0.95 nm, while the overpotential @10 mA/cm^2 current density (η_{10}) decreased from 256 to 210 mV. In addition, with the aid of ultrasound, the required time for expanding interlayer reduced to a great extent. As a result, the η_{10} further decreased to 203 mV [37]. Sirisomboonchai et al. [42] also fabricated NiO@NiFe-LDH composite with can enlarged interlayer distance of LDH by means of ultrasound. Even at high current densities (such as 50 and 100 mA/cm^2), the obtained composite can still perform a superior long-term stability.

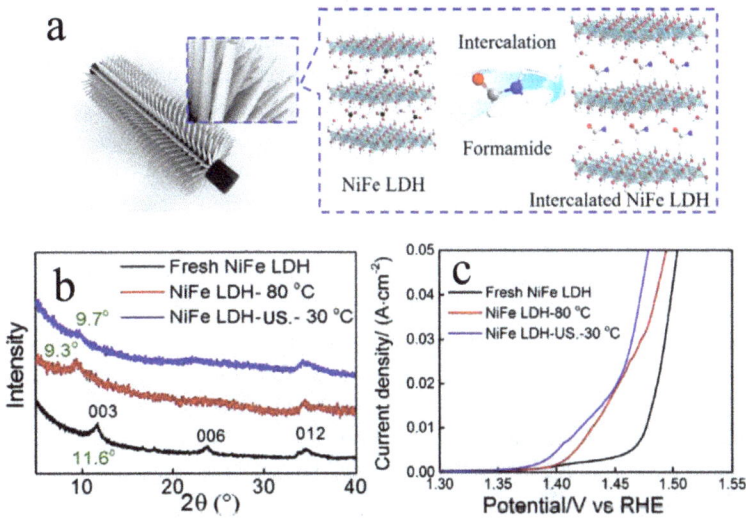

Fig. 4: (a)Schematic representation of in-situ intercalation process over NiFe LDH nanosheets; (b)XRD patterns of as-prepared NiFe LDH and intercalated NiFe LDHs at 80 °C and 30 °C with ultrasound; (c) Polarization curves (iR corrected) of as-prepared and intercalated NiFe LDHs. Reprinted with permission from ref. 37, copyright 2017, Elsevier.

Electrochemical Water Splitting: Materials and Applications   Materials Research Forum LLC
Materials Research Foundations **59** (2019) 37-58   doi: https://doi.org/10.21741/9781644900451-2

Other Fe-based LDHs OER catalysts, such as CoFe-LDH [38,39,43,44], LiFe-LDH [38], and NiFeAl [45], NiFeMn [46] and NiFeCo [47] ternary LDHs, also got extensively investigated in alkaline electrolytes. Particularly, preparation of ternary LDHs provided a strategy of the third transition metal doping to improve the OER property of LDH. For example, Lu et al. [46] fabricated a ternary NiFeMn-LDH, which exhibited superior OER property to NiFe-LDH and even commercial Ir/C catalysts. It is understood that this reinforced OER activity could be ascribed to the doping of Mn^{4+} into the laminate, which can facilitate modulating the electronic structure and improve the conductivity.

3.4   Other Fe-based composites

Recently, perovskite-type ABO_3 oxides (A: rare earth or alkaline earth element, B: transition metal element) are attracting special interest as an OER electrocatalyst in basic media due to their promising variable structures together with high intrinsic activities [7,48,49]. For Fe-based perovskite-type ABO_3 oxides, Zhu et al. [50] synthesized a series $La_{1-x}FeO_{3-\delta}$ (x=0.02, 0.05, 0.1) perovskite catalysts by a simply introducing A-site cation deficiency strategy. Among them, $La_{0.95}FeO_{3-\delta}$ showed the best OER activity. Herein, the enhanced OER activity compared with pristine $LaFeO_3$ should be attributed to Fe^{4+} species and oxygen vacancies in $La_{1-x}FeO_{3-\delta}$. Furthermore, the synergistic effect of various cations in the perovskite-type ABO_3 oxides could efficiently improve the electrocatalytic properties. Gui et al. [51] developed a Fe-Ni coupled perovskite family ($Pr_{0.5}Ba_{0.5}Fe_{1-x}Ni_xO_{3-\delta}$), in which the obtained $Pr_{0.5}Ba_{0.5}Fe_{0.75}Ni_{0.25}O_{3-\delta}$ had a lower overpotential (440 mV) than those of the pristine $Pr_{0.5}Ba_{0.5}FeO_{3-\delta}$ and IrO_2 electrocatalysts.

Besides, spinel compounds AB_2O_4 (A, B: metal ions) have good electric conductivity and stability in the alkaline solution, and they are also considered as promising candidates for OER [52]. Especially, Fe-based spinel oxides have identified to have tri-functional electrocatalytic activities in hydrogen evolution reaction (HER), oxygen reduction reaction(ORR) and OER [53]. For OER, Li et al. [54] found that the electrocatalytic activities of MFe_2O_4 are under a sequence of $CoFe_2O_4> CuFe_2O_4> NiFe_2O_4> MnFe_2O_4$. In addition, Silva et al. [55] fabricated 1D hollow MFe_2O_4 (M: Cu, Co, and Ni) fibers by the solution blow spinning method, and revealed the activities in the order of $CuFe_2O_4> CoFe_2O_4> NiFe_2O_4$.

In addition, other Fe-based non-oxide catalysts, such as carbides [56,57], nitrides [58], sulfides [18,24,59], phosphides [60, 61], selenides [2,62], have also been widely studied as the OER electrocatalyst. For these electrocatalysts, since their metallic nature can accelerate electron transport, they have high intrinsic electronic conductivity, which is beneficial for OER catalytic property [5]. Table 1 summarized the performances of various Fe-based OER catalysts.

Table 1. *Performances of various Fe-based OER catalysts.*

Catalyst	Electrolyte (KOH) Concentration	Substrate	Overpotential(s) [mV]	Tafel slope [mV/dec]	Ref.
Oxides					
Fe_2O_3/CNT	1 M	glassy carbon	η_{10}=410	62	[29]
Fe_2O_3	1 M	glassy carbon	η_{10}>850	126.39	[29]
Ni/Fe_3O_4	1 M		η_{20}=275	70	[15]
Fe-TiO_x	0.5 M (H_2SO_4)	Ti foam	η_1=260	126.2	[63]
$FeSO_y$	1 M	carbon cloth	η_{10}=370; η_{100}=481	93	[64]
Mn-Fe oxide	0.1 M	carbon cloth	$\eta_{11.5}$=770	80	[65]
Ni-Fe oxide	0.1 M	ITO	η_{10}=328; η_{100}= 420	42	[4]
Ni-Fe-O	1 M		η_{10}=244	39	[32]
Mo-Ni-Fe oxide	1 M	glassy carbon	η_{10}=231	39	[66]
Fe_3O_4@Co_9S_8/rGO	1 M		η_{10}=320	54.1	[17]
Fe_2O_3@Ni_2P/Ni(PO_3)$_2$	1 M	Ni foam	η_{50}=340; η_{100}=370	48.2	[30]
FeCoO	1 M	Ni foam	η_{10}=244	57	[33]
Co_3O_4/Co-Fe oxide	1 M		η_{10}=297	61	[67]
CuO@Fe-Co_3O_4	1 M	carbon rod	η_{10}=232		[68]
CoNi/$CoFe_2O_4$	1 M	Ni Foam	η_{1000}=360	45	[69]
$Na_{0.08}Ni_{0.9}Fe_{0.1}O_2$	1 M		η_{10}=260	44	[70]
Ni-Fe-O-P	1 M	glassy carbon	η_{10}=227; η_{20}=254	50	[62]
Ni-Fe-O-B	1 M	glassy carbon	η_{10}=243; η_{20}=261	53	[62]
Ni-Fe-O-S	1 M	glassy carbon	η_{10}=272; η_{20}=287	70	[62]
(oxy)hydroxides					
α-FeOOH	1 M	FTO	$\eta_{0.05}$=210		[9]
CNTs@FeOOH	1 M	carbon cloth	η_{10}=250	36	[11]
FeOOH/Co/FeOOH	1 M (NaOH)	Ni foam	η_{25}=~250	~32	[10]
FeOOH/CeO_2	1 M (NaOH)	Ni foam	$\eta_{31.3}$=250; $\eta_{140.2}$=350	92.3	[34]
$Ni_{50}Fe_{50}$OOH	0.1 M	rotating disk	η_{10}=~320	35	[13]
$Ni_{0.75}Fe_{0.25}$OOH	0.1 M	Si/Ti/Au	η_{10}=~375	-	[23]
NiOOH	0.1 M	Si/Ti/Au	η_{10}=~675	-	[23]
FeOOH	0.1 M	Si/Ti/Au	η_{10}=~560	-	[23]
P-CoFeOOH	1 M	glassy carbon	η_{10}=305	49.2	[71]
CoFeOOH	1 M	carbon paper	η_{10}=330	37	[72]
NiFeOOH	1 M	carbon paper	η_{10}=~270	-	[28]
FeOOH	1 M	carbon paper	η_{10}=~450	-	[28]
NiFeOH	1 M	Ni foam	η_{10}=210	31	[73]

Ni(Fe)OOH	1 M	Au	$\eta_8=250; \eta_{100}=300$	-	[74]
$Ni_{45}Fe_{55}OOH$	1 M	carbon	$\eta22.5=300$		[75]
FeOOH/NiFe LDHs	1 M	Ni foam	$\eta_{10}=208; \eta_{100}=247$	42	[35]
$\alpha\text{-}Co_{1-m}Fe_m(OH)_2$	1 M	glassy carbon	$\eta_{10}=295$	52	[76]
LDHs					
NiFe LDH	1 M	carbon rod	$\eta_{10}=210$	39	[37]
NiFe LDH-US.	1 M	carbon rod	$\eta_{10}=203$	42	[37]
NiFe-LDH	1 M	glassy carbon	$\eta_{10}=\sim280$	40	[77]
NiFe-LDH	1 M	Ni foam	$\eta_{10}=224$	52.8	[38]
CoFe-LDH	1 M	Ni foam	$\eta_{10}=288$	90.2	[38]
LiFe-LDH	1 M	Ni foam	$\eta_{10}=277$	104.0	[38]
$Ni_{2/3}Fe_{1/3}\text{-}GO$	1 M	glassy carbon	$\eta_{10}=230$	42	[78]
Ni-Fe-LDH	1 M	ErGO	$\eta_{10}=259$	39	[41]
CoFe LDH	1 M	glassy carbon	$\eta_{10}=300$	40	[44]
NiFeMn-LDH	1 M		$\eta_{20}=289$	47	[46]
NiFe-LDH	1 M		$\eta_{20}=401$	65	[46]
NiMn-LDH	1 M		$\eta_{20}=640$	70	[46]
NiFe-LDH	1 M	carbon paper	$\eta_{25}=310$	48	[47]
NiCoFe-LDH	1 M	carbon paper	$\eta_{25}=295; \eta_{200}=\sim300$	~35	[47]
$Ni_3FeAl_{0.91}\text{-}LDH$	1 M	Ni foam	$\eta_{20}=304$	57	[45]
$Cu(OH)_2@NiFe\text{-}LDH$	1 M	carbon paper	$\eta_{10}=283$	88	[40]
NiFe LDH-NS@DG10	1 M		$\eta_{10}=210$	52	[79]
CoFe-LDH	0.1 M	glassy carbon	$\eta_{10}=350$	49	[43]
$NiO@NiFe\text{-}LDH$	1 M	Ni foam	$\eta_{10}=265$	72	[42]
Others					
$CoFe_2O_4$	0.1 M	Ni foam	$\eta_5=408$	82.15	[54]
$CuFe_2O_4$	0.1 M	Ni foam	$\eta_5=450$	93.97	[54]
$NiFe_2O_4$	0.1 M	Ni foam	$\eta_5=467$	98.22	[54]
$MnFe_2O_4$	0.1 M	Ni foam	$\eta_5=520$	113.62	[54]
$CuFe_2O_4$	0.1 M	Ni foam	$\eta_{10}=360$	82	[55]
$CoFe_2O_4$	0.1 M	Ni foam	$\eta_{10}=414$	95	[55]
$NiFe_2O_4$	0.1 M	Ni foam	$\eta_{10}=433$	134	[55]
$CoFe_2O_4$	0.1 M	glass carbon	$\eta_{10}=380$	185.2	[53]
Fe-based film	0.1 M (PBS)	ITO	$\eta_7=220$	47	[80]
$Ni_{0.8}Fe_{0.2}$ nanosheets	1 M	stainless steel mesh	$\eta_{10}=206$	64	[81]
$CaCu_3Fe_4O_{12}$	0.1 M	rotating-disk	$\eta_{0.5}=310$	51	[49]
$Pr_{0.5}Ba_{0.5}Fe_{1-x}Ni_xO_{3-\delta}$	0.1 M	glassy carbon	$\eta_{10}=440$	73	[51]
Fe_3C	1 M	porous graphite	$\eta_{10}=299$	59.5	[56]
Ni_3FeN	1 M	glassy carbon	$\eta_{10}=280$	46	[82]
NiFeP	1 M	foils	$\eta_{10}=277$	-	[61]
Fe_2P	1 M	foils	$\eta_{10}=390$	-	[61]
NiFeP	1 M (NaOH)		$\eta_{10}=219$	32	[83]

$(Ni_xFe_{1-x})_2P$	1 M	Ni foam	$\eta_{20}=219$	57	[60]
FeP_4/CoP_{2-x}	1 M		$\eta_{10}=283$	41.4	[16]
$Fe_{0.1}-NiS_2$	1 M	Ti mash	$\eta_{100}=231$	43	[18]
Fe_7S_8	1 M		$\eta_{10}=270$	43	[24]
Ni-Fe-S	1 M	Ti plate	$\eta_{200}=321$		[84]
Ti-Fe sulfide	1 M	glassy carbon	$\eta_{10}=350$	55	[59]
$La_{0.95}FeO_{3-\delta}$	0.1 M	rotating disk	$\eta_{10}=410$	48	[50]
$Fe-N_x$	0.1 M	CNTs	$\eta_{10}=370$	82	[58]
Ni-Fe diselenide	1 M	glassy carbon	$\eta_{10}=240$	24	[2]
Ni-Fe-Se	1 M	glassy carbon	$\eta_{10}=216$	36	[62]

Conclusions and Outlook

In this chapter, the recent progress of Fe-based OER catalysts, including Fe oxides, Fe (oxy)hydroxides, Fe-based LDHs, Fe containing perovskites and spinel oxides is introduced. Table 1 summarized the performances of various Fe-based OER catalysts. Even though pure Fe-based catalysts are apparently inactive for OER due to their poor electrical conductivity; however, after incorporated with other elements, they show enhanced activities. Besides, hybridizing Fe-based catalysts with the conductive materials could substantially improve the electron/charge-transfer ability towards OER. Despite tremendous efforts being devoted to develop novel Fe-based materials with higher activity as well as stability, the detailed mechanism responsible for the enhanced activity resulted from the incorporation with other hetero-atoms is still difficult to clearly determine, and it is necessary to be considered by using theoretical and experimental ways in the following studies. Especially, besides development of novel Fe-based electrocatalyts, *in-situ* measurement of the intermediates on the surface of the electrocatalysts and *in-situ* observation of the ion and electron movements on the catalysts are important to understand the catalytic mechanisms.

References

[1]   L. Trotochaud, S.L. Young, J.K. Ranney, S.W. Boettcher, Nickel-iron oxyhydroxide oxygen-evolution electrocatalysts: the role of intentional and incidental iron incorporation, J. Am. Chem. Soc. 136 (2014) 6744-6753. https://doi.org/10.1021/ja502379c

[2]   J. Nai, Y. Lu, L. Yu, X. Wang, X.W.D. Lou, Formation of Ni-Fe mixed diselenide nanocages as a superior oxygen evolution electrocatalyst, Adv. Mater. 29 (2017). https://doi.org/10.1002/adma.201703870

[3]   L. Zhang, J. Xiao, H. Wang, M. Shao, Carbon-based electrocatalysts for hydrogen and oxygen evolution reactions, ACS Catal. 7 (2017) 7855-7865. https://doi.org/10.1021/acscatal.7b02718

[4]   J. Qi, W. Zhang, R. Xiang, K. Liu, H.Y. Wang, M. Chen, Y. Han, R. Cao, Porous nickel-iron oxide as a highly efficient electrocatalyst for oxygen evolution reaction, Adv.Sci. 2 (2015) 1500199. https://doi.org/10.1002/advs.201500199

[5]   F. Lu, M. Zhou, Y. Zhou, X. Zeng, First-row transition metal based catalysts for the oxygen evolution reaction under alkaline conditions: basic principles and recent advances, Small 13 (2017)1701931. https://doi.org/10.1002/smll.201701931

[6]   X. Li, S. Li, A. Yoshida, S. Sirisomboonchai, K. Tang, Z. Zuo, X. Hao, A. Abudula, G. Guan, Mn doped CoP nanoparticle clusters: an efficient electrocatalyst for hydrogen evolution reaction, Catal. Sci. Technol. 8 (2018) 4407-4412. https://doi.org/10.1039/C8CY01105B

[7]   Y. Chen, K. Rui, J. Zhu, S.X. Dou, W. Sun, Recent progress on nickel-based oxide/(oxy)hydroxide electrocatalysts for the oxygen evolution reaction, Chemistry 25 (2019) 703-713. https://doi.org/10.1002/chem.201802068

[8]   J. Wang, W. Cui, Q. Liu, Z. Xing, A.M. Asiri, X. Sun, Recent progress in cobalt-based heterogeneous catalysts for electrochemical water splitting, Adv. Mater. 28 (2016) 215-230. https://doi.org/10.1002/adma.201502696

[9]   W. Luo, C. Jiang, Y. Li, S.A. Shevlin, X. Han, K. Qiu, Y. Cheng, Z. Guo, W. Huang, J. Tang, Highly crystallized α-FeOOH for a stable and efficient oxygen evolution reaction, J. Mater. Chem. A 5 (2017) 2021-2028. https://doi.org/10.1039/C6TA08719A

[10]   J.X. Feng, H. Xu, Y.T. Dong, S.H. Ye, Y.X. Tong, G.R. Li, FeOOH/Co/FeOOH hybrid nanotube arrays as high-performance electrocatalysts for the oxygen evolution reaction, Angew. Chem. Int. Ed. Engl. 55 (2016) 3694-3698. https://doi.org/10.1002/anie.201511447

[11]   Y. Zhang, G. Jia, H. Wang, B. Ouyang, R.S. Rawat, H.J. Fan, Ultrathin CNTs@FeOOH nanoflake core/shell networks as efficient electrocatalysts for the oxygen evolution reaction, Mater. Chem. Front. 1 (2017) 709-715. https://doi.org/10.1039/C6QM00168H

[12]   Y. Wu, M. Chen, Y. Han, H. Luo, X. Su, M.T. Zhang, X. Lin, J. Sun, L. Wang, L. Deng, W. Zhang, R. Cao, Fast and simple preparation of iron-based thin films as highly efficient water-oxidation catalysts in neutral aqueous solution, Angew. Chem. Int. Ed. Engl. 54 (2015) 4870-4875. https://doi.org/10.1002/anie.201412389

[13]   M. Gorlin, P. Chernev, J. Ferreira de Araujo, T. Reier, S. Dresp, B. Paul, R. Krahnert, H. Dau, P. Strasser, Oxygen evolution reaction dynamics, faradaic charge efficiency, and the active metal redox states of Ni-Fe oxide water splitting electrocatalysts, J. Am. Chem. Soc. 138 (2016) 5603-5614. https://doi.org/10.1021/jacs.6b00332

[14]   L.J. Enman, M.B. Stevens, M.H. Dahan, M.R. Nellist, M.C. Toroker, S.W. Boettcher, Operando X-ray absorption spectroscopy shows iron oxidation is concurrent with oxygen evolution in cobalt-iron (oxy)hydroxide electrocatalysts, Angew. Chem. Int. Ed. Engl. 57 (2018) 12840-12844. https://doi.org/10.1002/anie.201808818

[15]   C. He, X. Kong, M. Jiang, X. Lei, Metal Ni-decorated Fe_3O_4 nanoparticles: A new and efficient electrocatalyst for oxygen evolution reaction, Mater. Lett. 222 (2018) 138-141. https://doi.org/10.1016/j.matlet.2018.03.142

[16]   Q. He, H. Xie, Z.u. Rehman, C. Wang, P. Wan, H. Jiang, W. Chu, L. Song, Highly defective Fe-based oxyhydroxides from electrochemical reconstruction for efficient oxygen evolution catalysis, ACS Energy Lett. 3 (2018) 861-868. https://doi.org/10.1021/acsenergylett.8b00342

[17]   J. Yang, G. Zhu, Y. Liu, J. Xia, Z. Ji, X. Shen, S. Wu, Fe_3O_4-decorated Co_9S_8 nanoparticles in situ grown on reduced graphene oxide: A new and efficient electrocatalyst for oxygen evolution reaction, Adv. Funct. Mater. 26 (2016) 4712-4721. https://doi.org/10.1002/adfm.201600674

[18]   N. Yang, C. Tang, K. Wang, G. Du, A.M. Asiri, X. Sun, Iron-doped nickel disulfide nanoarray: A highly efficient and stable electrocatalyst for water splitting, Nano Res. 9 (2016) 3346-3354. https://doi.org/10.1007/s12274-016-1211-x

[19]   X. Li, X. Hao, A. Abudula, G. Guan, Nanostructured catalysts for electrochemical water splitting: current state and prospects, J. Mater. Chem. A 4 (2016) 11973-12000. https://doi.org/10.1039/C6TA02334G

[20]   K.S. Exner, J. Anton, T. Jacob, H. Over, Ligand effects and their impact on electrocatalytic processes exemplified with the oxygen evolution reaction (OER) on $RuO_2(110)$, Chem. Electro. Chem. 2 (2015) 707-713. https://doi.org/10.1002/celc.201402430

[21]   Z. Zhang, D. Zhou, X. Bao, G. Huang, B. Huang, One-pot synthesis of Fe_2O_3/C by urea combustion method as an efficient electrocatalyst for oxygen evolution reaction, Int. J. Hydrogen Energy 44 (2019) 2877-2882. https://doi.org/10.1016/j.ijhydene.2018.12.034

[22]   M.S. Burke, L.J. Enman, A.S. Batchellor, S. Zou, S.W. Boettcher, Oxygen evolution reaction electrocatalysis on transition metal oxides and (oxy)hydroxides: activity trends and design principles, Chem. Mater. 27 (2015) 7549-7558. https://doi.org/10.1021/acs.chemmater.5b03148

[23]   D. Friebel, M.W. Louie, M. Bajdich, K.E. Sanwald, Y. Cai, A.M. Wise, M.J. Cheng, D. Sokaras, T.C. Weng, R. Alonso-Mori, R.C. Davis, J.R. Bargar, J.K. Norskov, A. Nilsson, A.T. Bell, Identification of highly active Fe sites in (Ni,Fe)OOH for electrocatalytic water splitting, J. Am. Chem. Soc. 137 (2015) 1305-1313. https://doi.org/10.1021/ja511559d

[24]   S. Chen, Z. Kang, X. Zhang, J. Xie, H. Wang, W. Shao, X. Zheng, W. Yan, B. Pan, Y. Xie, Highly active Fe sites in ultrathin pyrrhotite Fe_7S_8nanosheets realizing efficient electrocatalytic oxygen evolution, ACS Central Sci. 3 (2017) 1221-1227. https://doi.org/10.1021/acscentsci.7b00424

[25]   S. Zou, M.S. Burke, M.G. Kast, J. Fan, N. Danilovic, S.W. Boettcher, Fe (oxy)hydroxide oxygen evolution reaction electrocatalysis: intrinsic activity and the roles of electrical conductivity, substrate, and dissolution, Chem. Mater. 27 (2015) 8011-8020. https://doi.org/10.1021/acs.chemmater.5b03404

[26]   M.S. Burke, M.G. Kast, L. Trotochaud, A.M. Smith, S.W. Boettcher, Cobalt-iron (oxy)hydroxide oxygen evolution electrocatalysts: the role of structure and composition on activity, stability, and mechanism, J. Am. Chem. Soc. 137 (2015) 3638-3648. https://doi.org/10.1021/jacs.5b00281

[27]   M.S. Burke, S. Zou, L.J. Enman, J.E. Kellon, C.A. Gabor, E. Pledger, S.W. Boettcher, Revised oxygen evolution reaction activity trends for first-row transition-metal (oxy)hydroxides in alkaline media, J. Phys.Chem.Lett. 6 (2015) 3737-3742. https://doi.org/10.1021/acs.jpclett.5b01650

[28]   J.Y. Chen, L. Dang, H. Liang, W. Bi, J.B. Gerken, S. Jin, E.E. Alp, S.S. Stahl, Operando analysis of NiFe and Fe oxyhydroxide electrocatalysts for water oxidation: detection of Fe(4)(+) by mossbauer spectroscopy, J. Am. Chem. Soc. 137 (2015) 15090-15093. https://doi.org/10.1021/jacs.5b10699

[29]   H.A. Bandal, A.R. Jadhav, A.A. Chaugule, W.J. Chung, H. Kim, Fe_2O_3 hollow nanorods/CNT composites as an efficient electrocatalyst for oxygen evolution reaction, Electrochim. Acta 222 (2016) 1316-1325. https://doi.org/10.1016/j.electacta.2016.11.107

[30]   X. Cheng, Z. Pan, C. Lei, Y. Jin, B. Yang, Z. Li, X. Zhang, L. Lei, C. Yuan, Y. Hou, A strongly coupled 3D ternary $Fe_2O_3@Ni_2P/Ni(PO_3)_2$ hybrid for enhanced

electrocatalytic oxygen evolution at ultra-high current densities, J. Mater. Chem. A 7 (2019) 965-971. https://doi.org/10.1039/C8TA11223A

[31] M.W. Louie, A.T. Bell, An investigation of thin-film Ni-Fe oxide catalysts for the electrochemical evolution of oxygen, J. Am. Chem. Soc. 135 (2013) 12329-12337. https://doi.org/10.1021/ja405351s

[32] C. Dong, T. Kou, H. Gao, Z. Peng, Z. Zhang, Eutectic-derived mesoporous Ni-Fe-O nanowire network catalyzing oxygen evolution and overall water splitting, Adv. Energy Mater. 8 (2018) 1701347. https://doi.org/10.1002/aenm.201701347

[33] H.A. Bandal, A.R. Jadhav, A.H. Tamboli, H. Kim, Bimetallic iron cobalt oxide self-supported on Ni-Foam: An efficient bifunctional electrocatalyst for oxygen and hydrogen evolution reaction, Electrochim. Acta 249 (2017) 253-262. https://doi.org/10.1016/j.electacta.2017.07.178

[34] J.X. Feng, S.H. Ye, H. Xu, Y.X. Tong, G.R. Li, Design and synthesis of $FeOOH/CeO_2$heterolayered nanotube electrocatalysts for the oxygen evolution reaction, Adv. Mater. 28 (2016) 4698-4703. https://doi.org/10.1002/adma.201600054

[35] J. Chi, H. Yu, B. Qin, L. Fu, J. Jia, B. Yi, Z. Shao, Vertically aligned FeOOH/NiFe layered double hydroxides electrode for highly efficient oxygen evolution reaction, ACS Appl. Mater. Inter. 9 (2017) 464-471. https://doi.org/10.1021/acsami.6b13360

[36] A.M. Smith, L. Trotochaud, M.S. Burke, S.W. Boettcher, Contributions to activity enhancement via Fe incorporation in Ni-(oxy)hydroxide/borate catalysts for near-neutral pH oxygen evolution, Chem. Commun. 51 (2015) 5261-5263. https://doi.org/10.1039/C4CC08670H

[37] X. Li, X. Hao, Z. Wang, A. Abudula, G. Guan, In-situ intercalation of NiFe LDH materials: An efficient approach to improve electrocatalytic activity and stability for water splitting, J. Power Sources 347 (2017) 193-200. https://doi.org/10.1016/j.jpowsour.2017.02.062

[38] Z. Li, M. Shao, H. An, Z. Wang, S. Xu, M. Wei, D.G. Evans, X. Duan, Fast electrosynthesis of Fe-containing layered double hydroxide arrays toward highly efficient electrocatalytic oxidation reactions, Chem.Sci. 6 (2015) 6624-6631. https://doi.org/10.1039/C5SC02417J

[39] Y. Vlamidis, E. Scavetta, M. Gazzano, D. Tonelli, Iron vs aluminum based layered double hydroxides as water splitting catalysts, Electrochim. Acta 188 (2016) 653-660. https://doi.org/10.1016/j.electacta.2015.12.059

[40] X. Ma, X. Li, A.D. Jagadale, X. Hao, A. Abudula, G. Guan, Fabrication of $Cu(OH)_2$@NiFe-layered double hydroxid catalyst array for electrochemical water

splitting, Int. J. Hydrogen Energy 41 (2016) 14553-14561.
https://doi.org/10.1016/j.ijhydene.2016.05.174

[41]   X. Yu, M. Zhang, W. Yuan, G. Shi, A high-performance three-dimensional Ni–Fe layered double hydroxide/graphene electrode for water oxidation, J. Mater. Chem. A, 3 (2015) 6921-6928. https://doi.org/10.1039/C5TA01034A

[42]   S. Sirisomboonchai, S. Li, A. Yoshida, X. Li, C. Samart, A. Abudula, G. Guan, Fabrication of NiO microflake@NiFe-LDH nanosheet heterostructure electrocatalysts for oxygen evolution reaction, ACS Sustain. Chem. Eng. 7 (2018) 2327-2334. https://doi.org/10.1021/acssuschemeng.8b05088

[43]   F. Yang, K. Sliozberg, I. Sinev, H. Antoni, A. Bahr, K. Ollegott, W. Xia, J. Masa, W. Grunert, B.R. Cuenya, W. Schuhmann, M. Muhler, Synergistic effect of cobalt and iron in layered double hydroxide catalysts for the oxygen evolution reaction, Chem. Sus. Chem. 10 (2017) 156-165. https://doi.org/10.1002/cssc.201601272

[44]   P.F. Liu, S. Yang, B. Zhang, H.G. Yang, Defect-rich ultrathin cobalt-iron layered double hydroxide for electrochemical overall water splitting, ACS Appl. Mater. Inter. 8 (2016) 34474-34481. https://doi.org/10.1021/acsami.6b12803

[45]   H. Liu, Y. Wang, X. Lu, Y. Hu, G. Zhu, R. Chen, L. Ma, H. Zhu, Z. Tie, J. Liu, Z. Jin, The effects of Al substitution and partial dissolution on ultrathin NiFeAl trinary layered double hydroxide nanosheets for oxygen evolution reaction in alkaline solution, Nano Energy 35 (2017) 350-357. https://doi.org/10.1016/j.nanoen.2017.04.011

[46]   Z. Lu, L. Qian, Y. Tian, Y. Li, X. Sun, X. Duan, Ternary NiFeMn layered double hydroxides as highly-efficient oxygen evolution catalysts., Chem. Commun. 52 (2016) 908-911. https://doi.org/10.1039/C5CC08845C

[47]   Z. Lu, L. Qian, W. Xu, Y. Tian, M. Jiang, Y. Li, X. Sun, X. Duan, Dehydrated layered double hydroxides: Alcohothermal synthesis and oxygen evolution activity, Nano Res. 9 (2016) 3152-3161. https://doi.org/10.1007/s12274-016-1197-4

[48]   B. Han, A. Grimaud, L. Giordano, W.T. Hong, O. Diaz-Morales, L. Yueh-Lin, J. Hwang, N. Charles, K.A. Stoerzinger, W. Yang, M.T.M. Koper, Y. Shao-Horn, Iron-based perovskites for catalyzing oxygen evolution reaction, J. Phys. Chem. C 122 (2018) 8445-8454. https://doi.org/10.1021/acs.jpcc.8b01397

[49]   S. Yagi, I. Yamada, H. Tsukasaki, A. Seno, M. Murakami, H. Fujii, H. Chen, N. Umezawa, H. Abe, N. Nishiyama, S. Mori, Covalency-reinforced oxygen evolution reaction catalyst, Nat.Commun. 6 (2015) 8249. https://doi.org/10.1038/ncomms9249

[50]   Y. Zhu, W. Zhou, J. Yu, Y. Chen, M. Liu, Z. Shao, Enhancing electrocatalytic activity of perovskite oxides by tuning cation deficiency for oxygen reduction and

evolution reactions, Chem. Mater. 28 (2016) 1691-1697.
https://doi.org/10.1021/acs.chemmater.5b04457

[51]   L. Gui, Z. Huang, G. Li, Q. Wang, B. He, L. Zhao, Insights into Ni-Fe couple in perovskite electrocatalysts for highly efficient electrochemical oxygen evolution, Electrochim. Acta 293 (2019) 240-246. https://doi.org/10.1016/j.electacta.2018.10.033

[52]   N.T. Suen, S.F. Hung, Q. Quan, N. Zhang, Y.J. Xu, H.M. Chen, Electrocatalysis for the oxygen evolution reaction: recent development and future perspectives, Chem. Soc. Rev. 46 (2017) 337-365. https://doi.org/10.1039/C6CS00328A

[53]   S.M. Alshehri, A.N. Alhabarah, J. Ahmed, M. Naushad, T. Ahamad, An efficient and cost-effective tri-functional electrocatalyst based on cobalt ferrite embedded nitrogen doped carbon, J. Colloid Interface Sci. 514 (2018) 1-9.
https://doi.org/10.1016/j.jcis.2017.12.020

[54]   M. Li, Y. Xiong, X. Liu, X. Bo, Y. Zhang, C. Han, L. Guo, Facile synthesis of electrospun MFe_2O_4 (M = Co, Ni, Cu, Mn) spinel nanofibers with excellent electrocatalytic properties for oxygen evolution and hydrogen peroxide reduction, Nanoscale 7 (2015) 8920-8930. https://doi.org/10.1039/C4NR07243J

[55]   V.D. Silva, L.S. Ferreira, T.A. Simoes, E.S. Medeiros, D.A. Macedo, 1D hollow MFe_2O_4 (M=Cu, Co, Ni) fibers by solution blow spinning for oxygen evolution reaction, J. Colloid Interface Sci. 540 (2019) 59-65.
https://doi.org/10.1016/j.jcis.2019.01.003

[56]   Y. Zhang, J. Zai, K. He, X. Qian, Fe_3C nanoparticles encapsulated in highly crystalline porous graphite: salt-template synthesis and enhanced electrocatalytic oxygen evolution activity and stability, Chem. Commun. 54 (2018) 3158-3161.
https://doi.org/10.1039/C8CC01057A

[57]   M. Li, T. Liu, X. Bo, M. Zhou, L. Guo, S. Guo, Hybrid carbon nanowire networks with Fe–P bond active site for efficient oxygen/hydrogen-based electrocatalysis, Nano Energy 33 (2017) 221-228. https://doi.org/10.1016/j.nanoen.2017.01.026

[58]   P. Chen, T. Zhou, L. Xing, K. Xu, Y. Tong, H. Xie, L. Zhang, W. Yan, W. Chu, C. Wu, Y. Xie, Atomically dispersed iron-nitrogen species as electrocatalysts for bifunctional oxygen evolution and reduction reactions, Angew. Chem. Int. Ed. Engl. 56 (2017) 610-614. https://doi.org/10.1002/anie.201610119

[59]   J. Nai, Y. Lu, X.Y. Yu, Formation of Ti–Fe mixed sulfide nanoboxes for enhanced electrocatalytic oxygen evolution, J. Mater. Chem. A 6 (2018) 21891-21895.
https://doi.org/10.1039/C8TA02334D

[60]   J. Yu, G. Cheng, W. Luo, Hierarchical NiFeP microflowers directly grown on Ni foam for efficient electrocatalytic oxygen evolution, J. Mater. Chem. A 5 (2017) 11229-11235. https://doi.org/10.1039/C7TA02968C

[61]   C.G. Read, J.F. Callejas, C.F. Holder, R.E. Schaak, General strategy for the synthesis of transition metal phosphide films for electrocatalytic hydrogen and oxygen evolution, ACS Appl. Mater. Inter. 8 (2016) 12798-12803. https://doi.org/10.1021/acsami.6b02352

[62]   C. Xuan, J. Wang, W. Xia, J. Zhu, Z. Peng, K. Xia, W. Xiao, H.L. Xin, D. Wang, Heteroatom (P, B, or S) incorporated NiFe-based nanocubes as efficient electrocatalysts for the oxygen evolution reaction, J. Mater. Chem. A 6 (2018) 7062-7069. https://doi.org/10.1039/C8TA00410B

[63]   L. Zhao, Q. Cao, A. Wang, J. Duan, W. Zhou, Y. Sang, H. Liu, Iron oxide embedded titania nanowires-An active and stable electrocatalyst for oxygen evolution in acidic media, Nano Energy 45 (2018) 118-126. https://doi.org/10.1016/j.nanoen.2017.12.029

[64]   F. Yan, C. Zhu, S. Wang, Y. Zhao, X. Zhang, C. Li, Y. Chen, Electrochemically activated-iron oxide nanosheet arrays on carbon fiber cloth as a three-dimensional self-supported electrode for efficient water oxidation, J. Mater. Chem. A 4 (2016) 6048-6055. https://doi.org/10.1039/C6TA00456C

[65]   N. Bhandary, P.P. Ingole, S. Basu, Electrosynthesis of Mn-Fe oxide nanopetals on carbon paper as bi-functional electrocatalyst for oxygen reduction and oxygen evolution reaction, Int. J. Hydrogen Energy 43 (2018) 3165-3171. https://doi.org/10.1016/j.ijhydene.2017.12.102

[66]   Y. Chen, C. Dong, J. Zhang, C. Zhang, Z. Zhang, Hierarchically porous Mo-doped Ni–Fe oxide nanowires efficiently catalyzing oxygen/hydrogen evolution reactions, J. Mater. Chem. A 6 (2018) 8430-8440. https://doi.org/10.1039/C8TA00447A

[67]   X. Wang, L. Yu, B.Y. Guan, S. Song, X.W.D. Lou, Metal-organic framework hybrid-assisted formation of Co_3O_4/Co-Fe oxide double-shelled nanoboxes for enhanced oxygen evolution, Adv. Mater. 30 (2018)1801211. https://doi.org/10.1002/adma.201801211

[68]   X. Li, C. Li, A. Yoshida, X. Hao, Z. Zuo, Z. Wang, A. Abudula, G. Guan, Facile fabrication of CuO microcube@Fe-Co_3O_4 nanosheet array as a high-performance electrocatalyst for the oxygen evolution reaction, J. Mater. Chem. A 5 (2017) 21740-21749. https://doi.org/10.1039/C7TA05454H

[69]   S. Li, S. Sirisomboonchai, A. Yoshida, X. An, X. Hao, A. Abudula, G. Guan, Bifunctional CoNi/CoFe$_2$O$_4$/Ni foam electrodes for efficient overall water splitting at a

high current density, J. Mater. Chem. A 6 (2018) 19221-19230.
https://doi.org/10.1039/C8TA08223E

[70]　B. Weng, F. Xu, C. Wang, W. Meng, C.R. Grice, Y. Yan, A layered $Na_{1-x}Ni_yFe_{1-y}O_2$ double oxide oxygen evolution reaction electrocatalyst for highly efficient water-splitting, Energy Environ. Sci. 10 (2017) 121-128.
https://doi.org/10.1039/C6EE03088B

[71]　H. Xu, J. Wei, C. Liu, Y. Zhang, L. Tian, C. Wang, Y. Du, Phosphorus-doped cobalt-iron oxyhydroxide with untrafine nanosheet structure enable efficient oxygen evolution electrocatalysis, J. Colloid Interface Sci. 530 (2018) 146-153.
https://doi.org/10.1016/j.jcis.2018.06.073

[72]　M. Xiong, D.G. Ivey, Composition effects of electrodeposited Co-Fe as electrocatalysts for the oxygen evolution reaction, Electrochim. Acta 260 (2018) 872-881. https://doi.org/10.1016/j.electacta.2017.12.059

[73]　G. Dong, M. Fang, J. Zhang, R. Wei, L. Shu, Xiaoguang Liang, S. Yip, F. Wang, L. Guan, Z. Zheng, J.C. Ho, In situ formation of highly active Ni–Fe based oxygen-evolving electrocatalysts via simple reactive dip-coating, J. Mater. Chem. A 5 (2017) 11009-11015. https://doi.org/10.1039/C7TA01134B

[74]　A.S. Batchellor, S.W. Boettcher, Pulse-electrodeposited Ni–Fe (oxy)hydroxide oxygen evolution electrocatalysts with high geometric and intrinsic activities at large mass loadings, ACS Catal. 5 (2015) 6680-6689.
https://doi.org/10.1021/acscatal.5b01551

[75]　M. Gorlin, J. Ferreira de Araujo, H. Schmies, D. Bernsmeier, S. Dresp, M. Gliech, Z. Jusys, P. Chernev, R. Kraehnert, H. Dau, P. Strasser, Tracking catalyst redox states and reaction dynamics in Ni-Fe oxyhydroxide oxygen evolution reaction electrocatalysts: the role of catalyst support and electrolyte pH, J. Am. Chem. Soc. 139 (2017) 2070-2082. https://doi.org/10.1021/jacs.6b12250

[76]　H. Jin, S. Mao, G. Zhan, F. Xu, X. Bao, Y. Wang, Fe incorporated α-Co(OH)$_2$ nanosheets with remarkably improved activity towards the oxygen evolution reaction, J. Mater. Chem. A 5 (2017) 1078-1084. https://doi.org/10.1039/C6TA09959A

[77]　N. Han, F. Zhao, Y. Li, Ultrathin nickel–iron layered double hydroxide nanosheets intercalated with molybdate anions for electrocatalytic water oxidation, J. Mater. Chem. A 3 (2015) 16348-16353. https://doi.org/10.1039/C5TA03394B

[78]　W. Ma, R. Ma, C. Wang, J. Liang, X. Liu, K. Zhou, T. Sasaki, A superlattice of alternately stacked Ni–Fe hydroxide nanosheets and graphene for efficient splitting of water, ACS nano 9 (2015) 1977-1984. https://doi.org/10.1021/nn5069836

[79]   Y. Jia, L. Zhang, G. Gao, H. Chen, B. Wang, J. Zhou, M.T. Soo, M. Hong, X. Yan, G. Qian, J. Zou, A. Du, X. Yao, A heterostructure coupling of exfoliated Ni-Fe hydroxide nanosheet and defective graphene as a bifunctional electrocatalyst for overall water splitting, Adv. Mater. 29 (2017)1700017. https://doi.org/10.1002/adma.201700017

[80]   M. Chen, Y. Wu, Y. Han, X. Lin, J. Sun, W. Zhang, R. Cao, An iron-based film for highly efficient electrocatalytic oxygen evolution from neutral aqueous solution, ACS Appl.Mater.Inter. 7 (2015) 21852-21859. https://doi.org/10.1021/acsami.5b06195

[81]   M. Yao, N. Wang, W. Hu, S. Komarneni, Novel hydrothermal electrodeposition to fabricate mesoporous film of $Ni_{0.8}Fe_{0.2}$ nanosheets for high performance oxygen evolution reaction, Appl. Catal. B-Environ. 233 (2018) 226-233. https://doi.org/10.1016/j.apcatb.2018.04.009

[82]   X. Jia, Y. Zhao, G. Chen, L. Shang, R. Shi, X. Kang, G.I.N. Waterhouse, L.Z. Wu, C.H. Tung, T. Zhang, Ni_3FeN nanoparticles derived from ultrathin NiFe-layered double hydroxide nanosheets: An efficient overall water splitting electrocatalyst, Adv. Energy Mater. 6 (2016) 1502585. https://doi.org/10.1002/aenm.201502585

[83]   F. Hu, S. Zhu, S. Chen, Y. Li, L. Ma, T. Wu, Y. Zhang, C. Wang, C. Liu, X. Yang, L. Song, X. Yang, Y. Xiong, Amorphous metallic NiFeP: A conductive bulk material achieving high activity for oxygen evolution reaction in both alkaline and acidic media, Adv. Mater. 29 (2017)1606570. https://doi.org/10.1002/adma.201606570

[84]   J. Zhang, Y. Hu, D. Liu, Y. Yu, B. Zhang, Enhancing oxygen evolution reaction at high current densities on amorphous-like Ni-Fe-S ultrathin nanosheets via oxygen incorporation and electrochemical tuning, Adv. Sci. 4 (2017) 1600343. https://doi.org/10.1002/advs.201600343

Electrochemical Water Splitting: Materials and Applications   Materials Research Forum LLC
Materials Research Foundations **59** (2019) 59-96   doi: https://doi.org/10.21741/9781644900451-3

Chapter 3

Co-Based Electrocatalysts for Hydrogen-Evolution Reaction

Xiumin Li[1]*, Guoqing Guan[2]*

[1] School of Materials Science and Engineering, Zhengzhou University, Zhengzhou 450001, China

[2] Energy Conversion Engineering Lab, Institute of Regional Innovation, Hirosaki University, 2-1-3 Matsubara, Aomori 030-0813, Japan

xiuminli0516@zzu.edu.cn, guan@hirosaki-u.ac.jp

Abstract

Among the transition metals based electrocatalysts, cobalt based ones have been extensively studied for hydrogen evolution reaction (HER). Herein, the state-of-the-art of Co based electrocatalysts including Co metal, Co-N-C composites, Co based alloys, nitrides, phosphides, oxides, sulfides, selenides and binary nonmetal compounds for HER is introduced. The strategies for the synthesis of highly efficient Co based catalysts are summarized. Besides, the effective tactics including adjusting the bond strength of the catalyst-reactant, optimizing the electronic structure, and designing rational nanostructure for improving performance of electrocatalysts are reviewed and discussed.

Keywords

Hydrogen Evolution Reaction, Co-Based Material, Synthesis Tactic, Electrocatalysis, Water Splitting

Contents

1.   Introduction

Cobalt (Co)-based materials are widely distributed on the earth, and the Co-based catalysts have been extensively studied for anodic oxygen evolution reaction (OER) and cathodic hydrogen evolution reaction (HER) in water splitting. Co-based materials include nitride, phosphide, arsenide, arsenate, silicon phosphide, oxide, hydroxide oxide, hydroxide-containing water oxides, sulfide, selenide, antimonide, telluride, sulfoarsenide, sulfantimonide, selenium telluride, carbide, carbonate, silicate and so on, and some of them are considered as the most representative electrocatalysts among the transition metal-based electrocatalysts for HER. Several excellent review papers have been published for Co-based electrocatalysts during the past years [1-3]. It is considered that the unique facet structure, high surface area, and enhanced conductivity are generally in favor of improving the catalytic activity. Unfortunately, the performances of Co-based catalysts are still not over the noble metal ones. Thus, substantial endeavors are devoted to further improving the activity as well as the stability of the Co-based catalysts. Especially, a thorough understanding of the electrochemical HER mechanism on the catalysts is helpful to achieve insight into the existing issues. In addition, understanding the reaction rate determining step and offering directions for the synthesis and evaluation of electrocatalysts are also important.

The electrolysis HER process is considered as following two steps:

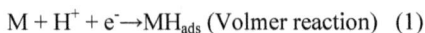

$M + H^+ + e^- \rightarrow MH_{ads}$ (Volmer reaction)   (1)

$MHads + H^+ + e^- \rightarrow H_2 + M$ (Heyrovsky reaction)   (2)

or

$MHads + MHads \rightarrow H_2 + 2M$ (Tafel reaction)   (3)

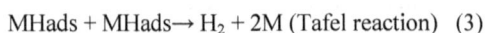

That is, the HER process performs via either the Volmer–Heyrovsky or the Volmer–Tafel mechanism [3]. During the investigations, valuable insights into the catalytic behaviors from the experimental results of current density, Tafel slope, Volcano plot, etc. can be obtained. Especially, the current density is widely used to evaluate the electrochemical

Electrochemical Water Splitting: Materials and Applications   Materials Research Forum LLC
Materials Research Foundations **59** (2019) 59-96   doi: https://doi.org/10.21741/9781644900451-3

reaction rate and catalytic activity by analysis of Butler–Volmer equation. Considering the concentrations of the reduced surface species are consistent with the bulk concentrations, the response current density can be calculated as

$$j = j_o[e^{-\alpha f\eta} - e^{(1-\alpha f\eta)}] \quad (4)$$

$$f = F/RT \quad (5)$$

where j_o refer to the exchange current density, α the transfer coefficient, and η overpotential. Moreover, the Tafel slope is used to analyze the catalytic kinetics and reveal the electrolysis mechanism. It is widely employed to get insight into the sensitivity of the applied potential and response current, which reveals the information of rate determining steps. In general, Tafel slopes of 120, 40 or 30 mV dec^{-1} can be revealed respectively when Volmer, Heyrovsky or Tafel is the rate determining step, respectively (Fig. 1 A-C) [4].

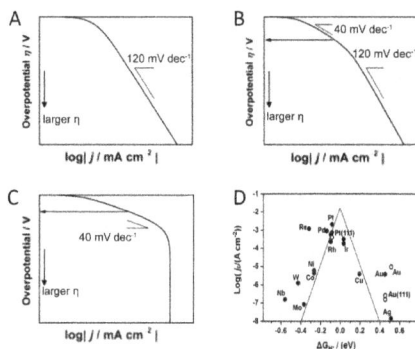

Fig. 1: *Simulative behavior of the Tafel slope for HER presuming (A) Volmer (B) Heyrovsky (C) or Tafel step as the rate-determining step [4]. (D)Volcano plot of exchange current density (j$_0$) as a function of DFT-calculated ΔG_{H*} on metals [5].*

Additionally, the HER activity of various catalysts can be graded according to the exchange current density and chemisorption energy (M–H binding energy) [5]. Herein, it is crucial to obtain an optimum M–H binding energy for HER. Especially, feeble binding energy is difficult to capture the reactants or intermediates, whereas the doughty one will impede the combination of the adsorbed H and the diffusion of the generated H$_2$. Volcano plot is a valuable technique to compare the exchange current density and Gibbs free energy change (ΔG_H) of different catalysts. Herein, density functional theory (DFT) is generally used to calculate the M–H hydrogen binding energy, in which a plane wave

Electrochemical Water Splitting: Materials and Applications   Materials Research Forum LLC
Materials Research Foundations **59** (2019) 59-96   doi: https://doi.org/10.21741/9781644900451-3

pseudo potential is applied to evaluate the ionic core, and the number of hydrogen species represents the surface coverage [6]. As shown in Fig. 1D, Pt catalyst has the highest j_o value and near zero negative ΔG_H in the acidic electrolyte, which supports the excellent HER activity. In contrast, the performance of pure Co metal for catalyzing HER is not satisfactory. However, after combining with other elements, the j_o and ΔG_H values of Co-based electrocatalysts could be greatly improved.

In this chapter, the research, and development progress of Co-based electrocatalysts including Co metal, Co-N-C composites, Co based alloys, nitrides, phosphides, oxides, sulfides selenides and binary nonmetal cobalt compounds for the cathodic HER in the water electrolysis is reviewed. Some recent strategies considered for the synthesis of highly efficient Co-based catalysts are summarized. Besides, based on the research achievements of the reaction kinetics and mechanism, a series of effective tactics, i. e. fine-adjusting the bond strength of catalyst–reactant, optimizing the catalyst electronic property, and designing a rational nanostructure to increase the active catalytic sites, for improving the performance of electrocatalysts are presented.

2.   Various Co-based electrocatalysts

2.1   Co metal, alloy, and their composites

To find potential low-cost catalysts as a replacement for high-cost noble metal ones, 31 alternative metals were evaluated for HER by using cyclic voltammetric measurement. It is reported that the order of their catalytic activity follows Ni > Mo > Co > W > Fe >Cu [7]. Unfortunately, the activity and/or stability of non-noble metals are always worse than the noble metals. Owing to nanotechnology, the influences of size, shape, morphology, dimensionality, support, oxidation state, confinement, composition, and inter particle distance on the inherent properties including activity, selectivity, stability, atom economy and energy efficiency of Co nanoparticle catalysts have been unveiled. For instance, Ma *et al.* [8] assembled hollow nanostructured cobalt particles as the HER electrocatalyst, which needed low overpotentials of 85 and 237 mV to support 10 and 100 mA cm^{-2} current densities, respectively, and stable operation for an extended period was also maintained. Recently, the HER over Co particles was investigated by operando and *in situ* X-ray absorption spectroscopy. The results indicated that the superficial Co was partly oxidized to Co^{2+}, but a fraction of which remained as the metallic outer shell, which is in favor of the good catalytic activity [9].

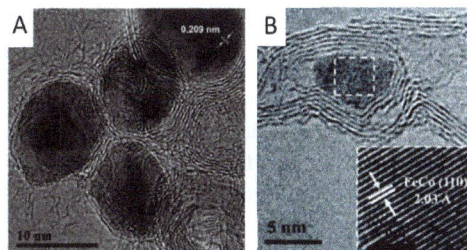

Fig. 2: *(A) TEM images of Co@N–C [15]. (B) TEM image of FeCo@NCNTs with the inset revealing the (110) plane of the FeCo material [16].*

In order to surmount the weaknesses of pure Co metal, a large family of Co-based composites were explored. Among of them, Co-N-C composites attract wide attentions. Co-N-C composites are generally generated from Co salts and N-rich organics by thermal treatment. Herein, it is considered that the N rich carbon nanotubes, N-doped graphene nanosheets, and other porous carbon are ideal supports to immobilize Co nanoparticle, which could reveal attractively synergistic effect between the introduced Co and the doped N, resulting in improved stability, electronic conductivity and the amount of exposed active sites. Recently, encapsulating of Co or Co alloy into N-rich carbon materials by developing efficient synthesis tactics have been widely investigated. Since Zhang et al. [10] firstly proposed single-atom catalysis (SACs) concept in 2011, SACs have attracted rising attention in electrocatalys is field due to the unique properties of high selectivity, high activity, maximum atom utilization efficiency, and low cost. To date, some typical work has been reported on the topic of introduction of a single Co atom into Co-N-C composites [11]. It is believed that the good performance is originated from the optimized electronic structure of the single atoms and the faster electron transfer between the single atoms and N-rich carbon material. Chen and Tour et al. [12] anchored single Co atoms on N-doped graphene (NG) by calcining graphene oxide and Co salts in Ar/NH3 atmosphere. Through X-ray absorption fine structure spectroscopy (EXAFS), high-angle annular dark field (HAADF) and electrochemical measurements, they found that the N doping of graphene played a crucial role for connecting single Co atoms, in which the formed Co-N bond acted as the catalytic active sites in HER. As a result, the single Co atom modified N-doped graphene showed very high catalytic acitivity for HER. Similarly, Zhang et al. [13] immobilized Co atoms on N-doped graphene, which was confimed by HAADF-STEM image. In this work, the EXAFS result indicated that the Co was bonded with N atom of graphene rather than C element. As a result, the

63

obtained catalyst only needed 146 mV overpotenial to maintain a current density of 10mA cm^{-2}.

Besides single Co atom, utilizition of Co nanoparticle for the engineering of Co-N-C composites is also attracting extensive attention. Herein, the moderate H absorption energy over Co metal plays a crucial role for HER. In addition, the protection of carbon shield extends the application of Co-N-C composites to a wide pH range. Recently, many novel strategies have been reported in capsuling, imbedding, or bonding Co metal to N-rich carbon materials. Typically, Asefa *et al.* [14] synthesized Co embedded N-rich CNTs as efficient and low-cost HER electrocatalyst by a two-step heating process. Especially, they calcined dicyandiamide and cobalt chloride at 500 °C, leading to functionalization of Co^{2+}on graphitic carbon nitride at first. Subsequently, the composite was pyrolyzed at 700 °C, and then it was etched in an acid solution to dislodge accessible metallic cobalt. Meanwhile, during the pyrolysis process, the in-situ generated Co metal could catalyze the conversion of carbon nitride to N-CNTs, which resulted in a nanotube capsuled Co particles microstructure. The obtained composite catalyst was utilized for HER at a wide pH of 0~14. It is considered that the Pt-close properties should be due to the N dopants and concomitant structural defects.

Moreover, Bao group [15-17] explored a series of methods to combine the Co and its alloy with N-rich carbon. For example, they prepared N-doped carbon encapsulated Co nanoparticles as an HER and OER bifunctional electrocatalyst by a heating cobalt-imidazole precursor (Fig. 2A), which revealed high activity and good stability for catalyzing HER in a wide pH range [15]. Besides, they also found the other route to fabricate CNTs shield covered FeCo bimetallic particles (Fig. 2B) [16]. Briefly, the chemical vapor deposition (CVD) method was employed for the formation of N-CNTs over FeCo catalysts supported by MgO substrate, and then the MgO substrate and those exposed metal were removed out. As expected, the achieved composite electrode had satisfactory. HER performance in acidic solution, which is much better than Co@NCNTs or Fe@NCNTs catalysts. Furthermore, they systematically investigated the significance of nitrogen dopant by DFT calculations. According to the Volcano principles, the HER catalyst with high activity generally has low free energy of H-adsorption near the thermoneutral one (ΔGH = 0). The calculation results showed that the ΔG_H values of FeCo@NCNTs and corresponding Co@NCNTs, N-CNTs and CNTs were −0.05, −0.15, 0.42 and 1.29 eV, respectively, indicating that the N dopant and encapsulated metal play crucial roles for improving the water splitting efficiency. In their further work, a facile method was demonstrated to fabricate CoNi alloy embedded N-doped graphene by employing ethylenediaminetetraacetate carbon source [17]. The achieved composite materials consisted of 1-3 layers of graphene shell and uniform CoNi nanoalloy core.

Electrochemical Water Splitting: Materials and Applications   Materials Research Forum LLC
Materials Research Foundations **59** (2019) 59-96   doi: https://doi.org/10.21741/9781644900451-3

This well-defined model is in favor of elucidating the role of Co-N-C composite catalysts for HER by evaluating theoretical calculations with the real system. Based on the DFT calculations, it can be concluded that the N doping and neighboring alloy particles can synergistically adjust the free energy of H-adsorption of graphene shells, enhancing the HER activity. Especially, the improvement of catalytic activity is attributed to the increase of graphene electro-negativity since the alloy and N element can synchronously contribute electrons to the closed carbon atoms, resulting in the activation of carbons in the graphene shell for H-adsorption. The last interesting work introduced here is that reported by Chen *et al.* [18]. They prepared N-doped graphene layers covered FeCo nanoalloy by direct heating MOF nanoparticles at 600°C in a nitrogen atmosphere. The obtained catalyst was very active and stable for HER application. Besides, the DFT calculations indicate that both the nitrogen dopants and unique heterostructure could promote the decrease of H-adsorption free energy.

It is an attractive topic to get HER catalysts with improved surface area and/or altered electronic structure by alloying and/or composition approaches to synthesize bi- or tri-metal material. Jakšić [19] found that the enhanced performance of alloy catalysts could be expounded by the theory of Brewer-Engel valence-bond. Especially, according to the interactive effect of hypo-hyper-d-electronic in Brewer-type inter metallic systems, the elements among the left half part of transition series in the periodic table of the elements endow the half-filled or empty d-electron orbitals. However, the elements distributed in the right half part own the interiorly paired d-electrons. Thus, the consummate cooperation of two kinds of elements may results in noble metal-like performance for electrolysis reaction. Based on this idea, the studies on the performance of Co-Mo alloys, such as hardness, valuable magnetic properties, high melting point, easier hydrogen evolution and good corrosion resistance have shown some interesting results. Especially, the work of Crooks *et al.* indicated that compositing small amount of Mo elements, Co or Ni based alloy could exhibit significantly improved HER activity [20] (Fig. 3). Recently, in order to explore the suitable combination of alloy, the group of Nørskov theoretically analyzed the HER activities of more than 700 metals and binary alloys via a computational screening procedure based on DFT [21]. They found that the metals such as Cu, Ni, Ru, Pd, Rh, Ag, Ir, Cd, and Pt were efficient candidates for alloying with Co. To date, many researchers have devoted to designing and analyzing the performance of Co-based alloy for HER. Most of the synthesis processes are similar with the preparation of the above Co-N-C composites generated from the thermal treatment of Co salts and N-rich organics or metal doped MOF precursors. For examples, CoFe, CoNi, CoCu, bimetal in layed N-doped carbon frameworks were prepared by using calcination treatment of bi-metal doped zeolitic imidazolate frameworks [22]. Due to synergistic advantages of M-

Electrochemical Water Splitting: Materials and Applications   Materials Research Forum LLC
Materials Research Foundations **59** (2019) 59-96   doi: https://doi.org/10.21741/9781644900451-3

N-C species, bi-metal adjustment, Co-based electroactive phases and porous structure, the alloy catalysts with optimum molar ratio always reveal better performances than the pure Co-based electrodes for HER. Besides binary alloys, Co containing ternary and quaternary alloys such as Ni-Co-Mo, Ni-Co-Ti, Ni-Co-Ti–Cu, Ni-Fe-Co, and Ni-Fe-Co-P have also been widely studied. For instance, Rosalbino *et al.* [23] alloyed Co–Ni with Y, Ce, Pr, and Er rare earth metals by the salt calcination way, and analyzed the good performances of the optimal $Co_{57.5}Ni_{36}Y_{6.5}$ and $Co_{57}Ni_{35}Ce_8$ electrodes. It is found that more obvious synergistic effect between Co–Ni and Y, Ce could be formed. As a result, the suitable recombination of $3d^8$-orbitals, d^7-orbitals of Co of Ni with d^1-orbitals of Y or Ce based on the theory of Brewer-Engel valence bond was considered to promote the hydrogen desorption during Heyrovsky and/or Tafel step. As shown in Fig. 4, Lee *et al.* [24] designed Pt-Ni-Co alloy nanocatalyst with a Cartesian-coordinate-like hexapod structure, which displayed a much better HER activity in alkaline solution than binary PtNi hexapods and Pt/C electrodes, with tenfold specific activity to Pt/C.

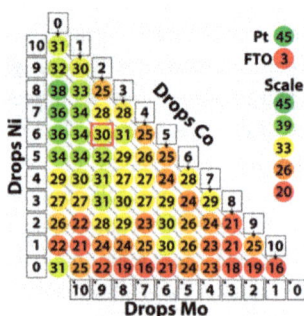

Fig. 3: *(A) Matrix plot revealing the number of Cr bands dissolved as a function of composition for bimetallic and trimetallic catalysts. The vertical axis is the drop number of the Ni precursor, while the diagonal axis represents Co, and the horizontal axis is for Mo component [20].*

For further enhancing the HER catalytic activity, grafting Co and its alloys on the catalytic and conductive materials is another efficient tactics for designing novel HER heterostructure catalysts. Especially, the assembly of three dimensional (3D) core-shell heterostructures attracts rising interest because it could expose a high surface area for the diffusions and transportations of the ions, electrons and generated gases efficiently in the catalysts, which is in favor of maximum utilization of the electrolysis materials. Recently, various 3D hetero-structure electrocatalysts have been developed, which showed enhanced performance for water electrolysis. Typically, Wang *et al.* [25] designed Co-CoO$_x$ loaded N-doped carbon sheets by one-pot thermal treatment method. The composite showed bifunctional catalytic activity for OER and HER. Herein, it is considered that the high HER efficiency of Co-CoO$_x$@CN hybrids could be due to the rational structure of the heterogeneous catalyst, the synergistic property of Co and CoO$_x$, the fine-adjusted H adsorption free energy and electron-donating effect near Co-N sites, and the protection of carbon shield.

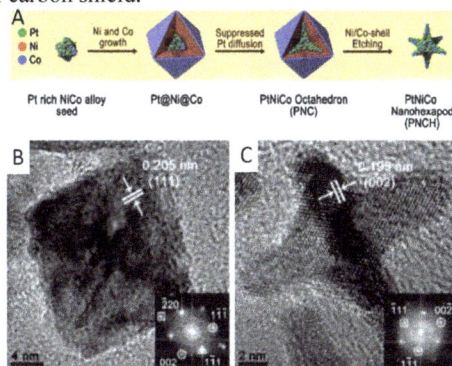

Fig. 4: *(A) Schematic diagram of the synthesis process of PtNiCo nanohexapod. (B) TEM with FFT images (insets) of (B) PtNiCo octahedron and (C) PtNiCo nanohexapod [24].*

2.2   Co nitrides

Metal nitrides materials are fabricated by integrating N atoms into the interval sites of the host metal. As such, the crystalline structure of parent metal can be finely modified by the intercalation of the N elements, resulting in the expansion of metal lattices, broadening metal d-band and increasing the distance between metal atoms [26]. Moreover, comparing with the parent metal, the contracted d-band could lead to a larger density of states near the Fermi level, which is favorable for finely adjusting its properties such as electronic conductivity, melting temperature and hardness [27]. Recently, metal nitride is believed as one of the potential HER electrocatalysts because of their good electrical conductivity, unique electronic structures, and good corrosion resistance [28,29].

Fig. 5 reveals the XRD data (left) of Co$_2$N, Co$_3$N, and Co$_4$N with the electron delocalization schematic [30]. Due to the d band center distance of Co$_x$N materials relatively far from the HER energy level comparing with the molybdenum nitrides, nickel nitrides and so on, few types of research focus on the HER behavior of cobalt nitrides [31,32]. However, this issue was resolved by adjusting the location of the d-band center by introducing V element [32]. For example, a Pt-like V-doped Co_4N catalyst was prepared recently and the HER mechanism was explored by the DFT calculations, ultraviolet photoemission spectroscopy (UPS) and synchrotron-based X-ray absorption near-edge structure (XANES). They found that the enhanced activity was triggered by shifting down the d-band center on neutralizing the electron density of Co by introducing V dopant. In addition, after V doping, the ΔG_H of V-Co_4N can be optimized from -0.56 eV to -0.25 eV which is close to the thermoneutral value, indicating that the V dopant could promote the H adsorption/desorption reaction. Besides, the d-band centers of V-Co_4N and Co_4N were calculated as -1.85 and -1.79 eV, respectively, explaining that the d-band center of V-Co_4N is far from the Fermi level. These results illustrate that the antibonding energy states can be decreased so as to weaken the interaction between the surface and adsorbate. Hence, the V-Co_4N catalyst revealed an improved HER catalytic activity Moreover, HER catalytic activity of cobalt nitride could be optimized by precisely regulating the electronic states and valence of the metal atoms by combination with a second metal element. To date, the enhanced HER performance of ternary Ni–Co, and Co–Fe nitrides was realized by the synergic effect arising from the bimetallic nitrides compared with the individual ones [29].

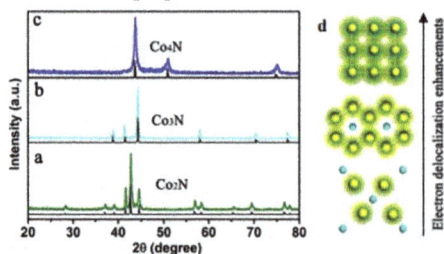

Fig. 5: *XRD patterns of a) Co_2N, b) Co_3N and c) Co_4N. d) diagram of the electron delocalization [30].*

In addition, it is reported that some single metal nitrides are hard to sustain a long-term electrochemical reaction stability in either alkaline or acidic condition [33]. In order to solve this issue, an effective way is to combine metal nitrides on various conductive and/or catalytic supports. For example, Guo's group [34] fabricated porous NiO/CoN electrocatalysts via *in-situ* nitrogenization of $NiCo_2O_4$, and found that the strongly interconnected interface of CoN and NiO domained the enhanced electrocatalytic stability and activity. Moreover, Qiao *et al.* [35] embedded Co/CoN composite nanoparticles in an N-rich carbon, which was derived from the Co^{2+} containing polymer in ammonia nitrogen source. The obtained Co/CoN composite showed good HER catalytic activity and long-term stability in both acidic and alkaline environments. Herein, the controllable contact interface of Co/CoN species is crucial for tuning the HER property.

To date, the most popular and feasible route for synthesizing of nano scaled nitrides is annealing the metal precursors in the existence of various N sources, e.g., NH_3, N_2, N_2H_4, urea and dicyanamide [29]. For the preparation of cobalt nitrides, NH_3 is a frequently used nitrogen source whose annealing temperature in the nitriding process is generally in the range of 400-1000°C. For instance, Co_3N, Co_2N, and Co_4N were synthesized by using β-$Co(OH)_2$, α-$Co(OH)_2$, and Co_3O_4 precursors, respectively, with NH_3 flow [30]. Besides, Fan *et al.* [36] synthesized CoN nanowires by using N_2 radio frequency plasma treatment, which needs markedly shorter reaction duration at room temperature compared to traditional NH_3 annealing at high temperature for several hours.

In addition, with the rising interest on Co based ternary metal nitrides for electrocatalysts, various efficient synthesis tactics of ternary metal nitrides have also been explored. For instance, Gajbhiye *et al.* [37] obtained $CoMoN_2$ nanoparticles from $CoMoO_4$ precursor with a nitriding process in the NH_3 atmosphere. Similarly, Khalifah *et al.* [28] synthesized cobalt molybdenum nitrides and further confirmed the stoichiometry of catalyst, i.e., $Co_{0.6}Mo_{1.4}N_2$, by pair distribution function and neutron diffraction measurement, which revealed high HER catalytic activity. In addition to the metal oxides, double hydroxide precursors were also utilized to prepare cobalt nitrides, which generally requires a low heating temperature. For example, porous Co_3FeN_x nanoparticle was derived from nitriding Co_3Fe double hydroxide in the flowing NH_3 atmosphere, which exhibited good HER and OER performances [38].

2.3   Co phosphides

The first research on transition metal phosphides (TMPs) can trace back to 18th century [39]. Unfortunately, no exciting application of them during the subsequent 200 years until the 1960s was reported. Now, TMPs have been widely applied in the fields of metallurgy, hydrodenitrogenation, hydrodesulfurization, lithium ion batteries and photocatalytic degradation, and so on. In 2005, Rodriguez and Liu [40] studied the property of Ni_2P (001) which was found to be similar to the [NiFe] hydrogenase based on DFT calculations, indicating that it could be a highly active HER catalyst. Recently, cobalt phosphide was confirmed as one member of the TMPs family with highly active HER activity. As shown in Fig.6 A [41], Co_2P has a structure as the Co_2Si, which includes edge-sharing CoP_4 tetrahedra and CoP_5 pentahedra, including 9 paratactic P atoms. In contrast, CoP, with B31 MnP type crystalline structure, was built by face-sharing CoP_6octahedra and edge-sharing PCo_6 trigonal prisms, forming a zig-zag chain along the b direction with a 2.70 Å P–P distance. Cobalt phosphides with MnP andCo_2Si type structures are both considered to be derived from the NiAs structure type, which have similar bond distances. Oppositely, comparing to Co_2P, the high-index surfaces of CoP endow an obvious exposure extent of Co atoms, which is related to the Co ratio in CoP and Co_2P materials.

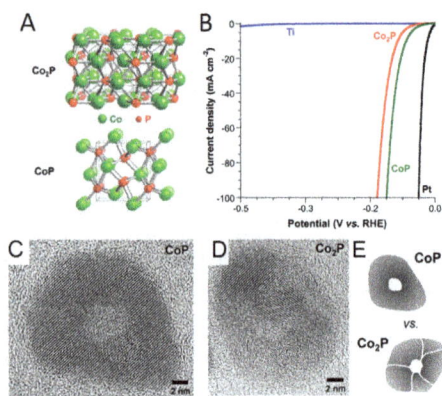

Fig. 6: *(A) Crystal structures of Co_2Si-type Co_2P (top) and MnP-type CoP (bottom). (B) Polarization curves of Co_2P and CoP electrodes, along with Pt mesh and bare Ti foil. (C-D)TEM images and corresponding schematic diagram of CoP and Co_2P particles [41].*

In 2005, single-crystalline CoP nanowires were prepared by the thermal-decomposition of cobalt acetylacetone and alkylphosphonic acid in the existence of trioctylphosphine oxide and hexadecylamine (Fig. 7 A) [42]. Since Sun et al. [43] reported the synthesis of CoP nanowire on carbon cloth by a gas–solid reaction method (Fig. 7 B), which exhibited good HER catalytic activity and long-term stability in a wide pH, a series of TMPs have been obtained for HER. To date, two popular synthesis routes are applied for the preparation cobalt phosphide electrocatalysts. The previously reported way is the solution-phase preparation with a phosphorous source of tri-n-octylphosphine (TOP). TOP is an efficient and versatile phosphorus source in various synthesis processes. Herein, P-C covalent bond of TOP will break at a high temperature of ~ 300°C, leading to

Electrochemical Water Splitting: Materials and Applications   Materials Research Forum LLC
Materials Research Foundations **59** (2019) 59-96   doi: https://doi.org/10.21741/9781644900451-3

the phosphorization of various precursors of metal, metal oxides, metal acetylacetonates, and metal carbonyl compounds. Due to the strong coordination effect of TOP and metal ions, only using a facile reaction process could generate a novel catalyst microstructure. For example, Yan *et al.* [44] synthesized CoP hollow nanoparticles, Co_2P nanorods and nanoparticles cumulated spheres by using the solution-phase synthesis with TOP phosphorus source and found that the composition and morphology of the obtained cobalt phosphides were highly dependent on the injection rate and reaction time.

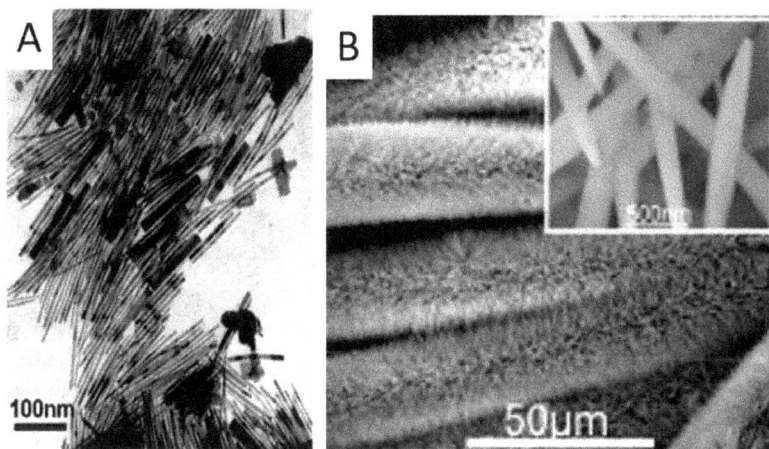

Fig. 7: (A) *TEM image of CoP nanowires [42]. (B) SEM images of CoPnanowire (inset: corresponding enlarge one) [43].*

Similarly, Shao-Horn's group [45] also prepared the CoP hollow nanoparticles by this method, which showed high activity for HER in both alkaline and acidic solutions. It is found that the applied potential increased with the decrease of the molar ratio of P/Co and with the appearance of (oxy) phosphate(s), and specific HER activity decreased. However, the aqueous insolubility and high decomposition temperature of TOP restricts the phosphorization process to carry out in organic solvent with high boiling-points, e.g., 1-octyl ether, squalene, and octadecylene, which resulted in a high corrosive and flammable reaction. Moreover, the phosphorization should be conducted instrict oxygen-free atmosphere. As a result, the tri-n-octylphosphine oxide (oxide of TOP, TOPO) and other phosphines (tri-phenylphosphine, etc.), are attempted to mix with TOP to provide the phosphorous source and cap molecules.

The second phosphorization method is the gas–solid reaction route, in which active PH_3 gas serves as the phosphorus source. Nevertheless, due to the exceeding toxicity of PH_3, some phosphorus sources, e.g., hypophosphites, are widely utilized for *in situ* generation of PH_3. During the annealing treatment, hypophosphites can decompose to PH_3 at above 250°C, which directly phosphorizes cobalt based metal, oxides, hydroxides, MOFs and so on to form cobalt phosphides. Moreover, the gas–solid reaction tactic is very applicable to maintain the morphology and dimension of the precursors [46]. In addition, another "gas–solid" reaction mode is explored for the synthesis of metal phosphates. That is, $Co_3(PO_4)_2$ precursor was deposited on substrates by hydrothermal method, which was then reduced to Co_2P by hydrogen. As such, toxic PH_3 can be avoided (Fig. 8) [47].

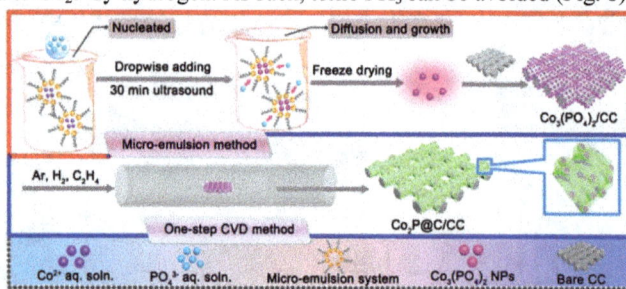

Fig. 8: *Schematic illustration of $Co_2P@C$ synthesis via a one-step CVD method using a $Co_3(PO_4)_2$ precursor [47].*

Other feasible ways have been developed for the synthesis of cobalt phosphide. For instances, the electrochemical cathodic reduction was utilized to deposit cobalt phosphide at room temperature [48]. Hydrothermal synthesis method can also be carried out at a temperature below 200°C to complete phosphorization, where red or white phosphorus and metal salts are used as phophorous source and cobalt source. It can promote the formation of various nanostructures, such as branched and hollow nanomaterials [49]. In addition, it is also feasible to fabricate cobalt phosphides with abundant structures by combining multiple synthetic routes.

To further improve the performance of cobalt phosphides, structural modification, foreign element doping, and compositing have been tried. Moreover, the rational design of microstructure or heterostructure to form a satisfying connection interface between the support and catalysts could promote the electron transition between the catalyst and the current collector. In addition, how to let the produced gas bubbles escape easily from the

Electrochemical Water Splitting: Materials and Applications   Materials Research Forum LLC
Materials Research Foundations **59** (2019) 59-96   doi: https://doi.org/10.21741/9781644900451-3

catalyst surface should also be considered. As demonstrated by Schaak et al. [50], highly branched CoP nanorods can be prepared by heating trioctylphosphine and cobalt acetylacetonate in the presence of trioctylphosphine oxide (Fig. 9 A). Herein, the spaces among the neighboring branches should facilitate the diffusion of electrolyte and generated gas.

Similarly, Sun *et al.* [51] reported a porous hedgehog-like CoP microspheres electrode with a 3D interconnected hierarchical structure, which also provided abundant spaces among the nanowire, and the interconnected configuration was also in favor of electron transfer (Fig. 9 B). As a result, the obtained CoP electrode needed 45 and 60 mV overpotentials to afford 10 mA cm^{-2} current density in acidic and alkaline solutions, respectively (Fig. 9 C). Besides, the design of super-hydrophilic and super-aerophobic nanostructure is also an efficient way to overcome the adhesion of produced gas bubbles. For example, Zhou *et al.* [52] prepared uniform CoP nanowires with good property of superhydrophilicity and superaerophobicity by a gas–solid reaction route, which efficiently evacuated the generated bubbles and assured full contact of the electrocatalyst and the electrolyte solution (Fig. 9 D).

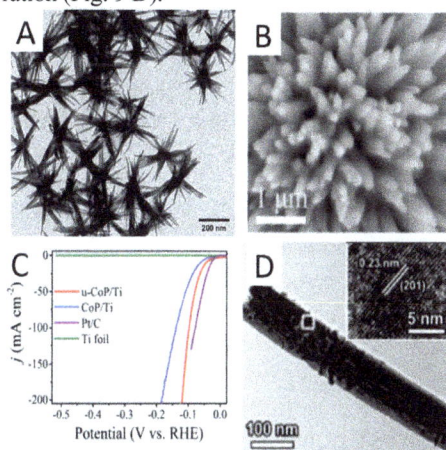

Fig. 9: *(A) TEM image of nanostructured CoP s. (B) SEM image of CoP. (C) Polarization curves of u-CoP/Ti, CoP/Ti, Pt/C/Ti, and Ti foil [51]. (D) TEM image of CoP nanowire with corresponding inserted HRTEM image [52].*

Electrochemical Water Splitting: Materials and Applications   Materials Research Forum LLC
Materials Research Foundations **59** (2019) 59-96   doi: https://doi.org/10.21741/9781644900451-3

It is well known that the electrolysis efficiency is highly dependent on the electrical conductivity and the active site dispersity. Good conductivity can distinctly reduce the applied voltage of the electrode, the overpotential and energy consumption. Meanwhile, the high dispersity of electrocatalysts can assure the adequate exposure of active catalytic sites, and enhance the surface area for the reaction per geometric area. To date, numerous works presented the composition of catalysts with conductive (N-rich) CNTs, graphene, carbon dots, etc., that will not only improve the catalyst dispersity and exposed extra HER active sites since the matrix of high surface area, but also increase the conductivity. Furthermore, the carbon material composition may also modulate the electrondensity to enhance the HER activity and stability of composite [17]. For instance, Liu *et al.* [53] loaded CoP and Co_2P on N doped CNTs materials to enhance the performance and considered that the enhancement should result from the high ratio of P:Co, strong interaction between CoP and CNTs, and the doping of N element in carbon. Moreover, Li *et al.* [47] wrapped carbon framework around Co_2P during the phosphorization process and found that the covered carbon not only improved the conductivity of composite and exposed more active sites of Co_2P for HER but also facilitated the rapid spread of formed H_2 bubbles.

Besides, adjusting the element ratio or doping the second element is also a useful route to increase the intrinsic catalyst activity. DFT calculations proved that P atom ratio is a crucial factor for HER activity of TMPs. Electronegative phosphorous atom will attract electrons from the metal atoms, which serves as a base active site for capturing positively charged proton. As such, the HER performance could be facilitated with the increase of the molar ratio of phosphorous. Schaak *et al.* [41] synthesized hollow Co_2P and CoP nanoparticles with equivalent morphology by setting the calcination temperature. As a result, CoP revealed better property than the Co_2P for the HER in acidic solution, identifying that CoP with more phosphorous molar ratio may provide more active sites (Fig. 6 B-E). Moreover, the second element doping is another efficient strategy to enhance the catalyst intrinsic activity by fine-adjustment of the electronic distribution. Many types of research have found that the foreign metal doping could optimize the thermos neutral ΔG_{H*} and adjust the interactions between the catalytic species and H atoms. Typically, as shown in Fig.10 a-c, Chen *et al.* [54] found that the ΔG_{H}^{*} of CoP (101) surface can be changed from -0.14 to -0.11 eV after the Mn doping. The electronic structure analysis proved that the Mn dopant could move the electrons to the adjacent phosphorous and Co atoms. By the Bader charge analysis, it is also found that the Co atom scan get 0.38 electrons from the nearby Mn atoms. Meanwhile, the interaction between H and Co atoms is weakened so that the absolute value of ΔG_{H*} is much closer to 0. As expected, Mn doped CoP exhibited a much better HER performance than the original CoP one (Fig. 10 d, e).

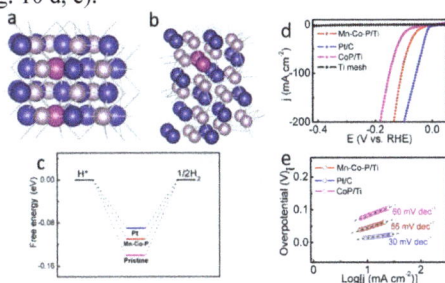

Fig. 10: *(a) Top and (b) side view Mn−Co−P(101) Gray, blue, and purple balls express the P, Co, and Mn atoms, respectively. (c) Free-energy diagram for HER on pristine Pt, CoP, and Mn−Co−P. (d) Polarization curves for CoP/Ti, Mn−Co−P/Ti, Pt/C on Ti mesh and bare Ti mesh,. (e) Tafel plots for CoP/Ti, Mn−Co−P/Ti, and Pt/C on Ti mesh [54].*

2.4   Co oxide

Due to the abundant and diversiform d-orbitals, cobalt oxides endow high activity for catalyzing water electrolysis [55, 56]. Cobalt oxides are considered as the promising low-cost OER catalysts for alkaline water electrolysis for decades, and the pure transition metal oxides are generally inactive for catalyzing HER due to the unbefitting hydrogen adsorption energy [57]. However, to wing the good stability within a wide range of electrochemical window in alkaline solution, considerable efforts have been devoted to overcome these barriers, and some efficient tactics are developed to synthesize the high active HER cobalt oxide catalysts. For alkaline water electrolysis, the oxidation state of Co atom in cobalt oxides is very important for the modulation of intrinsic catalytic activity. Especially, it is considered that the Co^{4+} formation due to the Co^{3+} oxidation in the electric field will promote OER which will play as active sites for OH^- adsorption. On the other hand, the reduction of Co^{3+} to Co^{2+} in an alkaline solution at the cathode is favorable for the HER since it can effectively adsorb H^+ ions. Therefore, cobalt oxide can also be used as HER electrocatalyst.

Due to the poor acid corrosion resistances of most cobalt oxides or cobaltites, it is still a challenge to fabricate cobalt oxide/cobaltite based HER catalysts used in acidic solutions. Interestingly, a layered mixed oxide of $Ca_3Co_4O_9$ with good acid corrosion resistance was successfully used for catalyzing HER in 0.5 M H_2SO_4 [58]. To date, several attractive cobalt oxide based catalysts have been evaluated for HER in alkaline solution. For

Electrochemical Water Splitting: Materials and Applications   Materials Research Forum LLC
Materials Research Foundations **59** (2019) 59-96   doi: https://doi.org/10.21741/9781644900451-3

example, Qiao *et al.* [59] introduced tensile strain into the surface of CoO nanorod, and converted the inactive material to an efficient electrocatalyst for HER. Based on the experimental data and theoretical calculations, the outstanding performance of the activated CoO catalyst was considered to be attributed to a large number of O-vacancies located on the outermost surface of catalysts, which can promote the water dissociation and weaken hydrogen adsorption energy toward the optimum state. As mentioned above, carbon materials are generally employed to enhance the catalyst conductivity, and the heteroatom doping of CoO_x on carbon also endows it with certain catalytic activity. For example, Jin *et al.* [25] synthesized cobalt-cobalt oxide/N-doped carbon hybrids ($CoO_x@CN$) as a HER and OER bifunctional catalyst for alkaline water electrolysis, which revealed low onset potential and good stability for HER. It is considered that the enhanced performance of $CoO_x@CN$ composite is attributed to the high conductivity of N-doped carbon, and the interactive function of Co metal and Co oxides.

Many studies exhibited that the bimetallic oxides have better intrinsic HER/OER property than the single metal oxides. In particular, it is found that the HER activity of bimetallic oxides can be easily improved by controllably modulating the electronic and valence states of the metal elements [60]. For instances, Gao *et al.* [61] applied hierarchical $NiCo_2O_4$ hollow micro cuboids for HER, which required 110 mV overpotential to support 10 mA cm^{-2} current density. Other attractive work was demonstrated by Wang *et al.* [62]. They prepared 3D hierarchical Co nanoparticles modified $Co_2Mo_3O_8$ nanosheets by thermal reduction of $CoMoO_4 \cdot nH_2O$ precursor. Interestingly, the dehydration and reduction process induced the formation of nanoporous morphology and Co vacancy defective structure, which resulted in a high HER catalytic activity.

Moreover, the potential of transition metal oxides to overcome sluggish HER kinetics was discussed by Markovic *et al. [63]* and it is considered that noble metal catalysts are adept at the attraction and recombination of the H intermediates, but the activity is limited by the prior step of water dissociation. On the contrary, the metal oxides are active for cracking the O-H bond, but inert at transferring the formed H_{ad} intermediates to hydrogen. Therefore, satisfactory HER catalyst can be fabricated by designing bifunctional metal oxide-metal systems, in which the catalytic specializations of metals and metal oxides are combined. Based on these considerations, various cobalt based oxides are composited with other HER catalysts including metals, sulfides, phosphides, and so on. Typically, as shown in Fig. 11, Liu group [64] synthesized Li intercalated 2D MoS_2 confined $Co(OH)_2$ nanoparticle composite, which remarkably promoted the H_2O splitting and revealed enhanced HER performance is comparing with the corresponding pure catalyst. Moreover, the DFT calculations indicated that the synergetic effect of

MoS_2 nanosheets and $Co(OH)_2$ nanoparticles should be attributed to the excellent HER activity.

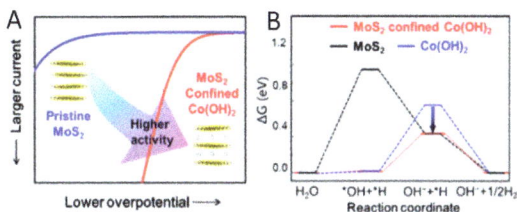

Fig. 11: *(A) HER activity of MoS_2 confined $Co(OH)_2$ and pure MoS_2. (B) Free energy diagrams of HER on the edges of $Co(OH)_2$, MoS_2, and MoS_2 confined $Co(OH)_2$ sample [64].*

2.5   Cobalt (Co) sulfides

CoS_2 is emerging as promising HER catalyst with superior HER activity to the similarly structured FeS_2 and NiS_2. To date, several feasible routes have been developed to synthesize transition metal chalcogenides. Especially, the hydrothermal method with steady temperatures and fast reaction kinetics is the most popular approach to prepare CoS_2 in a large scale [65, 66]. Moreover, the hydrothermal method is also a facile way to composite CoS_2 with other conductive and catalytic materials. Especially, it is found that compositing with carbon based materials such as CNTs, carbon fiber, graphene, and carbon dots could not only improve the conductivity and dispersability of CoS_2, but also provide more catalytic active sites. For instances, Ramakrishna *et al.* [67] fabricated an active 3D CoS_2/G-CNT with nano-heterostructure for HER by hydrothermal treatment with vacuum filtration method. They believed that high activity is attributed to the synergistic function of CoS_2, graphene, and CNTs. Addition to carbon materials, other HER catalysts have also been composited with Co sulfides by the hydrothermal method to reveal fascinating synergetic effect. Chen *et al.* [68] designed a hybrid core-shell nanomaterial of CoS_2 nanowire@MoS_2 nanosheet arrays by hydrothermal deposition combined with thermal sulfuration method, which exhibited a high HER activity with 87 mV overpotential to support 10 mA cm^{-2} current density.

Another popular synthesis approach of cobalt sulfides is the vapor deposition method, which includes sulfurization process over metal or metal oxide, thermal thiosalts decomposition and evaporation and recrystallization from powders [69]. This method can result in nanostructured chalcogenides with high quality on a large scale and the good interfaces between nanocatalysts and supports to promote charge transportation.

Electrochemical Water Splitting: Materials and Applications   Materials Research Forum LLC
Materials Research Foundations **59** (2019) 59-96   doi: https://doi.org/10.21741/9781644900451-3

Moreover, through modulation of the precursors ratio, the component of obtained catalysts can be adjusted to enhance the chemical and physical performance [70]. Stimulated by these merits, a lot of researches has been conducted to fabricate TMDs for HER via this way. For instance, CoS_2 nanowires and microwires were prepared by sulfurizing Co films, β-$Co(OH)_2$ microwires and $Co(OH)(CO_3)_{0.5} \cdot xH_2O$ nanowires in S vapor atmosphere [71].

Recently, an attractive method, the electrochemical deposition route which can easily deposit amorphous cobalt sulfide on conductive substrates at room temperature, has been used. Electrodeposition modes endow special advantages: a series of operational parameters, which can affect the formation and electrochemical performance of the grown materials, can be adjusted so as to achieve a good electrode [72]. It is found that the electrodeposition derived amorphous cobalt sulfides had high activity for HER, especially in neutral media [73]. However, the amorphous sulfide is unstable in acid solution and inert in alkaline solution.

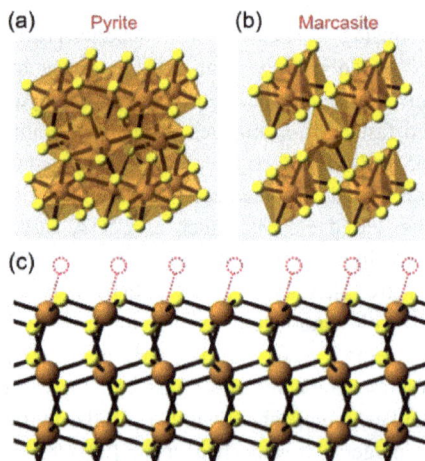

Fig. 12: *(a) pyrite-type CoS_2 structure, where orange and yellow balls represent Co and S atoms respectively. (b) marcasite-type CoS_2 structures. (c) Side-view of the nonpolar pyrite (100) surface [74].*

Current studies indicated that crystalline CoS_2 materials have become a strong competitor among precious metal-free HER catalysts. Fig. 12 shows the orthorhombic macarsite-type and cubic pyrite-type structures of crystalline cobalt sulfide where the Co atoms are

Electrochemical Water Splitting: Materials and Applications   Materials Research Forum LLC
Materials Research Foundations **59** (2019) 59-96   doi: https://doi.org/10.21741/9781644900451-3

octahedrally bonded to nearby S atoms [74]. The theoretical study proved that the barrier of H adsorption energy over Co atom is not remarkable [75]. Especially, CoS_2 and $CoSe_2$ materials possess intrinsically superior conductivity and stability than the commonly known MoS_2 and WS_2 so that they presented excellent catalytic activity without phase conversion [69]. Motivated by this, substantial effort has been devoted to the fabrication of novel crystalline CoS_2 catalysts to further enhance its performance for HER.

Generally, the improvement of CoS_2 electrocatalysts could be realized by the following three ways: (i) adjusting the bond strength of catalyst-reactant, (ii) optimizing the catalyst electronic property, and (iii) designing rational nanostructure to increase the surface area and active sites. For example, as revealed in Fig. 13, Jin et al. [71] designed metallic CoS_2 catalysts with three kinds of morphologies, i.e., nanowire, microwire and film. It is found that the microstructured CoS_2 had enlarged surface area and increased HER activity. Furthermore, the microstructure of obtained CoS_2 also facilitates the diffusion of produced gas bubbles from the catalyst surface.

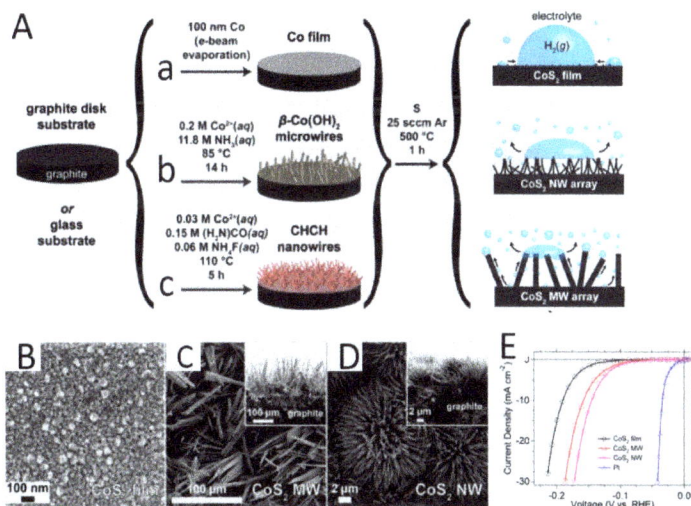

Fig. 13: *(A) Schematic diagram of the synthesis of CoS_2 film, microwire , or nanowire array, SEM images of (B) a polycrystalline CoS_2 film, (C) CoS_2 microwire (D) CoS_2 nanowire(with cross-sectional inserted SEM images) (E) Polarization curves of various CoS_2 electrodes for HER electrolysis [71].*

Besides, the electronic configuration of cobalt sulfides is tunable by chemical modifications such as strain engineering or doping. Especially, the doping is valuable to tune the free energy of H adsorption. As shown in Fig .14, DFT calculations proved that the second metal doping can efficiently activate the HER catalytic activity of CoS_2 [76]. Especially, the volcano plot indicated that Mn atom is among the superior dopant (Mn>Fe > Ni) for modulating the H adsorption behavior on the contiguous Co atoms. Moreover, the experimental study further confirmed that the Mn doping increased the catalytic activity and stability of CoS_2 nanowires. As a result, the overpotential @ 10 mA cm^{-2} was reduced from 187 mV with a Tafel slope of 87 mV dec^{-1} to 34 mV with a Tafel slope of 43 mV dec^{-1}.

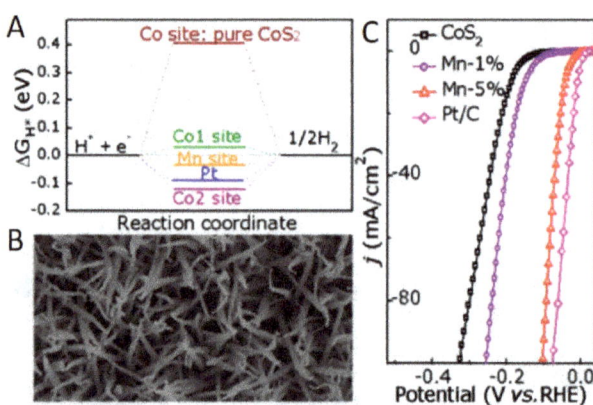

Fig. 14: *(A) H adsorption free-energy of CoS_2 and Mn- CoS_2. (B) SEM image of Mn-CoS_2 (C) Polarization curves of CoS_2, Mn- CoS_2, and Pt/C electrodes [76].*

Among the various methods for increasing HER catalytic activity of cobalt sulfides, bimetallic catalyst systems is also expected to exhibit the improved performance [77]. As other catalysts reviewed above, this enhancement is also caused by the fine-adjustment of the structural defects, conductivity, electrochemically active sites, and surface charge density. Irshad and Munichandraiah [78] considered that the bimetallic Co−Ni−S catalyst could combine the advantages of Ni−S and Co−S materials, and found that the obtained optimized Co−Ni−S electrodes had the superior HER activity than the corresponding monometallic sulfites, resulting from the higher electronic conductivity. Besides, it is reported that the enhanced activity of duplex metal sulfide can be expounded by the self-doping concept. For instance, Ni^{3+} of the material could offer additional electrons as the n-type doping, while the Co^{2+} could lead to additional holes as the p-type doping. Hence, the existence of Co^{2+} and Ni^{3+} in the Ni-Co-S resulted in a better conductivity and electrochemical activity, compared with the Ni−S and Co−S ones [79].

The selection of suitable support for fabricating heterogeneous catalysts is another way to optimize the catalyst-reactant bond strength. As discussed above, the conductivity of the catalyst can also be improved by combing sulfides with carbon materials [67]. In addition to the electrical conductivity, the free energy of hydrogen adsorption is also tunable by designing a composite catalyst with a heterostructure. Generally, the stronger the binding between the sulfide and the support is, the weaker the hydrogen adsorption is. For example, Ramakrishna and co-workers [80] fabricated porous Co_9S_8/WS_2 composites with nanoneedle, a nanobelt, and nanorhombus arrays on Ti foil, which presented a good catalyzing HER in the basic solution with 138 mV overpotential @10 mA cm^{-2}. This high activity was considered to be due to the unusual 3D structure and the synergistic function of WS_2 and Co_9S with a large electrochemically active surface area, efficient charge transport, and good electric conductivity. DFT calculations further indicated that the suitable energetics and kinetics for the H_2 formation and H adsorption lead to the enhanced HER catalytic activity.

2.6   Cobal selenides

Selenium and sulfur are congeners in group VIA of the periodic table, which shares several similar and discrepant properties. Since both have 6 electrons in the outer most shells, which generally determines the chemical properties of the compound, the cobalt selenide may show similar performance for catalyzing HER as the cobalt sulfide. However, the different number and configuration of intimal electrons of Se and S should generate some distinct characteristics. For instance, the cobalt selenide has a more obvious metallic property and lower ionization energy than sulfide [81].

Recently, the group of Cui [74] analyzed the HER performances of $FeSe_2$, $CoSe_2$, and $NiSe_2$, and found that the $CoSe_2$ had the highest activity. All of these metal dichalcogenides have orthorhombic macarsite-type or cubic pyrite-type crystalline, where the metal atoms are bonded with Se atoms octahedrally. As such, it is considered that the partially filled, e_g band of $CoSe_2$ should result in the excellent activity (similar to Fig. 15). Recently, Zhang's group [82] prepared polymorphic cobalt selenide with both cubic and orthorhombic crystal by annealing the electrodeposited cobalt selenide at 300 °C. As a result, the polymorphic $CoSe_2$ exhibited 70 mV onset overpotential with ~30 mV dec^{-1} Tafel slope, indicating a higher activity than the cubic $CoSe_2$, CoSe, and amorphous $CoSe_x$ (Fig. 15 A, B). Furthermore, Wu et al. [83] transformed the orthorhombic phase $CoSe_2$ to the cubic phase $CoSe_2$, and found that the electrical conductivity, H_2O adsorption energy and transformation efficiency of H_{ads} to H_2 are highly dependent on the phase of $CoSe_2$ material. As a result, comparing with the orthorhombic phase $CoSe_2$, the metallic cubic $CoSe_2$ electrocatalyst exhibited improved HER activity in an alkaline media.

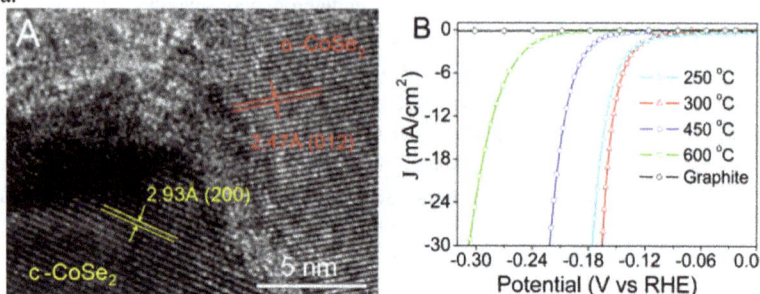

Fig. 15: *(A) TEM image of p-CoSe₂. (B) Polarization curves of CoSe_x (250 °C), p-CoSe₂ (300 °C), c-CoSe₂ (450 °C), CoSe (600 °C) and graphite [82].*

In particular, cobalt selenides have better stability in strong acidic media than cobalt sulfides, which provides a congenital advantage for electrolysis applications in the strong acids. To date, numerous nanostructured selenides are fabricated by using vapor deposition, hydrothermal synthesis, electrodeposition methods. Cui's group [84] prepared various $CoSe_2$ nanostructured materials. For example, $CoSe_2$ nanoparticle was fabricated by employing carbon black nanoparticle template. Meanwhile, they prepared another $CoSe_2$ nanoparticle with high surface area by selenizing cobalt oxide under Se vapor atmosphere, and such a 3D $CoSe_2$ nanomaterial need a low overpotential of 180 mV @100 mA cm^{-2} current density (Fig. 16 A and B). Besides, Lewis and co-workers[85] prepared amorphous CoSe in a polymeric Se matrix through the electrodeposition route in an aqueous solution of $Co(C_2H_3O_2)_2$ and SeO_2, and the prepared electrode needed 135 mV overpotential to afford 10 mA cm^{-2} in the acidic solution (Fig. 16 C and D). Moreover, Sun *et al.* [86] designed a $CoSe_2$ nanowire array via a hydrothermal selenization of a Co(OH)F nanowire array, which also revealed good HER catalytic activity and durability in the acidic media (Fig. 16 E and F).

Furthermore, $CoSe_2$ catalysts can be combined with other conductive and/or catalytic materials to improve its performance for HER. For instance, Yu *et al.* [87] reported a Ni/NiO/$CoSe_2$ composite in which Ni/NiO core–shell nanoparticle implanted on the surface of $CoSe_2$ nanobelts. The existence of $CoSe_2$ controlled the nucleation and generation of Ni. The obtained hybrid electrode provided more active sites for HER, in which the Ni core provided high conductivity, the thin NiO layer acted as Lewis acid to promote dissociation of the water molecule. Comparing with the corresponding single materials, better HER performance was achieved.

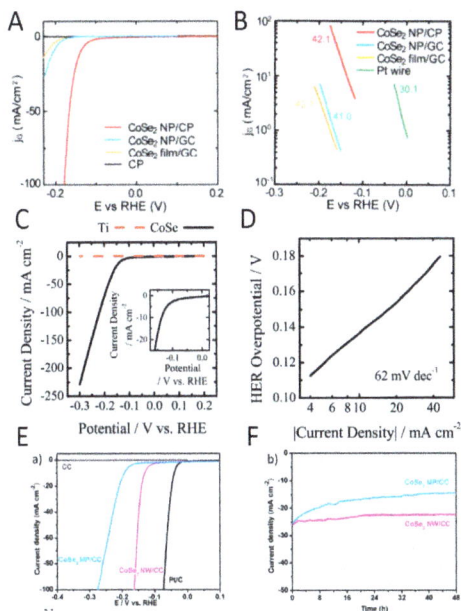

Fig. 16: *(A) Polarization curves of $CoSe_2$ nanoparticle/carbonfiber paper (CP) electrode, along with $CoSe_2$ nanoparticle/glassy carbon, $CoSe_2$ film/glassy carbon, and CP. (B) Corresponding Tafel plots in comparison with a Pt wire [84]. (C) Polarization curves of Tifoil and of a cobalt selenide electrodes (inset highlights behavior at low overpotentials). (D) Tafel plot derived from data in (C) [85]. (E) Polarization curves of $CoSe_2$ nanowire, $CoSe_2$microparticleand Pt/C [86].*

2.7   Binary nonmetal cobalt compounds

Currently, it is a rising hot topic to combining metal with bi-nonmetal element to controllably modify and engineer the electronic distribution for achieving the synergistic regulation of activity, conductivity, and stability of HER electrocatalysts [88]. For Co-based catalysts, the introduction of second nonmetal element into the host lattices could also cause obvious regulation of the electronic structure of the cobalt compound host and

Electrochemical Water Splitting: Materials and Applications   Materials Research Forum LLC
Materials Research Foundations **59** (2019) 59-96   doi: https://doi.org/10.21741/9781644900451-3

change the H adsorption free energy, leading to good HER catalytic properties. It is found that the introduction of the second nonmetal atom with smaller or larger radii could lead to the distortion of host lattices, which can affect the absorption energy of the reactants and intermediates. In addition, the nonmetal atom incorporation could also result in a smaller band gap than the parent material, and thus obtaining better conductivity with higher electrocatalytic activity [89]. For example, CoSP catalysts exhibited higher catalytic activity than CoS_2 or CoP with a low overpotential of 48 mV @ 10 mA cm^{-2}. Cobalt disulfide endows a pyrite-type crystalline, which generally maintains the same crystal form after P doping. Based on the first-principle calculations, the cubic structure CoS_2 has higher stability than the monoclinic one, however, the stability of monoclinic CoP_2 becomes higher as S is replaced by phosphorous, and the monoclinic phosphosulfides could be more stable in a wide range of x. Besides, the cubic $CoS_{2-x}P_x$ has higher stability for x≤1.0. It is considered that the enhanced chemical stability is attributed to the modified electron structure of CoSP materials after the substitute of P in CoS_2, which affect the chemical bonding between Co and S/P obviously [90].

The morphologies of catalysts are highly dependent on the preparation routes, which subsequently impact the HER catalytic performance. To date, the foreign element doping is the most popular synthetic tactic to obtain binary-nonmetal cobalt compound. Based on the incorporated species, phosphorization, sulfuration, selenization, and nitridation in the doping processes have been demonstrated. The gas/solid-phase doping is also a versatile route to synthesize bi-nonmetal cobalt compounds. Similar to the phosphide preparation, active PH_3 gas is widely used in phosphorization process.

Meanwhile, because of the extremely toxic and lethal, some substitutions such as hypophosphites and red phosphorus powders are always utilized, which can *in situ* produce PH_3 and phosphorus vapor. For example, phosphorous can be doped into the CoS_2 crystalline by using NaH_2PO_2 as the phosphorus source at 400°C. For the S and Se incorporation, sulphur and selenium powders are popular reagents to realize sulfuration and selenization reactions. One should notice that the post processing of tail gas for the solid/gas-phase doping is extremely necessary since most P, S, Se containing vapors are toxic and lethal. Another efficient synthetic approach is to use the mixed precursors in one-step for the formation of stoichiometric compositions. However, due to the different chemical reactivity of S, Se, and P sources, the incorporation conditions should be considered, and sometimes, it is necessary to separate the reaction regions of selenization, sulfuration, and phosphorization to achieve the optimum stoichiometric compositions.

Among the family of bi-nonmetal cobalt compounds, cobalt phosphosulphides are highly attractive electrocatalysts for the HER. Herein, the phosphorous sites are speculated to be valuable for H adsorption and H_2 desorption. Due to the different atomic radium and electronegativity of phosphorous and sulphur elements, phosphorous incorporation can induce the d-band center shift and electronic structure tuning of cobalt sulfide, which greatly influence the HER catalytic activity. The theoretical simulations suggest that substitution of sulphur by phosphorous will reduce the electron occupation in antibonding e_g* orbitals because less electron valence in P phosphorous than sulphur, which leads the chemical bonding enhancement between the ligands (S/P) and metal so that their catalytic and chemical stability can be improved in the HER process [90].

Moreover, by introducing P element, the surface properties of cobalt disulfide could be finely adjusted so that the HER activity is greatly enhanced with facilitated charge transfer and promoted proton adsorption [91]. DFT calculations proved that the ΔG_{H*} of the Co sites in CoPS is higher than that in CoS_2. In addition, after the spontaneous H adsorption at the open phosphorous sites, the ΔG_{H*} at the neighboring Co sites becomes almost thermo-neutral and spontaneous. Besides, comparing with CoS_2, including Co^{2+} octahedral and S_2^{2-} dumbbells, CoPS endows a lower lattice constant and consists of Co^{3+} octahedra and dumbbells with unform P^{2-} and S^- [92]. Since the P^{2-} ligands can donate more electrons than sulphur ligands, after H is adsorbed on the open phosphorous sites, the free energy of hydorgen adsorption at the neighbor Co sites becomes almost thermoneutral and spontaneous due to the spontaneous transition between Co^{2+} and Co^{3+} (Co^{3+} is reduced to Co^{2+} in the case of H adsorption on the neighboring phosphorous sites, and then Co^{2+} is oxidized back to Co^{3+} as the H is subsequently adsorbed on Co sites). Like this, the realized CoPS based electrode needs a low 48 mV overpotential to support a current density of 10 mAcm^2[92].

It should be noted that CoPS has semiconductor characteristics since it has one less valence electron comparing with CoS_2 [92]. In this case, it is crucial to enhance the property and stability of catalysts by the fabrication of robust contact interface between electrocatalyst and conductive support for laying an electrical transfer pathway with low resistance. Moreover, the composition of catalytic and conductive materials may also reveal synergistically drive inherent activity [93]. Besides phosphorous atom, as Se element was introduced into the cobalt sulfides, the obtained CoSSe also showed high activity for HER. Moreover, since the similarities of Se and sulphur, each of them can be doped into the other compound without the phase separation [94].

Cobalt phosphides can also serve as the host to accept foreign element. For example, Zhang et al. [95] introduced oxygen into CoP and MoP nanoparticles embedded in the layered graphene, and demonstrated that the incorporation of oxygen element into phosphides could improve the inherent conductivity, and increase the active sites by elongating the M-P bond for the HER and OER. DFT simulations indicated that oxygen doping in TMPs could realize a much higher DOS across the Fermi level, which indicates the improved inherent conductivity. EXAFS measurements showed a slightly positive shift after the oxygen doping in Mo-P or Co-P, with the surface structural disorder or coordinative unsaturation with oxygen atom existence [95]. All of these could result in a higher activity for the HER.

Conclusions and outlook

This chapter summarizes the most recent researches on Co based catalysts for hydrogen evolution reaction, especially the efficient tactics to improve their performance. It is generally considered that the catalytic activity could be enhanced by improving the surface area, fabricating unique facet structure, and enhancing the conductivity. However, there are still several challenges to obtain HER electrocatalyst with high performance for the practical process. Although water electrolysis is one of the simplest reactions, its fundamental understanding at the molecular level is still not complete. Spectroscopic measurements such as Raman spectroscopy and X-ray absorption spectroscopy are always employed to in-situ study the interfacial structure during the electrolysis process. However, it should be noted that the measurement accuracy is seriously disturbed by the Faradic currents and hydrogen bubbles during the HER process.

Moreover, the theoretical prediction is also lacking in the current HER catalyst development. DFT calculation is a powerful rising tool to design and predict the catalyst performance, which is also valuable to get insight into the influences of crystal structure, electronic state, and constitution on the HER performance of the catalyst. These methods could also guide us to fabricate and optimize Co based catalysts with high performance.

In order to meet the requirements of industrial application, the performance of Co based catalysts should be further improved. As an effective way, nanostructuring the catalyst surface to increase catalytic active sites is the most popular strategy to increase the catalytic activity. When the material size is adjusted from macroscale to nanoscale, which will not only extends the electrochemically active surface area dramatically but also probably introduce structural disorders or defects. For example, Guan et al. [96] fabricated $Cu@Fe-Co(OH)_2$ core-shell structure by using a one-step electrodeposition method, which showed high catalytic performance. The crystallinity and crystal structure of Co based catalysts affects their HER performance. Numerous works have proven that alloying, compositing, doping, and defect introduction could enhance the catalytic activity by optimizing the catalyst-reactant bond strength, altering the electronic property of catalyst. In addition, the stability of Co based catalysts also need more improvements since cobalt metal alloy, chalcogenides, phosphides, carbides, and nitrides are all prone to be oxidized gradually in the air or aerated solution, forming an oxide, phosphate, and inert surface layers, which is even worse in alkaline solution. Therefore, optimizing the crystal structure or compositing it with anticorrosion materials (e.g., carbon materials) should be helpful. It is expected the stability can be maintained on the scale of months, years or decades. All in all, as other electrocatalysts, many encouraging efforts have been achieved for the Co based catalysts, but it is still necessary to decrease the fabrication cost and improve the activity and stability in the future study.

References

[1]   J. Wang, W. Cui, Q. Liu, Z. Xing, A.M. Asiri, X. Sun, Recent progress in cobalt-based heterogeneous catalysts for electrochemical water splitting, Adv. Mater. 28 (2016) 215-230. https://doi.org/10.1002/adma.201502696

[2]   M. Zeng, Y. Li, Recent advances in heterogeneous electrocatalysts for the hydrogen evolution reaction, J. Mater. Chem. A 3 (2015) 14942-14962. https://doi.org/10.1039/C5TA02974K

[3]   X. Li, X. Hao, A. Abudula, G. Guan, Nanostructured catalysts for electrochemical water splitting: current state and prospects, J. Mater. Chem. A 4 (2016) 11973-12000. https://doi.org/10.1039/C6TA02334G

[4]   T. Shinagawa, A.T. Garcia-Esparza, K. Takanabe, Insight on Tafel slopes from a microkinetic analysis of aqueous electrocatalysis for energy conversion, Sci. Reports 5 (2015) 13801. https://doi.org/10.1038/srep13801

[5]   S. Trasatti, Work function, electronegativity, and electrochemical behaviour of metals: III. Electrolytic hydrogen evolution in acid solutions, J. Electroanal. Chem. Interfacial Electrochem. 39 (1972) 163-184. https://doi.org/10.1016/S0022-0728(72)80485-6

[6]   J.K. Nørskov, T. Bligaard, A. Logadottir, J. Kitchin, J.G. Chen, S. Pandelov, U. Stimming, Trends in the exchange current for hydrogen evolution, J. Electrochem. Soc.152 (2005) J23-J26. https://doi.org/10.1149/1.1856988

[7]   M. Miles, M. Thomason, Periodic variations of overvoltages for water electrolysis in acid solutions from cyclic voltammetric studies, J. Electrochem. Soc. 123 (1976) 1459-1461. https://doi.org/10.1149/1.2132619

[8]   B. Liu, L. Zhang, W. Xiong, M. Ma, Cobalt-nanocrystal-assembled hollow nanoparticles for electrocatalytic hydrogen generation from neutral-pH water, Angew. Chem. Int. Ed. 55 (2016) 6725-6729. https://doi.org/10.1002/anie.201601367

[9]   B. Lassalle-Kaiser, A. Zitolo, E. Fonda, M. Robert, E. Anxolabéhère-Mallart, In situ observation of the formation and structure of hydrogen-evolving amorphous cobalt electrocatalysts, ACS Energy Lett. 2 (2017) 2545-2551. https://doi.org/10.1021/acsenergylett.7b00789

[10]   B. Qiao, A. Wang, X. Yang, L.F. Allard, Z. Jiang, Y. Cui, J. Liu, J. Li, T. Zhang, Single-atom catalysis of CO oxidation using Pt_1/FeO_x, Nature Chem. 3 (2011) 634. https://doi.org/10.1038/nchem.1095

[11]   H. Liu, X. Peng, X. Liu, Single atom catalysts for the hydrogen evolution reaction, Chem. Electro. Chem. 5 (2018) 2963-2974. https://doi.org/10.1002/celc.201800507

[12]   H. Fei, J. Dong, M. J. Arellano-Jiménez, G. Ye, N. D. Kim, E. L. Samuel, Z. Peng, Z. Zhu, F. Qin, J. Bao, M. J. Yacaman, P. M. Ajayan, D. Chen, J. M. Tour,Atomic cobalt on nitrogen-doped graphene for hydrogen generation, Nature Commun. 6 (2015) 8668. https://doi.org/10.1038/ncomms9668

[13]   Y. Zhang, W. Li, L. Lu, W. Song, C. Wang, L. Zhou, J. Liu, Y. Chen, H. Jin, Y. Zhang, Tuning active sites on cobalt/nitrogen doped graphene for electrocatalytic hydrogen and oxygen evolution, Electrochim. Acta 265 (2018) 497-506. https://doi.org/10.1016/j.electacta.2018.01.203

[14]   X. Zou, X. Huang, A. Goswami, R. Silva, B.R. Sathe, E. Mikmeková, T. Asefa, Cobalt-embedded nitrogen-rich carbon nanotubes efficiently catalyze hydrogen evolution reaction at all pH values, Angew. Chem. Int. Ed. 53 (2014) 4372-4376. https://doi.org/10.1002/anie.201311111

[15]   J. Wang, D. Gao, G. Wang, S. Miao, H. Wu, J. Li, X. Bao, Cobalt nanoparticles encapsulated in nitrogen-doped carbon as a bifunctional catalyst for water electrolysis, J. Mater. Chem. A 2 (2014) 20067-20074. https://doi.org/10.1039/C4TA04337E

[16]   J. Deng, P. Ren, D. Deng, L. Yu, F. Yang, X. Bao, Highly active and durable non-precious-metal catalysts encapsulated in carbon nanotubes for hydrogen evolution reaction, Energy Environ. Sci. 7 (2014) 1919-1923. https://doi.org/10.1039/C4EE00370E

[17]   J. Deng, P. Ren, D. Deng, X. Bao, Enhanced electron penetration through an ultrathin graphene layer for highly efficient catalysis of the hydrogen evolution reaction, Angew. Chem. Int. Ed. 54 (2015) 2100-2104. https://doi.org/10.1002/anie.201409524

[18]   Y. Yang, Z. Lun, G. Xia, F. Zheng, M. He, Q. Chen, Non-precious alloy encapsulated in nitrogen-doped graphene layers derived from MOFs as an active and durable hydrogen evolution reaction catalyst, Energy Environ. Sci. 8 (2015) 3563-3571. https://doi.org/10.1039/C5EE02460A

[19]   M. Jakšić, Advances in electrocatalysis for hydrogen evolution in the light of the Brewer-Engel valence-bond theory, Int. J. Hydrogen Energy 12 (1987) 727-752. https://doi.org/10.1016/0360-3199(87)90090-5

[20]   S.E. Fosdick, S.P. Berglund, C.B. Mullins, R.M. Crooks, Evaluating electrocatalysts for the hydrogen evolution reaction using bipolar electrode arrays: bi- and trimetallic combinations of Co, Fe, Ni, Mo, and W, ACS Catal. 4 (2014) 1332-1339. https://doi.org/10.1021/cs500168t

[21]   J. Greeley, T.F. Jaramillo, J. Bonde, I. Chorkendorff, J.K. Nørskov, Computational high-throughput screening of electrocatalytic materials for hydrogen evolution, Nature Mater. 5 (2006) 909. https://doi.org/10.1038/nmat1752

[22]   X. Feng, X. Bo, L. Guo, CoM (M= Fe, Cu, Ni)-embedded nitrogen-enriched porous carbon framework for efficient oxygen and hydrogen evolution reactions, J. Power Sources 389 (2018) 249-259. https://doi.org/10.1016/j.jpowsour.2018.04.027

[23]   F. Rosalbino, S. Delsante, G. Borzone, E. Angelini, Electrocatalytic behaviour of Co–Ni–R (R= Rare earth metal) crystalline alloys as electrode materials for hydrogen evolution reaction in alkaline medium, Int. J. Hydrogen Energy 33 (2008) 6696-6703. https://doi.org/10.1016/j.ijhydene.2008.07.125

[24]   A. Oh, Y.J. Sa, H. Hwang, H. Baik, J. Kim, B. Kim, S.H. Joo, K. Lee, Rational design of Pt–Ni–Co ternary alloy nanoframe crystals as highly efficient catalysts toward the alkaline hydrogen evolution reaction, Nanoscale 8 (2016) 16379-16386. https://doi.org/10.1039/C6NR04572C

[25]   H. Jin, J. Wang, D. Su, Z. Wei, Z. Pang, Y. Wang, In situ cobalt–cobalt oxide/N-doped   carbon hybrids as superior bifunctional electrocatalysts for hydrogen and oxygen evolution, J. Am. Chem. Soc. 137 (2015) 2688-2694. https://doi.org/10.1021/ja5127165

[26]   J. Houston, G. Laramore, R.L. Park, Surface electronic properties of tungsten, tungsten carbide, and platinum, Science 185 (1974) 258-260. https://doi.org/10.1021/ja5127165

[27]   S. Dong, X. Chen, X. Zhang, G. Cui, Nanostructured transition metal nitrides for energy storage and fuel cells, Coord. Chem. Rev. 257 (2013) 1946-1956. https://doi.org/10.1016/j.ccr.2012.12.012

[28]   B. Cao, G.M. Veith, J.C. Neuefeind, R.R. Adzic, P.G. Khalifah, Mixed close-packed cobalt molybdenum nitrides as non-noble metal electrocatalysts for the hydrogen evolution reaction, J. Am. Chem. Soc. 135 (2013) 19186-19192. https://doi.org/10.1021/ja4081056

[29]   N. Han, P. Liu, J. Jiang, L. Ai, Z. Shao, S. Liu, Recent advances in nanostructured metal nitrides for water splitting, J. Mater. Chem.A 6 (2018) 19912-19933. https://doi.org/10.1039/C8TA06529B

[30] P. Chen, K. Xu, Y. Tong, X. Li, S. Tao, Z. Fang, W. Chu, X. Wu, C. Wu, Cobalt nitrides as a class of metallic electrocatalysts for the oxygen evolution reaction, Inorg. Chem. Front. 3 (2016) 236-242. https://doi.org/10.1039/C8TA06529B

[31] Z. Chen, Y. Ha, Y. Liu, H. Wang, H. Yang, H. Xu, Y. Li, R. Wu, In situ formation of cobalt nitrides/graphitic carbon composites as efficient bifunctional electrocatalysts for overall water splitting, ACS Appl. Mater. interfaces 10 (2018) 7134-7144. https://doi.org/10.1021/acsami.7b18858

[32] Z. Chen, Y. Song, J. Cai, X. Zheng, D. Han, Y. Wu, Y. Zang, S. Niu, Y. Liu, J. Zhu, Tailoring the d-band centers enables Co_4N nanosheets to be highly active for hydrogen evolution catalysis, Angew. Chem. Int. Ed. 57 (2018) 5076-5080. https://doi.org/10.1002/anie.201801834

[33] Y. Zhong, X. Xia, F. Shi, J. Zhan, J. Tu, H.J. Fan, Transition metal carbides and nitrides in energy storage and conversion, Adv. Sci.3 (2016) 1500286. https://doi.org/10.1002/advs.201500286

[34] J. Yin, Y. Li, F. Lv, Q. Fan, Y.Q. Zhao, Q. Zhang, W. Wang, F. Cheng, P. Xi, S. Guo, NiO/CoN Porous nanowires as efficient bifunctional catalysts for Zn–air batteries, ACS nano 11 (2017) 2275-2283. https://doi.org/10.1021/acsnano.7b00417

[35] M. Fan, Y. Zheng, A. Li, K. Li, H. Liu, Z.A. Qiao, Janus CoN/Co cocatalyst in porous N-doped carbon: toward enhanced catalytic activity for hydrogen evolution, Catal. Sci. Technol. 8 (2018) 3695-3703. https://doi.org/10.1039/C8CY00571K

[36] Y. Zhang, B. Ouyang, J. Xu, G. Jia, S. Chen, R.S. Rawat, H.J. Fan, Rapid synthesis of cobalt nitride nanowires: Highly efficient and low-cost catalysts for oxygen evolution, Angew. Chem. Int. Ed. 55 (2016) 8670-8674. https://doi.org/10.1002/anie.201604372

[37] S. Bhattacharyya, S. Kurian, S. Shivaprasad, N. Gajbhiye, Synthesis and magnetic characterization of $CoMoN_2$ nanoparticles, J. Nanoparticle Res. 12 (2010) 1107-1116. https://doi.org/10.1007/s11051-009-9639-5

[38] Y. Wang, D. Liu, Z. Liu, C. Xie, J. Huo, S. Wang, Porous cobalt–iron nitride nanowires as excellent bifunctional electrocatalysts for overall water splitting, Chem. Commun. 52 (2016) 12614-12617. https://doi.org/10.1039/C6CC06608A

[39] S. Carenco, D. Portehault, C. Boissiere, N. Mezailles, C. Sanchez, Nanoscaled metal borides and phosphides: recent developments and perspectives, Chem. Rev. 113 (2013) 7981-8065. https://doi.org/10.1021/cr400020d

[40] P. Liu, J.A. Rodriguez, Catalysts for hydrogen evolution from the [NiFe] hydrogenase to the Ni_2P (001) surface: the importance of ensemble effect, J. Am. Chem. Soc. 127 (2005) 14871-14878. https://doi.org/10.1021/ja0540019

[41]   J.F. Callejas, C.G. Read, E.J. Popczun, J.M. McEnaney, R.E. Schaak, Nanostructured Co_2P electrocatalyst for the hydrogen evolution reaction and direct comparison with morphologically equivalent CoP, Chem. Mater. 27 (2015) 3769-3774. https://doi.org/10.1021/acs.chemmater.5b01284

[42]   Y. Li, M.A. Malik, P. O'Brien, Synthesis of single-crystalline cop nanowires by a one-pot metal− organic route, J. Am. Chem. Soc. 127 (2005) 16020-16021. https://doi.org/10.1021/ja055963i

[43]   J. Tian, Q. Liu, A. M. Asiri, X. Sun, Self-supported nanoporous cobalt phosphide nanowire arrays: An efficient 3D hydrogen-evolving cathode over the wide range of pH 0−14, J. Am. Chem. Soc. 136(2014) 7587−7590. https://doi.org/10.1021/ja503372r

[44]   D. Yang, J. Zhu, X. Rui, H. Tan, R. Cai, H.E. Hoster, D.Y. Yu, H.H. Hng, Q. Yan, Synthesis of cobalt phosphides and their application as anodes for lithium ion batteries, ACS appl. mater. interfaces 5 (2013) 1093-1099. https://doi.org/10.1021/am302877q

[45]   D.H. Ha, B. Han, M. Risch, L. Giordano, K.P. Yao, P. Karayaylali, Y. Shao-Horn, Activity and stability of cobalt phosphides for hydrogen evolution upon water splitting, Nano Energy 29 (2016) 37-45. https://doi.org/10.1016/j.nanoen.2016.04.034

[46]   X. Yang, A.Y. Lu, Y. Zhu, M.N. Hedhili, S. Min, K.W. Huang, Y. Han, L.J. Li, CoP nanosheet assembly grown on carbon cloth: A highly efficient electrocatalyst for hydrogen generation, Nano Energy 15 (2015) 634-641. https://doi.org/10.1016/j.nanoen.2015.05.026

[47]   C. Ye, M.Q. Wang, G. Chen, Y.H. Deng, L.J. Li, H.Q. Luo, N.B. Li, One-step CVD synthesis of carbon framework wrapped Co_2P as a flexible electrocatalyst for efficient hydrogen evolution, J. Mater. Chem. A 5 (2017) 7791-7795. https://doi.org/10.1039/C7TA00592J

[48]   Y.P. Zhu, Y.P. Liu, T.Z. Ren, Z.Y. Yuan, Self-supported cobalt phosphide mesoporous nanorod arrays: a flexible and bifunctional electrode for highly active electrocatalytic water reduction and oxidation, Adv. Funct. Mater. 25 (2015) 7337-7347. https://doi.org/10.1002/adfm.201503666

[49]   Y. Lv, X. Wang, T. Mei, J. Li, J. Wang, Reduced graphene oxide-supported cobalt phosphide nanoflowers via in situ hydrothermal synthesis as pt-free effective electrocatalysts for oxygen reduction reaction, Nano 13 (2018) 1850047. https://doi.org/10.1142/S1793292018500479

[50]   E.J. Popczun, C.W. Roske, C.G. Read, J.C. Crompton, J.M. McEnaney, J.F. Callejas, N.S. Lewis, R.E. Schaak, Highly branched cobalt phosphide nanostructures

for hydrogen-evolution electrocatalysis, J. Mater. Chem. A 3 (2015) 5420-5425. https://doi.org/10.1039/C4TA06642A

[51]   D. Zhou, L. He, W. Zhu, X. Hou, K. Wang, G. Du, C. Zheng, X. Sun, A.M. Asiri, Interconnected urchin-like cobalt phosphide microspheres film for highly efficient electrochemical hydrogen evolution in both acidic and basic media, J. Mater. Chem.A 4 (2016) 10114-10117. https://doi.org/10.1039/C6TA03628G

[52]   C. Lyu, J. Zheng, R. Zhang, R. Zou, B. Liu, W. Zhou, Homologous Co_3O_4|CoP nanowires grown on carbon cloth as a high-performance electrode pair for triclosan degradation and hydrogen evolution, Mater. Chem. Front. 2 (2018) 323-330. https://doi.org/10.1039/C7QM00533D

[53]   Q. Liu, J. Tian, W. Cui, P. Jiang, N. Cheng, A.M. Asiri, X. Sun, Carbon nanotubes decorated with CoP nanocrystals: A highly active non-noble-metal nanohybrid electrocatalyst for hydrogen evolution, Angew. Chem. Int. Ed. 53 (2014) 6710-6714. https://doi.org/10.1002/anie.201404161

[54]   T. Liu, X. Ma, D. Liu, S. Hao, G. Du, Y. Ma, A.M. Asiri, X. Sun, L. Chen, Mn doping of CoP nanosheets array: An efficient electrocatalyst for hydrogen evolution reaction with enhanced activity at all pH values, ACS Catal. 7 (2016) 98-102. https://doi.org/10.1021/acscatal.6b02849

[55]   Z. Zhuang, W. Sheng, Y. Yan, Synthesis of monodispere Au@ Co_3O_4 core-shell nanocrystals and their enhanced catalytic activity for oxygen evolution reaction, Adv. mater. 26 (2014) 3950-3955. https://doi.org/10.1002/adma.201400336

[56]   G. Mattioli, P. Giannozzi, A. Amore Bonapasta, L. Guidoni, Reaction pathways for oxygen evolution promoted by cobalt catalyst, J. Am. Chem. Soc. 135 (2013) 15353-15363. https://doi.org/10.1021/ja401797v

[57]   M. Gong, W. Zhou, M.C. Tsai, J. Zhou, M. Guan, M.C. Lin, B. Zhang, Y. Hu, D.Y. Wang, J. Yang, Nanoscale nickel oxide/nickel heterostructures for active hydrogen evolution electrocatalysis, Nature commun. 5 (2014) 4695. https://doi.org/10.1038/ncomms5695

[58]   C.S. Lim, C.K. Chua, Z. Sofer, O. E. Jankovský, M. Pumera, Alternating misfit layered transition/alkaline earth metal chalcogenide $Ca_3Co_4O_9$ as a new class of chalcogenide materials for hydrogen evolution, Chem. Mater. 26 (2014) 4130-4136. https://doi.org/10.1021/cm501181j

[59]   T. Ling, D. Y. Yan, H. Wang, Y. Jiao, Z. Hu, Y. Zheng, L. Zheng, J. Mao, H. Liu, X. W. Du, M. Jaroniec, S. Qiao, Activating cobalt (II) oxide nanorods for efficient electrocatalysis by strain engineering, Nature Commun. 8 (2017) 1509. https://doi.org/10.1038/s41467-017-01872-y

[60]   Y. Yan, B.Y. Xia, B. Zhao, X. Wang, A review on noble-metal-free bifunctional heterogeneous catalysts for overall electrochemical water splitting, J. Mater. Chem. A 4 (2016) 17587-17603. https://doi.org/10.1039/C6TA08075H

[61]   X. Gao, H. Zhang, Q. Li, X. Yu, Z. Hong, X. Zhang, C. Liang, Z. Lin, Hierarchical $NiCo_2O_4$ hollow microcuboids as bifunctional electrocatalysts for overall water-splitting, Angew. Chem. 128 (2016) 6398-6402. https://doi.org/10.1002/ange.201600525

[62]   M. Zang, N. Xu, G. Cao, Z. Chen, J. Cui, L. Gan, H. Dai, X. Yang, P. Wang, Cobalt molybdenum oxide derived high-performance electrocatalyst for the hydrogen evolution reaction, ACS Catal. 8 (2018) 5062-5069. https://doi.org/10.1021/acscatal.8b00949

[63]   R. Subbaraman, D. Tripkovic, D. Strmcnik, K.C. Chang, M. Uchimura, A.P. Paulikas, V. Stamenkovic, N.M. Markovic, Enhancing hydrogen evolution activity in water splitting by tailoring Li^+-Ni $(OH)_2$-Pt interfaces, Science 334 (2011) 1256-1260. https://doi.org/10.1126/science.1211934

[64]   Y. Luo, X. Li, X. Cai, X. Zou, F. Kang, H.M. Cheng, B. Liu, Two-dimensional MoS_2 confined Co(OH)$_2$ electrocatalysts for hydrogen evolution in alkaline electrolytes, ACS nano 12 (2018) 4565-4573. https://doi.org/10.1021/acsnano.8b00942

[65]   S.J. Bao, C.M. Li, C.X. Guo, Y. Qiao, Biomolecule-assisted synthesis of cobalt sulfide nanowires for application in supercapacitors, J. Power Sources 180 (2008) 676-681. https://doi.org/10.1016/j.jpowsour.2008.01.085

[66]   J. Li, X. Zhou, Z. Xia, Z. Zhang, J. Li, Y. Ma, Y. Qu, Facile synthesis of CoX (X= S, P) as an efficient electrocatalyst for hydrogen evolution reaction, J. Mater. Chem. A 3 (2015) 13066-13071. https://doi.org/10.1039/C5TA03153B

[67]   S. Peng, L. Li, X. Han, W. Sun, M. Srinivasan, S.G. Mhaisalkar, F. Cheng, Q. Yan, J. Chen, S. Ramakrishna, Cobalt sulfide nanosheet/graphene/carbon nanotube nanocomposites as flexible electrodes for hydrogen evolution, Angew. Chem. 126 (2014) 12802-12807. https://doi.org/10.1002/ange.201408876

[68]   J. Huang, D. Hou, Y. Zhou, W. Zhou, G. Li, Z. Tang, L. Li, S. Chen, MoS_2 nanosheet-coated CoS_2 nanowire arrays on carbon cloth as three-dimensional electrodes for efficient electrocatalytic hydrogen evolution, J. Mater. Chem. A 3 (2015) 22886-22891. https://doi.org/10.1039/C5TA07234D

[69]   F. Wang, T.A. Shifa, X. Zhan, Y. Huang, K. Liu, Z. Cheng, C. Jiang, J. He, Recent advances in transition-metal dichalcogenide based nanomaterials for water splitting, Nanoscale 7 (2015) 19764-19788. https://doi.org/10.1039/C5NR06718A

[70]   J. Mann, Q. Ma, P.M. Odenthal, M. Isarraraz, D. Le, E. Preciado, D. Barroso, K. Yamaguchi, G. von Son Palacio, A. Nguyen, 2-Dimensional transition metal dichalcogenides with tunable direct band gaps: $MoS_{2(1-x)}Se_{2x}$ monolayers, Adv. Mater. 26 (2014) 1399-1404. https://doi.org/10.1002/adma.201304389

[71]   M.S. Faber, R. Dziedzic, M.A. Lukowski, N.S. Kaiser, Q. Ding, S. Jin, High-performance electrocatalysis using metallic cobalt pyrite (CoS_2) micro-and nanostructures, J. Am. Chem. Soc. 136 (2014) 10053-10061. https://doi.org/10.1021/ja504099w

[72]   X. Li, G. Guan, X. Du, A.D. Jagadale, J. Cao, X. Hao, X. Ma, A. Abudula, Homogeneous nanosheet Co_3O_4 film prepared by novel unipolar pulse electro-deposition method for electrochemical water splitting, RSC Advances 5 (2015) 76026-76031. https://doi.org/10.1039/C5RA12822F

[73]   Y. Sun, C. Liu, D.C. Grauer, J. Yano, J.R. Long, P. Yang, C.J. Chang, Electrodeposited cobalt-sulfide catalyst for electrochemical and photoelectrochemical hydrogen generation from water, J. Am. Chem. Soc. 135 (2013) 17699-17702. https://doi.org/10.1021/ja4094764

[74]   D. Kong, J.J. Cha, H. Wang, H.R. Lee, Y. Cui, First-row transition metal dichalcogenide catalysts for hydrogen evolution reaction, Energy Environ. Sci. 6 (2013) 3553-3558. https://doi.org/10.1039/c3ee42413h

[75]   Y. Zheng, Y. Jiao, M. Jaroniec, S.Z. Qiao, Advancing the electrochemistry of the hydrogen-evolution reaction through combining experiment and theory, Angew. Chem. Int. Ed. 54 (2015) 52-65. https://doi.org/10.1002/anie.201407031

[76]   J. Zhang, Y. Liu, C. Sun, P. Xi, S. Peng, D. Gao, D. Xue, Accelerated hydrogen evolution reaction in CoS_2 by transition-metal doping, ACS Energy Lett. 3 (2018) 779-786. https://doi.org/10.1021/acsenergylett.8b00066

[77]   Q. Lu, G.S. Hutchings, W. Yu, Y. Zhou, R.V. Forest, R. Tao, J. Rosen, B.T. Yonemoto, Z. Cao, H. Zheng, Highly porous non-precious bimetallic electrocatalysts for efficient hydrogen evolution, Nature commun. 6 (2015) 6567. https://doi.org/10.1038/ncomms7567

[78]   A. Irshad, N. Munichandraiah, Electrodeposited nickel–cobalt–sulfide catalyst for the hydrogen evolution reaction, ACS Appl. Mater. Interfaces 9 (2017) 19746-19755. https://doi.org/10.1021/acsami.6b15399

[79]   X. Li, Q. Li, Y. Wu, M. Rui, H. Zeng, Two-dimensional, porous nickel–cobalt sulfide for high-performance asymmetric supercapacitors, ACS Appl. Mater. Interfaces 7 (2015) 19316-19323. https://doi.org/10.1021/acsami.5b05400

[80]   S. Peng, L. Li, J. Zhang, T. L. Tan, T. Zhang, D. Ji, X. Han, F. Cheng, S. Ramakrishna, Engineering Co_9S_8/WS_2 array films as bifunctional electrocatalysts for efficient water splitting, J. Mater. Chem. A 5 (2017) 23361-23368. https://doi.org/10.1039/C7TA08518D

[81]   X. Zou, Y. Zhang, Noble metal-free hydrogen evolution catalysts for water splitting, Chem. Soc. Rev. 44 (2015) 5148-5180. https://doi.org/10.1039/C4CS00448E

[82]   H. Zhang, B. Yang, X. Wu, Z. Li, L. Lei, X. Zhang, Polymorphic $CoSe_2$ with mixed orthorhombic and cubic phases for highly efficient hydrogen evolution reaction, ACS Appl. Mater. Interfaces 7 (2015) 1772-1779. https://doi.org/10.1021/am507373g

[83]   P. Chen, K. Xu, S. Tao, T. Zhou, Y. Tong, H. Ding, L. Zhang, W. Chu, C. Wu, Y. Xie, Phase-transformation engineering in cobalt diselenide realizing enhanced catalytic activity for hydrogen evolution in an alkaline medium, Adv. Mater. 28 (2016) 7527-7532. https://doi.org/10.1002/adma.201601663

[84]   D. Kong, H. Wang, Z. Lu, Y. Cui, $CoSe_2$ nanoparticles grown on carbon fiber paper: an efficient and stable electrocatalyst for hydrogen evolution reaction, J. Am. Chem. Soc. 136 (2014) 4897-4900. https://doi.org/10.1021/ja501497n

[85]   A.I. Carim, F.H. Saadi, M.P. Soriaga, N.S. Lewis, Electrocatalysis of the hydrogen-evolution reaction by electrodeposited amorphous cobalt selenide films, J. Mater. Chem. A 2 (2014) 13835-13839. https://doi.org/10.1039/C4TA02611J

[86]   Q. Liu, J. Shi, J. Hu, A.M. Asiri, Y. Luo, X. Sun, $CoSe_2$ nanowires array as a 3D electrode for highly efficient electrochemical hydrogen evolution, ACS Appl. Mater. Interfaces 7 (2015) 3877-3881. https://doi.org/10.1021/am509185x

[87]   Y.F. Xu, M.R. Gao, Y.R. Zheng, J. Jiang, S.H. Yu, Nickel/nickel (II) oxide nanoparticles anchored onto cobalt (IV) diselenide nanobelts for the electrochemical production of hydrogen, Angew. Chem. 125 (2013) 8708-8712. https://doi.org/10.1002/ange.201303495

[88]   R. Ye, P. del Angel-Vicente, Y. Liu, M.J. Arellano-Jimenez, Z. Peng, T. Wang, Y. Li, B.I. Yakobson, S.H. Wei, M.J. Yacaman, High-performance hydrogen evolution from MoS2 (1–x) P x solid solution, Adv. Mater. 28 (2016) 1427-1432. https://doi.org/10.1002/adma.201504866

[89]   J. Xie, J. Zhang, S. Li, F. Grote, X. Zhang, H. Zhang, R. Wang, Y. Lei, B. Pan, Y. Xie, Controllable disorder engineering in oxygen-incorporated MoS_2 ultrathin nanosheets for efficient hydrogen evolution, J. Am. Chem. Soc. 135 (2013) 17881-17888. https://doi.org/10.1021/ja408329q

[90]   W. Liu, E. Hu, H. Jiang, Y. Xiang, Z. Weng, M. Li, Q. Fan, X. Yu, E.I. Altman, H. Wang, A highly active and stable hydrogen evolution catalyst based on pyrite-

structured cobalt phosphosulfide, Nature commun. 7 (2016) 10771.
https://doi.org/10.1038/ncomms10771

[91]   C. Ouyang, X. Wang, S. Wang, Phosphorus-doped CoS_2 nanosheet arrays as ultra-efficient electrocatalysts for the hydrogen evolution reaction, Chem. Commun. 51 (2015) 14160-14163. https://doi.org/10.1039/C5CC05541E

[92]   M. Cabán-Acevedo, M.L. Stone, J. Schmidt, J.G. Thomas, Q. Ding, H.C. Chang, M.L. Tsai, J.H. He, S. Jin, Efficient hydrogen evolution catalysis using ternary pyrite-type cobalt phosphosulphide, Nature mater. 14 (2015) 1245. https://doi.org/10.1038/nmat4410

[93]   M.S. Faber, S. Jin, Earth-abundant inorganic electrocatalysts and their nanostructures for energy conversion applications, Energy Environ. Sci. 7 (2014) 3519-3542. https://doi.org/10.1039/C4EE01760A

[94]   K. Liu, F. Wang, K. Xu, T.A. Shifa, Z. Cheng, X. Zhan, J. He, $CoS_{2x}Se_{2(1-x)}$ nanowire array: An efficient ternary electrocatalyst for the hydrogen evolution reaction, Nanoscale 8 (2016) 4699-4704. https://doi.org/10.1039/C5NR07735D

[95]   G. Zhang, G. Wang, Y. Liu, H. Liu, J. Qu, J. Li, Highly active and stable catalysts of phytic acid-derivative transition metal phosphides for full water splitting, J. Am. Chem. Soc. 138 (2016) 14686-14693. https://doi.org/10.1021/jacs.6b08491

[96]   X. Li, C. Li, A. Yoshida, X. Hao, Z. Zuo, Z. Wang, A. Abudula, G. Guan, Facile fabrication of CuO microcube@ $Fe–Co_3O_4$ nanosheet array as a high-performance electrocatalyst for the oxygen evolution reaction, J. Mater. Chem. A 5 (2017) 21740-21749. https://doi.org/10.1039/C7TA05454H

Electrochemical Water Splitting: Materials and Applications
Materials Research Foundations **59** (2019) 97-124

Materials Research Forum LLC
doi: https://doi.org/10.21741/9781644900451-4

Chapter 4

Metal Free Catalysts for Water Splitting

Paramita Karfa, Kartick Chandra Majhi, Rashmi Madhuri*

Department of Applied Chemistry, Indian Institute of Technology (Indian School of Mines), Dhanbad, Jharkhand 826 004, India

Abstract

Water splitting through electrolysis with the use of renewable sources is a highly versatile method of energy conversion producing hydrogen, which is a very clean energy carrier. Development of earth abundant catalyst with long term stability and high efficiency is of current research interest which can replace noble metal catalyst like Pt, Ir. Catalyst prepared through metal precursors like metal oxides, metal selenides, metal phosphides suffer from high expenses, poor durability, vulnerability to gas poisoning, limitations in electrolytic medium, formation of by-product which are not environmental friendly, low selectivity. Researcher have synthesized a new class of material i.e., metal free catalyst which acquires some of the best properties like large surface area, variety of shape and size and high robustness. Among the metal free catalysts, the carbon based catalyst was studied and researched more expansively. Research is going on full steam to overcome different key challenges like catalyst stability, high Faradaic efficiency and to explore further future opportunities in this exhilarating field.

Keywords

Water Splitting, Oxygen Evolution Reaction (OER), Hydrogen Evolution Reaction (HER), Metal Free Catalyst, Graphene, Carbon Nanotube, Graphitic Carbon Nitride

Contents

1.   Introduction

Exhaustive utilization of fossil fuels, petroleum, and coal causes high grade pollution, global warming, anthropogenic losses, acid rains, health hazards and large energy crisis with depletion of these traditional sources of energy [1]. Widespread research is going on for the development of clean, sustainable and renewable energy sources like geothermal, wind, solar energy as well as an environmental compatible energy carrier to meet the crisis and become a strong alternative to fossil fuels [2]. Because of considerable energy density and low molecular weight, causing no pollution, hydrogen has become popular in recent time as the supreme energy carrier [3]. Hydrogen in elemental form cannot be directly used as hydrogen fuel disparate of other traditional fuel. Therefore, we need a renewable clean energy source for their production [4].

Now-a-days, photochemical, photo electrochemical, and electrochemical water splitting has gained lots of importance towards the production of hydrogen [5-7]. Among these, electrochemical water splitting is the promising technology to produce hydrogen gas in large quantity at low cost safely and cleanly [8]. Electrochemical water splitting involves two half-cell reactions, at cathode reduction of water takes place, which generates hydrogen and termed as hydrogen evolution reaction (HER). While at the anode oxidation reaction takes place, which produces oxygen gas and reaction is termed as oxygen evolution reaction (OER) [9]. The total set up of water splitting showing both HER and OER using electrocatalyst both at anode and cathode is shown in Fig. 1. Hydrogen produced in the above electrolysis process can be stocked up and used for power generation in fuel cells, turbines, as well as in transportation. While oxygen produced in the anode is used for chemical fuel cells, metal air batteries, etc. [10].

Electrochemical Water Splitting: Materials and Applications   Materials Research Forum LLC
Materials Research Foundations **59** (2019) 97-124   doi: https://doi.org/10.21741/9781644900451-4

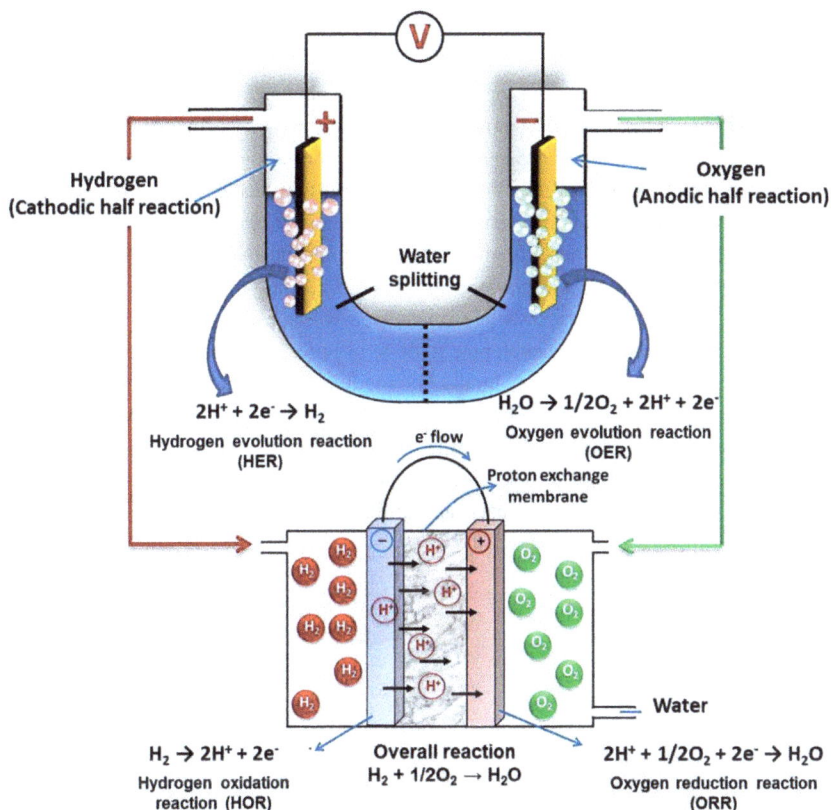

Fig. 1: *Illustration showing total set up of water splitting with electrocatalyst modified anode and cathode [46].*

Although the electrochemical water splitting looks simpler in the outer surface, it is a tedious job to be performed because of sluggish kinetics of HER and OER. Therefore, efficient electrocatalyst is needed to accelerate both the reaction kinetics and make the reaction easier [11]. Some scarcely available metals in earth-crust like platinum, ruthenium, iridium are the state of the art material used as benchmark catalyst for water splitting because of low overpotential value, but their towering cost, low availability, easy gas poisoning, non-biocompatibility restrain their use in large scale industrial purposes. In recent time, plentiful efforts have been made to prepare competent catalyst without the use of the above-mentioned precious metal for water splitting [12,13].

Researchers are working day and night to design the new age catalyst for water splitting, which must have a large number of active sites for electrolysis of hydrogen and oxygen, large specific surface area, tailored nano-structured morphology, low overpotential value, low onset potential, and low Tafel slope [14]. The most studied catalyst illustrated in Fig. 2, other than noble metals, for water splitting, is metal phosphides, metal sulphides, transition metal chalcogenides, polymer composites, carbon-based materials, boron-based materials, metal selenides, and their composites [15]. Catalytic reactions in recent days are subjugated by transition metals. However, their role for overall water splitting is still not well explored. Because development of a bifunctional catalyst for overall water splitting i.e. HER and OER is easy in alkaline medium but much challenging in an acidic medium because of the mild stability of this catalyst in lower pH value. Although, metal based catalysts for overall water splitting are most studied in the literature, but they possess certain disadvantages also like low intrinsic conductivity, stumpy durability, vulnerable to gas poisoning, high costs, susceptible to corrosion, etc. [16].

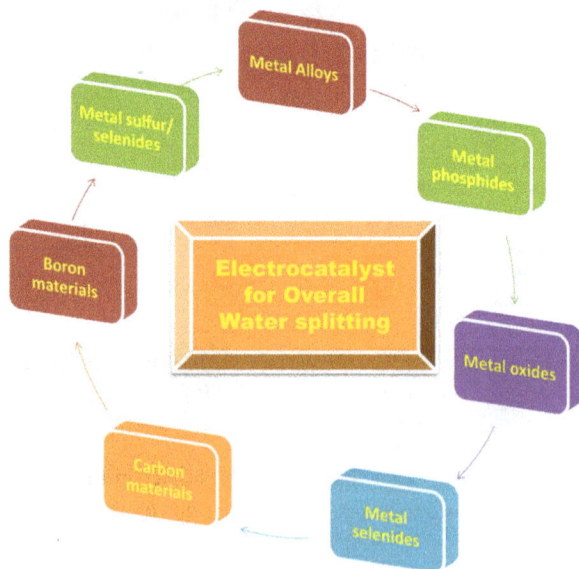

Fig. 2: *Diagram showing popularly used efficient electrocatalyst reported in the literature for overall water splitting.*

Therefore, to overcome these problems, the best alternative option is the use of metal free catalyst which can be easily available, affordable, and sustainable. With the use of metal based catalysts, we have limited option towards the engineering of active sites, structure, basal planes, etc. But, metal free catalysts as a contemporary catalyst to metal based catalyst has good electrical conductivity, diverse morphology, and structural arrangement, high thermal stability, easily tunable band gap, high mechanical strength, easily availability and low weight, which make them a challenging catalyst towards water splitting [17,18]. Most commonly, carbon based material like graphene, carbon nitride, carbon nanotubes and boron, silicon based materials are popularly used in overall water splitting [19]. Here, in this chapter, we have focused on the role of these nanomaterials towards overall water splitting, but before that, the basic chemistry behind the OER and HER reactions is necessary to be studied. In the next sections, we have discussed first the fundamentals of OER and HER.

1.1   Hydrogen evolution reaction (HER)

Hydrogen fuel came to attention in the year 1874 by the researcher Jules Verne, when hydrogen was used as energy storage medium and transportation fuel [20]. Sources of hydrogen production include biomass, natural gas, gasification of coal, from the gasification of coal and a natural gas large amount of hydrogen can be produced, but it causes global warming and a large amount of pollution due to the carbonaceous by-product [21]. Production of hydrogen through biomass is the cleanest way, but their yield is poor. Water, on the other hand, is abundant and limitless on earth. Therefore, it could be a probable alternate clean and sustainable source of hydrogen [22]. In general reaction via thermal process, production of hydrogen from water splitting requires a very high temperature of about 2000 °C. Thus, electrochemical water splitting is more favorable than thermal water splitting with the help of an efficient catalyst [23].

As we are aware that among the two reactions involved in water splitting, one of the reactions is hydrogen evolution reaction which occurs at cathode and leads to the formation of hydrogen gas through two electron pathways. The overall main reaction can be represented as: $2H^+ + 2e^- \rightarrow H_2$. However, this is not the actual reaction that occurs in the presence of catalysts. The reaction representation and steps get modified based on the choice of catalysts as well as the pH of the medium. For example, the hydrogen evolution reactions proceed through three reactions step in the acidic medium [24]. The reactions in the acidic medium are as follows:

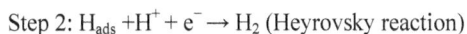

Step 1: $H_3O^+ + e^- \rightarrow H_{ads} + H_2O$ (Volmer reaction)

Step 2: $H_{ads} + H^+ + e^- \rightarrow H_2$ (Heyrovsky reaction)

Step 3: $H_{ads} + H_{ads} \rightarrow H_2$ (Tafel reaction)

In the acidic medium H_3O^+ takes an electron from the electrolyte to produce a surface adsorbed hydrogen atom, which is designated as the Volmer reaction step. The second step actually consists of two processes, i.e., desorption of the hydrogen atom from the surface of the catalyst and their combination with a proton to form one molecule of hydrogen gas. This reaction step is termed as Heyrovsky reaction step. Another step, involved in the HER, is Tafel reaction step, where two adsorbed hydrogen combine each other to form one hydrogen molecule [25].

Similar to the acidic medium, HER also proceeds through three step process in basic medium. The reactions are as follows:

Step 1: $H_2O + e^- \rightarrow OH^- + H_{ads}$ (Volmer reaction)

Step 2: $H_{ads} + H_2O + e^- \rightarrow OH^- + H_2$ (Heyrovsky reaction)

Step 3: $H_{ads} + H_{ads} \rightarrow H_2$ (Tafel reaction)

In basic medium water molecule take one electron from the electrolyte to form surface adsorbed hydrogen atom species [26]. In the second reaction step, the same as an acidic medium it can follow any one of the Heyrovsky pathways or Tafel pathway and results in the formation of hydrogen gas. A small difference in Heyrovsky pathway occurs in the basic medium; here the adsorbed hydrogen atom reacts with one water molecule to form one hydrogen gas molecule, while the Tafel pathway remains the same in both acidic and basic medium [27].

The energy conversion process of HER can also be represented in terms of Gibbs free energy also, i.e. $\Delta H^0 = \Delta G^0 + T\Delta S^0$, where ΔG^0 = Change in Gibbs free energy, ΔS^0 = Change in entropy, $T\Delta S^0$ = Change in heat energy. Second step, which is the main step in the HER process, i.e. adsorbed hydrogen on the catalyst gets adsorbed or removed, can be easily done if a catalyst possesses $\Delta G^0 = 0$. It can be treated as a criterion to select the good catalyst for HER, which has comparable activity to that of noble metals like Pt, Ir, etc. [28].

Similar to change in Gibbs free energy, Tafel slope is another important factor, which has to be considered while selecting the catalyst for HER. Tafel slope is a crucial kinetically manageable factor in the hydrogen evolution reaction for better understanding of catalyst performances and the pace of reaction [29]. The Tafel slope be able to denoted through: $\eta = \alpha + \frac{2.3\,RT}{\alpha nF} \log (j)$

Here, η denotes the overpotential value, j designates current density, and the Tafel slope can be represented as the slope of the linear plot between overpotential and current density i.e.$\frac{2.3\ RT}{\alpha nF}$. According to the literature, low Tafel slope means the catalyst has better performances towards HER [30]. In addition, by observing the Tafel slope, rate determining step during HER process can also be marked. For example, Tafel slope of 120 mVsec^{-1} means catalyst follows the volmer reaction step as the rate determining step. When the Tafel slope is equal to 30 mVsec^{-1} and 40 mVsec^{-1}, Tafel and Heyrovsky pathway is considered as the rate determining, respectively [31]. Table 1 [32-41] gives the summary of carbon-based catalyst used for hydrogen evolution reaction.

Table 1: *Summary of metal free carbon based catalyst used for hydrogen evolution reaction (HER) in 0.5 M H$_2$SO$_4$.*

S. N.	Electrode material	Overpotential*	Tafel slope (mV dec^{-1})	References
1.	HOPG	0.64 (0.1 mA cm^{-2})	122	32
2.	GGNR	183 (10 mA cm^{-2})	49	33
3.	SG-P	178 (10 mA cm^{-2})	86	34
4.	HT-AFNG	350 10 mA cm^{-2})	113	35
5.	I-G	245 (10 mA cm^{-2})	96	36
6.	NHC	185 (10 mA cm^{-2})	56.7	37
7.	N and S co-doped porous carbon	-	57.4	38
8.	B-SuG	430 (10 mA cm^{-2})	99	39
9.	Activated carbon nanotubes	220 (10 mA cm^{-2})	71.3	40
10.	N, P doped carbon	204 (10 mA cm^{-2})	58.4	41

*at the current density in bracket, HOPG = highly oriented pyrolytic graphite; GGNR= graphene/graphene nanoribbon aerogel; SG-P = S-doped Plasma-etched graphene; HT-AFNG= highly torn amine functionalized nitrogen doped graphene; I-G= I-doped graphene; NHC= nitrogen doped hexagonal carbon; B-SuG= Boron-substituted graphene.

1.2   Oxygen evolution reaction (OER)

The other counterpart of HER and another important reaction of overall water splitting is the oxygen evolution reaction, which takes place at the anode and is a intrinsically slow multiple electrons and proton transfer reaction [42]. OER is four electron transfer process, proceeds through O-O bond formation, which is a sluggish reaction process. The thermodynamic potential to achieve oxygen evolution reaction is +1.23 V. So, the catalyst which can overcome this overpotential value can split water into hydrogen and oxygen. Similar to HER, the catalyst used for OER must have low cost, made up of abundant earth material, have long term stability in a corrosive environment and have low overpotential value (nearly equal to +1.23V), large surface area, high porosity, and plentiful active sites for oxygen transport [43].

The well-known catalyst for OER in the literature is RuO_2 and IrO_2, but they are not easily available because of their low abundances in the earth. Some of the metal based catalyst and metal free catalyst popularly used for OER include metal nanoparticles, metal oxides, metal selenides, and carbon based material [44]. The composites of metals and non-metals are the new generation material used as a catalyst for OER and are expected to increase the rate of the reaction. OER can be performed in all pH electrolytes like acidic, basic, or neutral but have different reaction mechanism according to the different medium. The OER mechanism based on the pH of the medium are as follows [45]:

1. In basic medium: $4OH^- \longrightarrow 2H_2O + O_2 + 4e^-$

2. In acidic medium: $2H_2O \longrightarrow 4H^+ + O_2 + 4e^-$

During the reaction process, several intermediates like O*, OH* and HOO* were also formed and the control of reaction intermediate like formation and binding to the electrode surface determines the efficiency of the OER catalyst. However, OER reaction products, by-products, and intermediates are all environmentally benign and can be used for real time application [46]. The reaction of OER proceeds through intermediate steps have different mechanism proposed by different researchers. Some of them are discussed below: Tuckerman et al. have proposed a mechanism of OER with an intermediate step in the year 1995 and are as follows [47]:

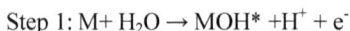

Step 1: $M + H_2O \rightarrow MOH^* + H^+ + e^-$

Step 2: $MOH^* \rightarrow MOH$

According to them, water molecule reacts with adsorbed reaction intermediate to form a hydrogen bond between the reaction intermediate and the hydroxyl group. In the next step, discharge of hydroxide, molecule takes place with electron transfer fascilating movement of hydroxyl from the electrolyte to surface. MOH* is the surface bound hydroxyl group of high energy. However, Shen et al. [48] have proposed another pathway for OER, in which different intermediates were formed over GaN surface and the reactions are as follows:

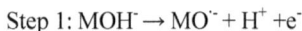

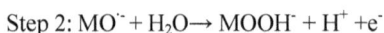

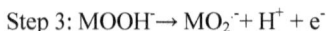

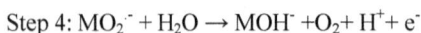

Step 1: $MOH^- \rightarrow MO^{-} + H^+ + e^-$

Step 2: $MO^{-} + H_2O \rightarrow MOOH^- + H^+ + e^-$

Step 3: $MOOH^- \rightarrow MO_2^{-} + H^+ + e^-$

Step 4: $MO_2^{-} + H_2O \rightarrow MOH^- + O_2 + H^+ + e^-$

According to the above method, the cation like Ga accepts hydroxide ions on their surface and the nitrogen anion accepts protons acting as bases which results in the hydroxylated surface. The author suggests that the reaction starts at the surface attached hydroxide ion rather than at the active sites stating that the active sites are unstable when vacant.

Therefore, it is very clear from HER and OER mechanism study that by kinetics optimization, OER will no longer remain as the obstacles for fuel energy conversion. Similarly, the use of hydrogen as sustainable fuel obtained from water splitting is also not a problem, if a suitable catalyst for overall water splitting is found. In order to find the best catalyst for both HER/OER or overall water splitting, it is needed that prior to their use, we must be aware of the intrinsic activity of the catalyst, their design, active sites as well as mechanistic pathways and therefore, extensively study their fundamentals by theoretical as well as experimental approach [49]. Table 2 [50-59] gives the review of carbon catalyst used for the oxygen evolution reaction.

Table 2: Summary of some of the metal free carbon based catalyst used for oxygen evolution reaction (OER).

S. N.	Electrode material	Electrolyte	Onset Potential	Tafel slope (mV dec^{-1})	Ref.
1.	N, S-GCNT	1 M KOH	1.61 V (vs. R.H.E)	94	50
2.	SWCNT@NPC	0.1 M KOH	1.68 V (vs. R.H.E)	78	51
3.	CC@VG	1 M KOH	1.28 V (vs. R.H.E)	56	52
4.	N-CCs	0.1 M NaOH	1.69 V (vs. R.H.E)	-	53
5.	g-C$_3$N$_4$/graphene	0.1 M KOH	0.58 V (vs. Ag/AgCl)	68	54
6.	Acidic oxidized CC	0.1 M KOH	0.32 V (vs. Ag/AgCl)	82	55
7.	NG-CNT	0.1 M KOH	0.12 V (vs. Ag/AgCl)	141	56
8.	N-MCF/MGF	0.1 M KOH	-	67	57
9.	NPMC	0.1 M KOH	0.79 (vs. Ag/AgCl)	-	58
10.	N-HC@G$_{-x}$	0.1 M KOH	1.44 (vs. R.H.E)	88.1	59

CC@VG=graphene grown on carbon cloth; NG-CNT=N,O-dual doped graphene-CNT; N-MCF/MGF= N-doped mesoporous carbon/graphene framework; NPMC=Mesoporous carbon foam co-doped with nitrogen and phosphorus; N-HC@G$_{-x}$=N-doped hollow carbon (N-HC) layer on GO substrate; SWCNT@NPC= single-wall carbon nanotube on N,P doped carbon; g-C$_3$N$_4$= graphitic carbon nitride; N-CCs= N-doped carbon nanocages.

Electrochemical Water Splitting: Materials and Applications   Materials Research Forum LLC
Materials Research Foundations **59** (2019) 97-124   doi: https://doi.org/10.21741/9781644900451-4

2.   Factors affecting the efficiency of electrochemical water splitting

Here, we have discussed some of the fundamental properties and factors which have to be considered, before selecting any material as electrocatalyst for overall water splitting. Some very basic parameters, which may affect the efficiency of electrochemical water splitting, are [60]:

1. Nature of catalysts, i.e. crystalline or amorphous;

2. The band gap of the material used;

3. Size and dimensionality of the catalyst; and

4. pH dependency of the catalyst

It is considered that crystalline material has higher efficiency than that of amorphous material because of considerable charge transfer properties and increase in density of defects. Size and dimensionality also affect water splitting, like there are 0D, 1D, 2D, and 3D materials with the small size used as a catalyst for water splitting and their efficiency vary with change in their size from micro to nanometer as well as zero to three dimensions. Like, small sized material with high surface area assists easy transportation of charge carriers and facilitates hydrogen generation [61]. Low band gap or tuneable band gap of the catalyst can increase their efficiency towards HER. Similarly, as the electrochemical cells for water splitting operate in various pH medium, the evolution of hydrogen and oxygen very much depends on the pH of the electrolyte. So, it can be concluded that catalysts which can show high stability and high corrosion resistance in all electrolytes without weakening of electrode surface, due to ion migration are opted as an efficient catalyst for electrochemical water splitting.

3.   Electrochemical matrices used for determining talent of the catalyst

Water splitting through two half-cell reactions was first discovered by Van Troostwijik and Deiman in 1789 [62]. But after such a long period, we are still at a similar position, where large overpotential value is still required for anodic OER and cathodic HER. The valid cell set up used in the laboratory for water splitting is shown in the image captured by camera Fig. 3. As HER and OER are both electrochemically uphill reactions, suitable electrocatalyst is needed to reduce the overpotential value required for HER/OER, which is actually the difference between thermodynamic cell potential and applied cell potential. In general, linear sweep voltammetry (LSV) and cyclic voltammetry (CV) studies were performed with infra-red (IR) compensation to eliminate the resistance over potential and to obtain the polarization curve. The polarization curve recorded by LSV can be used to calculate the accurate over potential, and current density obtained using the particular catalyst or catalyst modified electrodes. Overpotential value is the main parameter to evaluate the efficiency of the catalyst and it is calculated at geometric current density of 10 mA cm^2[63].

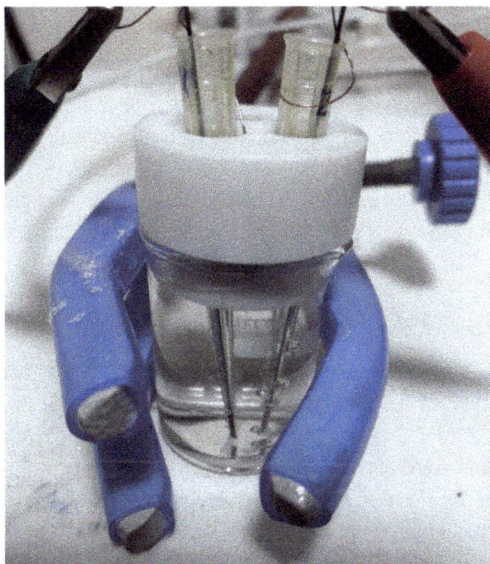

Fig. 3: *Camera picture of commonly used electrochemical cell set up used in the laboratory for water splitting study.*

One of the quantitative parameters used for the calculation of electron transfer on the electrode surface is the ratio of gas evolved (O_2 and H_2) during water splitting. While the ratio of experimental and theoretical values of gas evolution are used to calculate Faradaic efficiency (FE) of the catalysts. A better catalyst provides us nearly 100% Faradaic efficiency [64].

Turn over frequency (TOF) is another parameter showing the intrinsic catalytic property of the catalyst per active site on a particular potential and the molecule of hydrogen gas evolved per active sites per unit time. With evaluating the TOF value, the better catalyst can be prepared, which will give high H_2 and O_2 production on the active sites. TOF value calculation for the heterogeneous catalyst is very difficult, but, if the catalyst possesses homogeneous chemical composition, symmetrical structure, homogeneous active sites density, TOF value can be calculated very precisely. High TOF value does not mean that the catalyst is best for water splitting, the catalytic efficiency evolution depends on other parameters also (as defined above). Catalysts possessing high value of TOF having a large number of active sites per unit area of the electrode can be said as the efficient catalyst which can replace noble metal as a catalyst for water splitting [65].

Another important auxiliary tool towards the search of proficient catalyst for HER and OER is double layer capacitance (C_{dl}), which is used to estimate the electrochemically active surface area of the prepared catalyst. The electrochemically active surface area of the catalyst is proportional to the capacitance value of the catalyst. C_{dl} can be calculated by cyclic voltammetry method, using a non-faradaic current region of the CV runs recorded at different scan rates. The current density of anodic and cathodic non-faradaic regions (i.e., J_A and J_C, respectively) were used to calculate ΔJ (= J_A-J_C), which is plotted against the scan rate. The slope of the linear graph between ΔJ versus scan rate will give the C_{dl} value. A higher value of double layer capacitance means the catalyst has higher electrochemical surface area [66].

Stability of the catalyst for water splitting can be examined by taking cyclic voltammetry runs for multiple cycles like 5000 cycles at a particular scan rate. After the multiple cyclic runs, LSV runs were also recorded to observe the change in onset potential and over potential value of the polarization curve. If there is no change in the onset of potential and overpotential value in both CV and LSV runs, the catalyst can be considered as stable over a large number of cycles and can be used for long-term study. To check the stability of the catalyst in harsh condition or wide pH range, similarly, CV and LSV runs can be recorded in various pH electrolytes. No change in current density and overpotential value confers the catalyst to be a robust one [67].

On the different track, long term chronoamperometric run was also recorded, where constant stable current at particular overpotential for long hours illustrate that the high end stability of the catalyst. Storage stability was also tested to confer the behavior of catalyst if they are stored for a long time. For this study, catalyst or catalyst modified electrodes were stored for several months, and after particular time intervals, their LSV or CV runs were recorded. If the catalyst shows good stability, no considerable change in current density and overpotential value was observed after even the storage of several months [68].

4.　Electrocatalysts for overall water splitting

Electrocatalyst to be used for water splitting must have considerable efficiency in comparison to the benchmark noble metal catalysts. Several metals based or metal free

Electrochemical Water Splitting: Materials and Applications   Materials Research Forum LLC
Materials Research Foundations **59** (2019) 97-124   doi: https://doi.org/10.21741/9781644900451-4

catalysts has been reported in the literature till date for overall water splitting, however, metal free catalysts have always been considered as the better alternative. Metal free catalysts are usually carbon based materials, which provide high surface area and portability to design new structures and morphologies. Herein, we have compiled the performances of the metal free catalyst used to date in various researches towards water splitting. Metal free catalysts are environment friendly, have plentiful reserves, show high electron conductivity, and good catalytic performances. Metal free catalysts are new generation material, which is reasonably priced compared to metal counterparts, have renewable precursor or source, needs a very easy process of synthesis, resistance to corrosion, and have high selectivity. On the other hand, metal based catalyst always leads to the detrimental by product and give threat to the environment. Additionally, based on the limited availability of metal precursors, metal-based catalysts are not commercially feasible for energy application in a wide range. Thus, by designing suitable morphology with better composition, engineered band gap, and unique property metal free catalyst can be developed as a high-performance catalyst for overall water splitting. The next section discusses some of the metal free catalyst used for electrochemical H_2 and O_2 evolution in overall water splitting.

5.   Carbon based metal free catalyst

This type of catalysts is considered as one of the ideal catalysts for water splitting. Carbon is one of the copious elements in the earth crust; therefore, it is of great demand in ecology and human economy [69]. Carbon based material holds unique properties like high thermal and electrical conductivity, mechanical strength, diverse dimensionality and morphology, exceptional surface property, and considerable opto-electronic property [70]. Illustration showing both HER and OER for carbon doped with various element is shown in Fig. 4. Long ago, glassy carbon and activated carbon with simple molecular structure are both used as a catalyst for electrochemical reactions. These days carbon material discovered so far has a strange molecular structure with advance features, which recommend new opportunities meant for the expansion of carbon based metal free catalyst with improved catalytic properties.

Depending upon the arrangement of the carbon atom, it can be easily classified into three kinds, i.e. graphite, diamond, and amorphous carbon. Graphite is the best studied and stable form of carbon with sp^2 hybridization and possesses covalent bonding among the carbon atoms, where the layers are attached to each other by weak van der Waals forces. However, in diamond, one carbon atom is attached to four other carbon atoms, and they are very transparent and hard [71]. The other popular structural analogues of carbon-based nanomaterials are fullerene, carbon nanotube, and graphene. To increase and modify the electrical conductivity of these carbon-based materials or nanomaterials, introduction of several heteroatoms like nitrogen, sulfur, boron, phosphorus onto the surface was also done, which modulates the electronic configuration and make them a suitable candidate to be used as a remarkable alternative catalyst for water splitting. Heteroatom doping influences the energy levels of valance orbital for the neighboring carbon atom, which helps in hydrogen adsorption [72]. After getting an influential result from single hetero atom doping, multiple hetero-atom doping on carbon nanomaterial has also become popularized and showed excellent behavior towards HER and OER. Other than these traditional and popular carbonaceous materials, some of the newly discovered carbon based catalysts like carbon nitride (g-C_3N_4) are also gaining interest towards HER/OER or overall water splitting. They have interesting electronic, magnetic, photonic properties with a symmetrical structure which paved their way towards new generation electrocatalyst.

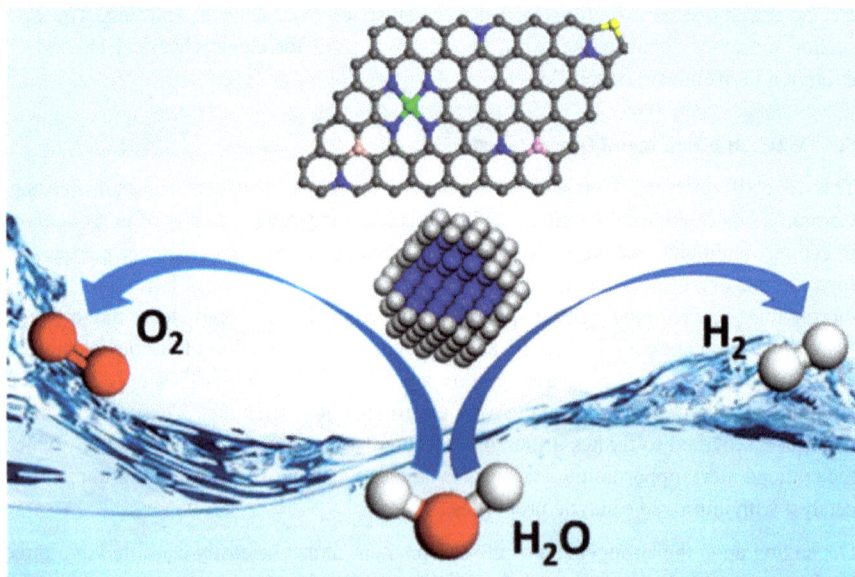

Fig. 4: *Illustration presenting overall water splitting using different element doped carbon-based catalyst [70].*

5.1 Graphene based electrocatalysts for water splitting

Graphene is one of the standard 2D materials having single layered carbon atom configured in a hexagonal honeycomb like arrangement containing a large number of benzene rings. Structure of 2D graphene was vividly discovered by Wallace in the year 1947 [73] but initially synthesized by Geim in layered form through mechanical exfoliation method. Geim and Novoselov were awarded a noble prize for this innovation. Graphene comprises of sp^2 hybridized carbon atom with alternate single and double bond [74]. Conjugated structure, p and π orbital overlapping of the carbon atom arises due to the alternative single and double bond, which also gives easy electron transfer.

Graphene possesses a stable and crystalline structure with exciting properties like high surface area, high optical transparency, high catalytic properties, high electronic properties, high mechanical strength, and thermal conductivity. Because of these properties, graphene has a considerable number of applications in fuel cells, energy storage devices, energy conversion devices, and battery [75]. Graphene acquiring the above unique properties can be used as an electrode material for electrochemical water splitting. For example, Lai et al. [76] have synthesized oxygen, nitrogen, phosphorus tri-doped porous graphitic carbon on oxidized carbon cloth (ONPPGC/OCC) and used it as a catalyst for overall water splitting. The precursors used by the group for the synthesis of the catalyst are phytic acid, aniline, and OCC. The prepared 3D catalyst shows water splitting in both acidic and neutral medium. For OER the catalyst shows an overpotential value of 410 mV with a Tafel slope of 83 mV dec^{-1}. For HER, ONPPGC/OCC catalyst shows an overpotential value of 446 mV with a Tafel slope of 154 mV dec^{-1}.

Similarly, Zhang and his group [77] have also synthesized a multifunctional catalyst for electrochemical water splitting using phosphorus, nitrogen, and fluorine tri-doped graphene. The catalyst shows OER onset potential of +1.62 V with a Tafel slope of 136 mV dec^{-1}. For HER, the catalyst shows onset potential value of -0.4 V and an overpotential value of +0.52 V vs Ag/AgCl. The catalyst was also used as a real time application in zinc air battery.

Zhang et al. [78] have synthesized a highly efficient laser induced graphene (LIG) electrode for HER and OER showed in Fig. 5. The overpotential value of the proposed catalyst towards HER is found to be 214 mV with Tafel slope of 54 mV dec^{-1} and for OER the overpotential of 380 mV and Tafel slope 49 mV dec^{-1}. Jia and his partners [79] have synthesized 2D graphene material having carbon defects by removing nitrogen from nitrogen containing precursor porous organic framework material (PAF-40). In OER experiment the onset potential was found to be +1.57 V at 10 mA cm^{-2} current density with lower Tafel slope of 97 mV dec^{-1}. For HER activity the onset potential is measured to be -0.15 V at 10 mA cm^{-2} current density in 0.5 M H$_2$SO$_4$. Yue et al. [80] have developed nitrogen and fluorine dual doped graphene nanosheets (NFPGNS) of porous nature through chemical etching method. The overall onset potential for water splitting is found to be +1.6 V. While, for OER the particular onset potential value is found to be +1.53 V. For HER the onset potential is found to be 120 mV. The Tafel slope for HER and OER for NFPGNS is calculated to be 109 mV dec^{-1} and 78 mV dec^{-1}, respectively.

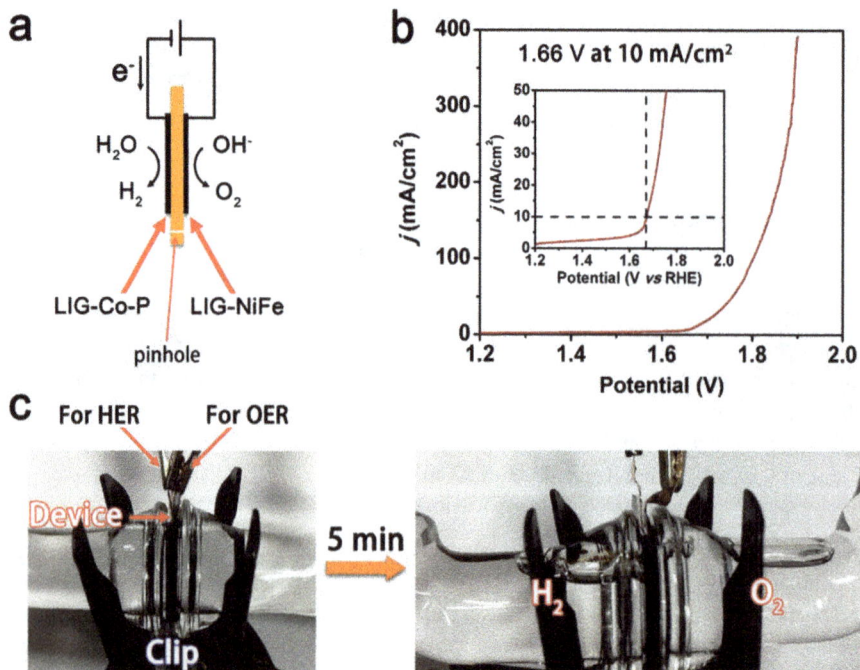

Fig. 5: *Diagram of performances of laser induced graphene (LIG): (a) illustration showing the set up of water splitting, (b) LSV plot of LIG showing overall water splitting (with inset showing zoomed version), (c) Photograph showing bubbles formation in the device after generation of hydrogen and oxygen gas evolution, after 5 minutes of electrolysis [78].*

5.2 Carbon nanotube based electrocatalysts for water splitting

Carbon nanotubes (CNTs) have a flawless cylindrical structure composed of layers of graphene, and it is of nanometer size and have nanometer diameter. The carbon network in CNTs have honeycomb like the arrangement, similar to that of graphene sheets. The

Electrochemical Water Splitting: Materials and Applications   Materials Research Forum LLC
Materials Research Foundations **59** (2019) 97-124   doi: https://doi.org/10.21741/9781644900451-4

quasi one dimensional morphology provides it with amazing electronic and mechanical property [81]. Iijima in 1991 first discovered the cylindrical structure of carbon. He observed that nanotubes are made up of several graphitic cells and called it multiwalled carbon nanotube (MWCNT), which had 0.34 nm diameter [82]. After two years of this discovery, Iijima and Ichihashi with another researcher Bethune have synthesized single walled carbon nanotube (SWCNT), which has single layer structure rolled in tube shape [83]. SWCNT and MWCNT can be prepared through catalytic growth, arc discharge, and laser ablation methods. CNT have symmetrical structure and possess high electrical conductivity, high chemical stability, and can be used as a high performance electrocatalyst.

CNT can be easily functionalized or can be easily combined to form a composite with other material, due to the presence of conjugated π-electrons [84]. CNT is used as a profound carbon-based catalyst separately as OER and HER, more than OER, carbon nanotube draw attention as hydrogen evolution catalyst. One of its examples is shown in Fig. 6, where incorporation of MoP nanoparticle on the carbon nanotube is exhibited. Some of the CNT based metal free catalysts are also used in electrochemical water splitting and are discussed below. Davodi et al. [85] have synthesized nitrogen doped CNTs through cost effective synthesis procedure. The prepared N-CNTs showed better performances towards HER and OER, owing to the presence of pyridinic N-sites. For OER the N-CNT shows an overpotential value of 320 mV at 10 mA cm^{-2} current density. For HER the catalyst shows an overpotential value of 340 mV at 10 mA cm^{-2}in 0.1 M KOH. The catalyst shows long term stability after constant water electrolysis.

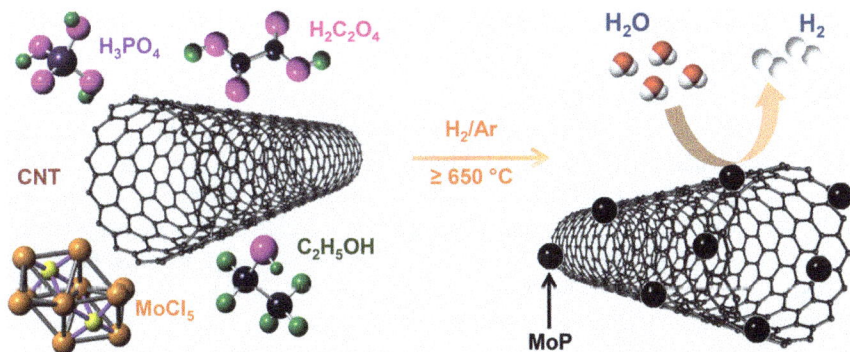

Fig. 6: *Diagram showing the synthesis and performances of CNT decorated with MoP for hydrogen evolution reaction [84].*

Qu et al. [86] have synthesized sulfur and nitrogen dual doped carbon nanotube through in situ reaction with high sulfur doping concentration and examined it for electrochemical water splitting. The electrolyte used for electrochemical water splitting is 1.0 M KOH solution. The onset overpotential value for the catalyst towards OER is found to be +1.59 V with a Tafel slop of 56 mV dec^{-1}. From HER study, it is observed that N, S-CNT shows a cathodic potential of -0.40 V at 5 mA cm^{-2} current density and a Tafel slope of 133 mV dec^{-1}. The catalyst shows long term durability for about 20 hours showing constant current in alkaline medium.

Ali and coworkers [87] have synthesized template assisted free standing multiwalled carbon nanotube as water splitting catalyst. The catalyst was prepared through chemical vapor deposition (CVD) using a silicon substrate. The prepared catalyst shows the onset potential of +1.60 V for OER and a current density of 0.89 mA cm^{-2}. The author states that the prepared catalyst would be a better replacement of Pt for overall water splitting. Zhang et al. [88] have prepared N-doped cotton cloth by carbonization method. For the nitrogen doping, polyacrylonitrile (PANI) was used as a precursor, and the prepared catalyst was used for overall water splitting. The small overpotential value of 351 mV was obtained for OER with a Tafel slope of 88 mV dec^{-1}. For HER overpotential value was observed to be 233 mV with a Tafel slope of 135 mV dec^{-1}. Cheng et al. [89] have synthesized pristine carbon nanotube having 2-7 walls with a diameter of 2-4 nm and explored its application for OER in water splitting. From the LSV run in 1.0 M KOH, the onset potential was observed to be +1.75 V at 5.9±1.7 mA cm^{-2} current density and Tafel slope of 60 mV dec^{-1}.

5.3   Graphitic carbon nitride (g-C$_3$N$_4$) based electrocatalysts for overall water splitting

Now a days graphitic carbon nitride (g-C$_3$N$_4$) have challenged the assembly of catalyst used for water splitting and is used as a promising catalyst for electrochemical water splitting. Graphitic carbon nitride has certain unique properties like high mechanical strength, high chemical durability, corrosion resistance, considerable hardness, and stability in different pH [90]. Graphitic carbon contains s-triazine units and has structural similarity to that of graphite. Presence of triazine group and its C-N bond, sp^2 hybridization, π-electrons and hexagonal structure affix exceptional physiochemical properties to these materials. Graphitic carbon has a different crystalline configuration like α-C$_3$N$_4$, β-C$_3$N$_4$, γ-C$_3$N$_4$, cubic C$_3$N$_4$, or zinc blende structure. Because of the presence of additional electron on nitrogen atoms and a smaller number of hydrogen atoms, more surface defects are present in the graphitic carbonnitride materials, which causes enhancement in their catalytic activity. Graphitic carbon nitride can be synthesized from nitrogen rich precursor through thermal condensation of urea, cyanamide, and/or melamine [91].

Zhong et al. [92] have reported graphitic carbon nitride based water splitting electrode. They have synthesized water splitting catalyst by making composite of graphitic carbon nitride carbon dot settled on graphene shown in Fig. 7. However, the authors only illustrated HER property of the catalyst having a low overpotential value of 110 mV with a Tafel slope of 53 mV dec^{-1}. Other than these, metal free carbon, nitrogen catalyst was synthesized by Zheng et al. [93] using composite of graphitic carbon nitride and N doped graphene ($C_3N_4@NG$), which possess unique molecular and electronic structure. HER overpotential value is obtained to be 240 mV at 10 mA cm^{-2} in 0.5 M H_2SO_4. It is also claimed that the prepared catalyst can be used as a bifunctional catalyst for water splitting.

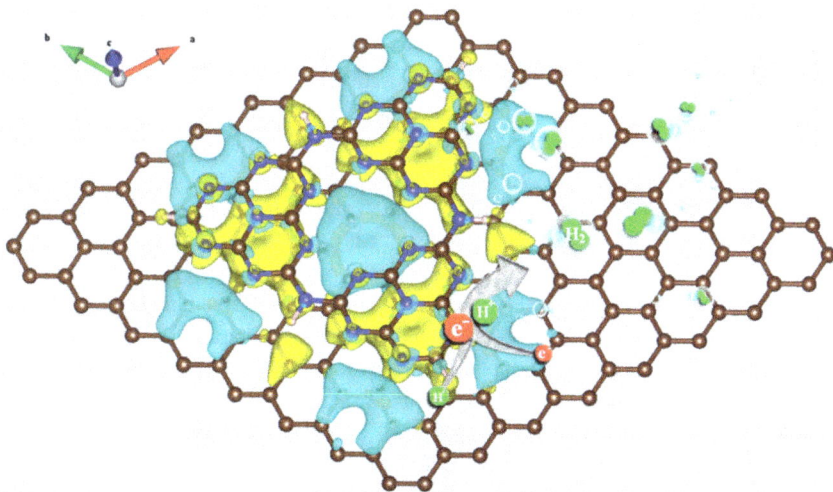

Fig. 7: *Graphical representation showing the performances of graphitic carbon nitride decorated with carbon dot over HER in water splitting [92].*

6.   Future aspects and outlook

Due to the increase in consumption of global energy and perilous impact of conventional energy resources on environment causes a high menace to flora and fauna. The electrochemical production of hydrogen and oxygen generate electricity from the fuel cell technology through electrochemical water splitting, which virtually causes no pollution.

Hence intensive research must be done to discover a new class of material which must have high catalytic performances and easy adsorption of H_2 and O_2 with high charge transfer for water splitting. It can be seen that metal free catalyst carbon based material shows low overpotential value with nearly zero onset potential, low Tafel slope, high stability, and robustness. A large number of articles are mostly observed for the doping of nitrogen and sulfur, inadequate articles are there for boron doping or other non-metal doping, as boron and other lower electronegativity elements create charge sites easy for adsorption of H_2 and O_2 for water splitting. Research effort must be enhanced for the synthesis of three dimensional carbon nanostructures which have a porous structure, high surface area, easy transport properties which can become a breakthrough in the field of water splitting. For metal free catalyst research is only limited to carbon based catalyst and there are seldom articles for the study of other nonmetals for water splitting like boron, silicon, and phosphorus as the principal element. As carbon based materials have shown great performances but knowledge in the area of mechanism and real active sites, atomic level understanding of those materials is still lacking. High resolution spectroscopic technique for the advanced theoretical study is required to explore those intrinsic properties and mechanism is required. Most of the research for metal free catalyst are seen to exhibit only HER or only OER and ORR, there is a limited number of articles in both HER and OER, so effort must be given for exploration of bifunctional catalyst for both hydrogen evolution and oxygen evolution reaction. Real time application like the use of the metal free catalyst as one of the high performances electrode for proton exchange membrane fuel cell (PEMFC), zinc air battery, etc. need to draw attention more for research purposes. The durability of the metal free catalyst is another challenging area of electrocatalyst which is of high concern in the scientific community. Harsh electrolytic environment causes aggregation, phase transfer, bleaching out of the catalyst, so preparation of metal free catalyst with improved stability for practical application is of great demand. Continual research in the field of water splitting catalyst will indubitably revolutionize the way in which the prospect of energy system be supposed to develop for better economy of fuel with decrease in environmental degradation and reduction of dependence on tradition energy sources.

Author declaration

Ms. Karfa has given the major contribution in writing to this book chapter along with drawing the figures and tables, requiring copyright permission, etc.

Reference

[1] M. K. Hubbert, Energy from fossil fuels, Science 109 (1949) 103-109. https://doi.org/10.1126/science.109.2823.103

[2] J.A Turner, A realizable renewable energy future, Science 285 (1999) 687-689.
https://doi.org/10.1126/science.285.5428.687

[3] K. Mazloomi, C. Gomes, Hydrogen as an energy carrier: Prospects and challenges,
Renew. Sust. Energ. Rev. 16 (2012) 3024-3033.
https://doi.org/10.1016/j.rser.2012.02.028

[4] A. Züttel, A. Remhof, A. Borgschulte, O. Friedrichs, Hydrogen: the future energy
carrier, Philos. Trans. Royal Soc. A 368, (2010) 3329-3342.
https://doi.org/10.1098/rsta.2010.0113

[5] S.U.M Khan, M. Al-Shahry, W.B. Ingler, Efficient photochemical water splitting by a
chemically modified n-TiO$_2$, Science 297 (2002) 2243-2245.
https://doi.org/10.1126/science.1075035

[6] T. Hisatomi, J. Kubota, K. Domen, Recent advances in semiconductors for
photocatalytic and photoelectrochemical water splitting, Chem. Soc. Rev. 43 (2014)
7520-7535. https://doi.org/10.1039/C3CS60378D

[7] J. Wang, W. Cui, Q. Liu, Z. Xing, A.M. Asiri, X. Sun, Recent progress in cobalt-
based heterogeneous catalysts for electrochemical water splitting. Adv. Mater. 28
(2016) 215-230. https://doi.org/10.1002/adma.201502696

[8] X. Li, X. Hao, A. Abudula, G. Guan, Nanostructured catalysts for electrochemical
water splitting: Current state and prospects, J. Mater. Chem. A (2016) 11973-12000.
https://doi.org/10.1039/C6TA02334G

[9] X. Zou, Z. Xiaoxin, Y. Zhang, Noble metal-free hydrogen evolution catalysts for
water splitting. Chem. Soc. Rev. 44 (2015) 5148-5180.
https://doi.org/10.1039/C4CS00448E

[10]   D. Kong, J.J. Cha, H. Wang, H.R. Lee, Y. Cui, First-row transition metal
dichalcogenide catalysts for hydrogen evolution reaction, Energy. Environ. Sci. 6
(2013) 3553-3558. https://doi.org/10.1039/c3ee42413h

[11]   D.R. Gamelin, Water splitting: Catalyst or spectator, Nat Chem. 4 (2012) 965.
https://doi.org/10.1038/nchem.1514

[12]   J.D Benck, T.R. Hellstern, J. Kibsgaard, P. Chakthranont, T.F. Jaramillo,
Catalyzing the hydrogen evolution reaction (HER) with molybdenum sulfide
nanomaterials, ACS Catal. 4, (2014) 3957-3971. https://doi.org/10.1021/cs500923c

[13]   N-T Suen, S-F Hung, Q Quan, N. Zhang, Y-J. Xu, H.M. Chen, Electrocatalysis
for the oxygen evolution reaction: recent development and future perspectives, Chem.
Soc. Rev. 46 (2017) 337-365. https://doi.org/10.1039/C6CS00328A

[14]   C.C.L McCrory, S. Jung, J.C. Peters, T.F. Jaramillo. Benchmarking heterogeneous
electrocatalysts for the oxygen evolution reaction. J. Am. Chem. Soc. 135 (2013)
16977-16987. https://doi.org/10.1021/ja407115p

[15] T. Reier, M. Oezaslan, P. Strasser, Electrocatalytic oxygen evolution reaction (OER) on Ru, Ir, and Pt catalysts: a comparative study of nanoparticles and bulk materials, ACS Catal. 2 (2012) 1765-1772. https://doi.org/10.1021/cs3003098

[16] A.T. Swesi, J. Masud, M. Nath, Nickel selenide as a high-efficiency catalyst for oxygen evolution reaction, Energy. Environ. Sci. Science 9 (2016) 1771-1782. https://doi.org/10.1039/C5EE02463C

[17] E. Fabbri, A. Habereder, K. Waltar, R. Kötz, T.J. Schmidt, Developments and perspectives of oxide-based catalysts for the oxygen evolution reaction, Catal. Sci. Technol.4 (2014) 3800-3821. https://doi.org/10.1039/C4CY00669K

[18] M.S. Burke, L.J. Enman, A.S. Batchellor, S. Zou, S.W. Boettcher, Oxygen evolution reaction electrocatalysis on transition metal oxides and (oxy) hydroxides: activity trends and design principles, Chem. Mater. 27 (2015) 7549-7558. https://doi.org/10.1021/acs.chemmater.5b03148

[19] Y. Zheng, Y. Jiao, L.H. Li, T. Xing, Y. Chen, M. Jaroniec, S.Z. Qiao, Toward design of synergistically active carbon-based catalysts for electrocatalytic hydrogen evolution, ACS nano 8 (2014) 5290-5296. https://doi.org/10.1021/nn501434a

[20] M.T.M. Koper, Hydrogen electrocatalysis: A basic solution, Nat. Chem. 5 (2013): 255. https://doi.org/10.1038/nchem.1600

[21] J.A. Turner, Sustainable hydrogen production, Science 305 (2004) 972-974. https://doi.org/10.1126/science.1103197

[22] J.D. Holladay, J. Hu, D.L. King, Y. Wang, An overview of hydrogen production technologies, Catal. Today 139 (2009) 244-260. https://doi.org/10.1016/j.cattod.2008.08.039

[23] I.K. Kapdan, F.Kargi, Bio-hydrogen production from waste materials, Enzyme Microb. Technol. 38 (2006) 569-582. https://doi.org/10.1016/j.enzmictec.2005.09.015

[24] Y. Shi, B. Zhang, Recent advances in transition metal phosphide nanomaterials: synthesis and applications in hydrogen evolution reaction, Chem. Soc. Rev. 45 (2016) 1529-1541. https://doi.org/10.1039/C5CS00434A

[25] J.D. Benck, T.R. Hellstern, J. Kibsgaard, P. Chakthranont, T.F. Jaramillo, Catalyzing the hydrogen evolution reaction (HER) with molybdenum sulfide nanomaterials, ACS Catal. 4 (2014) 3957-3971. https://doi.org/10.1021/cs500923c

[26] Y. Li, H. Wang, L. Xie, Y. Liang, G. Hong, H. Dai, MoS_2 nanoparticles grown on graphene: an advanced catalyst for the hydrogen evolution reaction, J. Am. Chem. Soc. 133 (2011) 7296-7299. https://doi.org/10.1021/ja201269b

[27] P. Xiao, M. AlamSk, L. Thia, X. Ge, R.J. Lim, J-Y Wang, K.H. Lim, X. Wang, Molybdenum phosphide as an efficient electrocatalyst for the hydrogen evolution reaction,Energy. Environ. Sci. 7 (2014) 2624-2629. https://doi.org/10.1039/C4EE00957F

[28]   Y. Ito, W. Cong, T. Fujita, Z. Tang, M. Chen, High catalytic activity of nitrogen and sulfur co-doped nanoporous graphene in the hydrogen evolution reaction, Angew. Chem. Int. Ed. 127 (2015) 2159-2164. https://doi.org/10.1002/ange.201410050

[29]   Y. Shi, B. Zhang, Recent advances in transition metal phosphide nanomaterials: Synthesis and applications in hydrogen evolution reaction, Chem. Soc. Rev. 45 (2016) 1529-1541. https://doi.org/10.1039/C5CS00434A

[30]   S. Lu, Z. Zhuang, Electrocatalysts for hydrogen oxidation and evolution reactions, Sci. China Mater. 59 (2016) 217-238. https://doi.org/10.1007/s40843-016-0127-9

[31]   D. Chialvo, MR Gennero, A. C. Chialvo, Hydrogen evolution reaction: analysis of the Volmer-Heyrovsky-Tafel mechanism with a generalized adsorption model, J. Electroanal. Chem. 372 (1994) 209-223. https://doi.org/10.1016/0022-0728(93)03043-O

[32]   A. Xie, N. Xuan, K. Ba, Z. Sun, Pristine graphene electrode in hydrogen evolution reaction, ACS Appl. Mater. Interfaces 9 (2017) 4643-4648. https://doi.org/10.1021/acsami.6b14732

[33]   Z. Sun, W. Fan, T. Liu, Graphene/graphene nanoribbon aerogels as tunable three-dimensional framework for efficient hydrogen evolution reaction, Electrochim. Acta 250 (2017) 91-98. https://doi.org/10.1016/j.electacta.2017.08.009

[34]   Y. Tian, Z. Wei, X. Wang, S. Peng, X. Zhang, W-M Liu, Plasma-etched, S-doped graphene for effective hydrogen evolution reaction, Int. J. Hydrog. Energy 42 (2017) 4184-4192. https://doi.org/10.1016/j.ijhydene.2016.09.142

[35]   B. Deng, D. Wang, Z. Jiang, J. Zhang, S. Shi, Z-J. Jiang, M. Liu, Amine group induced high activity of highly torn amine functionalized nitrogen doped graphene as the metal-free catalyst for hydrogen evolution reaction, Carbon 138 (2018) 169-178. https://doi.org/10.1016/j.carbon.2018.06.008

[36]   K. Chu, F. Wang, X-L. Zhao, X-P Wei, Xin-wei Wang, Y. Tian, One-step and low-temperature synthesis of iodine-doped graphene and its multifunctional applications for hydrogen evolution reaction and electrochemical sensing, Electrochim. Acta 246 (2017) 1155-1162. https://doi.org/10.1016/j.electacta.2017.07.001

[37]   Y. Liu, H. Yu, X. Quan, S. Chen, H. Zhao, Y. Zhang, Efficient and durable hydrogen evolution electrocatalyst based on nonmetallic nitrogen doped hexagonal carbon, Sci. Rep. 4 (2014) 6843. https://doi.org/10.1038/srep06843

[38]   X. Liu, W. Zhou, L. Yang, L. Li, Z. Zhang, Y. Ke, S. Chen, Nitrogen and sulfur co-doped porous carbon derived from human hair as highly efficient metal-free electrocatalysts for hydrogen evolution reactions, J. Mater. Chem. A 3 (2015)10135-10135. https://doi.org/10.1039/C5TA90086G

[39]   B.R Sathe, X. Zou, T. Asefa, Metal-free B-doped graphene with efficient electrocatalytic activity for hydrogen evolution reaction, Catal. Sci. Technol. 4 (2014) 2023-2030. https://doi.org/10.1039/C4CY00075G

[40]   W. Cui, Q. Liu, N. Cheng, A. M. Asiri, X. Sun, Activated carbon nanotubes: a highly-active metal-free electrocatalyst for hydrogen evolution reaction, Chem. Commun. 50 (2014) 9340-9342. https://doi.org/10.1039/C4CC02713B

[41]   L Wei, H.E.Karahan, K. Goh, W. Jiang, D. Yu, Ö. Birer, R. Jiang, Y. Chen, A high-performance metal-free hydrogen-evolution reaction electrocatalyst from bacterium derived carbon, J. Mater. Chem. A 3 (2015) 7210-7214. https://doi.org/10.1039/C5TA00966A

[42]   M. Tahir, L. Pan, F. Idrees, X. Zhang, L. Wang, J-J Zou, Z.L Wang, Electrocatalytic oxygen evolution reaction for energy conversion and storage: a comprehensive review, Nano Energy 37 (2017)136-157. https://doi.org/10.1016/j.nanoen.2017.05.022

[43]   J.A. Koza, Z. He, A.S. Miller, J.A. Switzer, Electrodeposition of crystalline Co_3O_4. A catalyst for the oxygen evolution reaction, Chem. Mater. 24 (2012) 3567-3573. https://doi.org/10.1021/cm3012205

[44]   N. Mamaca, E. Mayousse, S. Arrii-Clacens, T. W. Napporn, K. Servat, N. Guillet, K. B. Kokoh, Electrochemical activity of ruthenium and iridium based catalysts for oxygen evolution reaction, Appl. Catal. B. 111 (2012)376-380. https://doi.org/10.1016/j.apcatb.2011.10.020

[45]   I.C. Man, H.Y. Su, F.C. Vallejo, H.A. Hansen, J.I. Martínez, N.G. Inoglu, J. Kitchin, T.F. Jaramillo, J.K. Norskov, J. Rossmeisl, Universality in oxygen evolution electrocatalysis on oxide surfaces, Chem. Cat. Chem. 3 (2011) 1159-1165. https://doi.org/10.1002/cctc.201000397

[46]   V. Vij, S. Sultan, A.M. Harzandi, A. Meena, J.N. Tiwari, W.G. Lee, T. Yoon, K.S. Kim, Nickel-based electrocatalysts for energy-related applications: oxygen reduction, oxygen evolution, and hydrogen evolution reactions, ACS Catal.7 (2017) 7196-7225. https://doi.org/10.1021/acscatal.7b01800

[47]   M. Tuckerman, K. Laasonen, M. Sprik, M. Parrinello, Ab initio molecular dynamics simulation of the solvation and transport of hydronium and hydroxyl ions in water. J. Chem. Phys. 103(1995) 150-161. https://doi.org/10.1063/1.469654

[48]   X. Shen, Y.A. Small, J. Wang, P.B. Allen, M.V. Fernandez-Serra, M.S. Hybertsen, J.T.Muckerman, Photocatalytic water oxidation at the GaN (1010) – water interface, J. Phys. Chem. C114(2010) 13695-13704. https://doi.org/10.1021/jp102958s

[49]   J. Yang, D. Wang, X. Zhou, C. Li, A theoretical study on the mechanism of photocatalytic oxygen evolution on BiVO4 in aqueous solution, Chem. Eur. J 19(2013) 1320-1326. https://doi.org/10.1002/chem.201202365

[50] J. Zhang, X. Song, P. Li, Z. Wu, Y. Wu, S. Wang, X. Liu, Sulfur, nitrogen co-doped nanocomposite of graphene and carbon nanotube as an efficient bifunctional electrocatalyst for oxygen reduction and evolution reactions, J. Taiwan Inst. Chem. Eng. 93 (2018) 336-341. https://doi.org/10.1016/j.jtice.2018.07.040

[51] J-C. Li, P.X. Hou, M. Cheng, C. Liu, H.M. Cheng, M. Shao, Carbon nanotube encapsulated in nitrogen and phosphorus co-doped carbon as a bifunctional electrocatalyst for oxygen reduction and evolution reactions, Carbon 139 (2018) 156-163. https://doi.org/10.1016/j.carbon.2018.06.023

[52] B, Zhang, E. Zhang, S. Wang, Y. Zhang, Z. Ma, Y.Qiu, Bifunctional oxygen electrocatalyst derived from photochlorinated graphene for rechargeable solid-state Zn-air battery, J. Colloid Interface Sci. 543 (2019) 84-95. https://doi.org/10.1016/j.jcis.2019.02.044

[53] N. Jia, Q. Weng, Y. Shi, X. Shi, X. Chen, P. Chen, Z. An, Yu Chen, N-doped carbon nanocages: Bifunctional electrocatalysts for the oxygen reduction and evolution reactions, Nano Res. 11 (2018)1905-1916. https://doi.org/10.1007/s12274-017-1808-8

[54] J. Tian, Q. Liu, A.M. Asiri, K.A. Alamry, X. Sun, Ultrathin graphitic C_3N_4nanosheets/graphene composites: efficient organic electrocatalyst for oxygen evolution reaction, Chem. Sus. Chem. 7 (2014) 2125-2130. https://doi.org/10.1002/cssc.201402118

[55] N. Cheng, Q. Liu, J. Tian, Y. Xue, A.M. Asiri, H. Jiang, Y. He, X. Sun, Acidically oxidized carbon cloth: a novel metal-free oxygen evolution electrode with high catalytic activity, Chem. Commun. 51 (2015) 1616-1619. https://doi.org/10.1039/C4CC07120D

[56] S. Chen, J. Duan, M. Jaroniec, S.Z.Qiao, Nitrogen and oxygen dual-doped carbon hydrogel film as a substrate-free electrode for highly efficient oxygen evolution reaction, Adv. Mater. 26 (2014), 2925-2930. https://doi.org/10.1002/adma.201305608

[57] C. Zhang, B. Wang, X. Shen, J. Liu, X. Kong, S.S.C Chuang, D. Yang, A. Dong, Zhenmeng Peng, A nitrogen-doped ordered mesoporous carbon/graphene framework as bifunctional electrocatalyst for oxygen reduction and evolution reactions, Nano Energy 30 (2016)503-510. https://doi.org/10.1016/j.nanoen.2016.10.051

[58] J. Zhang, Z. Zhao, Z. Xia, L. Dai, A metal-free bifunctional electrocatalyst for oxygen reduction and oxygen evolution reactions, Nat. Nanotechnol. 10 (2015) 444. https://doi.org/10.1038/nnano.2015.48

[59] J. Sun, S.E. Lowe, L. Zhang, Y. Wang, K. Pang, Y. Wang, Y. Zhong, Ultrathin nitrogen-doped holey carbon@ graphene bifunctional electrocatalyst for oxygen reduction and evolution reactions in alkaline and acidic media, Angew. Chem. Int. Ed. 130(2018)16749-16753. https://doi.org/10.1016/j.jpowsour.2008.01.070

[60]   W. Schmittinger, A. Vahidi, A review of the main parameters influencing long-term performance and durability of PEM fuel cells, J. Power Sources. 180 (2008) 1-14. https://doi.org/10.1016/j.jpowsour.2008.01.070

[61]   A. Romo-Negreira, D.J. Cott, S.D. Gendt, K. Maex, M.M. Heyns, P.M. Vereecken, Electrochemical tailoring of catalyst nanoparticles for CNT spatial-dimension control, J. Electrochem. Soc. 157 (2010) K47-K51. https://doi.org/10.1149/1.3280245

[62]   T. Shinagawa, K.Takanabe, Towards versatile and sustainable hydrogen production through electrocatalytic water splitting: electrolyte engineering, Chem. Sus. Chem. 10 (2017)1318-1336. https://doi.org/10.1002/cssc.201601583

[63]   J. Heyrovský, A theory of overpotential. Recueil des TravauxChimiques des Pays-Bas 46 (1927) 582-585. https://doi.org/10.1002/recl.19270460805

[64]   M. Görlin, P. Chernev, J. Ferreira de Araújo, T. Reier, S. Dresp, B. Paul, R. Krähnert, H. Dau, P.Strasser, Oxygen evolution reaction dynamics, faradaic charge efficiency, and the active metal redox states of Ni–Fe oxide water splitting electrocatalysts, J. Am. Chem. Soc. 138(2016)5603-5614. https://doi.org/10.1021/jacs.6b00332

[65]   J.R. Hollenbeck, C.R. Williams, Turnover functionality versus turnover frequency: A note on work attitudes and organizational effectiveness, J. Appl. Psychol. 71 (1986) 606. https://doi.org/10.1037//0021-9010.71.4.606

[66]   M. Lukaszewski, M. Soszko, A. Czerwiński, Electrochemical methods of real surface area determination of noble metal electrodes–an overview, Int. J. Electrochem. Sci. 11 (2016) 4442-4469. https://doi.org/10.20964/2016.06.71

[67]   D. Astruc, ed. Nanoparticles and catalysis. John Wiley & Sons, (2008). https://doi.org/10.1002/9783527621323

[68]   J.D. Benck, Z. Chen, L.Y. Kuritzky, A.J. Forman, T.F. Jaramillo, Amorphous molybdenum sulfide catalysts for electrochemical hydrogen production: insights into the origin of their catalytic activity, ACS Catal. 2 (2012) 1916-1923. https://doi.org/10.1021/cs300451q

[69]   C. Hu, L. Dai, Carbon-based metal-free catalysts for electrocatalysis beyond the ORR, Angew. Chem. Int. Ed. 55 (2016) 11736-11758. https://doi.org/10.1002/anie.201509982

[70]   L. Zhang, J. Xiao, H. Wang, M. Shao, Carbon-based electrocatalysts for hydrogen and oxygen evolution reactions, ACS Catal. 7 (2017)7855-7865. https://doi.org/10.1021/acscatal.7b02718

[71]   L. Dai, D.W. Chang, J.B. Baek, W. Lu, Carbon nanomaterials for advanced energy conversion and storage, Small 8(2012)1130-1166. https://doi.org/10.1002/smll.201101594

[72]   Y. Zhao, R. Nakamura, K. Kamiya, S. Nakanishi, K. Hashimoto, Nitrogen-doped carbon nanomaterials as non-metal electrocatalysts for water oxidation, Nat. Commun. 4 (2013) 2390. https://doi.org/10.1038/ncomms3390

[73]   H. Chen, M.B. Müller, K.J. Gilmore, G.G. Wallace, D. Li, Mechanically strong, electrically conductive, and biocompatible graphene paper, Adv. Mater. 20(2008) 3557-3561. https://doi.org/10.1002/adma.200800757

[74]   A.K. Geim, K.S. Novoselov, The rise of graphene, In Nanoscience and Technology: A Collection of Reviews from Nature Journals (2010) 11-19. https://doi.org/10.1142/9789814287005_0002

[75]   A.H.C Neto, F. Guinea, N.M.R Peres, K.S. Novoselov, A.K. Geim, The electronic properties of graphene, Rev. Mod. Phys. 81 (2009) 109. https://doi.org/10.1103/RevModPhys.81.109

[76]   J. Lai, S. Li, F. Wu, M.Saqib, R. Luque, G. Xu, Unprecedented metal-free 3D porous carbonaceous electrodes for full water splitting, Energy Environ. Sci. 9 (2016) 1210-1214. https://doi.org/10.1039/C5EE02996A

[77]   J. Zhang, L. Dai, Nitrogen, Phosphorus, and Fluorine Tri-doped Graphene as a Multifunctional Catalyst for Self-Powered Electrochemical Water Splitting, Angew. Chem. Int. Ed 55 (2016)13296-13300. https://doi.org/10.1002/anie.201607405

[78]   J. Zhang, C. Zhang, J. Sha, H. Fei, Y. Li, J.M. Tour, Efficient water-splitting electrodes based on laser-induced graphene, ACS Appl. Mater. Interfaces 9 (2017) 26840-26847. https://doi.org/10.1021/acsami.7b06727

[79]   Y. Jia, L. Zhang, A. Du, G. Gao, J. Chen, X. Yan, C. L. Brown, X. Yao, Defect graphene as a trifunctional catalyst for electrochemical reactions, Adv. Mater. 28 (2016) 9532-9538. https://doi.org/10.1002/adma.201602912

[80]   X. Yue, S. Huang, J. Cai, Y. Jin, P.K. Shen, Heteroatoms dual doped porous graphene nanosheets as efficient bifunctional metal-free electrocatalysts for overall water-splitting, J. Mater. Chem. A 5 (2017) 7784-7790. https://doi.org/10.1039/C7TA01957B

[81]   T.W. Ebbesen, P.M. Ajayan, Large-scale synthesis of carbon nanotubes, Nature 358 (1992) 220. https://doi.org/10.1038/358220a0

[82]   S. Iijima, T. Ichihashi, Single-shell carbon nanotubes of 1-nm diameter, Nature 363(1993)603. https://doi.org/10.1038/363603a0

[83]   X. Lin, X.K. Wang, V.P. Dravid, R.P.H. Chang, J.B. Ketterson, Large scale synthesis of single-shell carbon nanotubes, Appl. Phys. Lett. 64 (1994) 181-183. https://doi.org/10.1063/1.111525

[84]   A. Adam, M.H. Suliman, H. Dafalla, A.R. Al-Arfaj, M.N. Siddiqui, M. Qamar, Rationally dispersed molybdenum phosphide on carbon nanotubes for the hydrogen

evolution reaction, ACS Sustain. Chem. Eng.6 (2018) 11414-11423.
https://doi.org/10.1021/acssuschemeng.8b01359

[85]   F. Davodi, M. Tavakkoli, J. Lahtinen, T. Kallio, Straightforward synthesis of nitrogen-doped carbon nanotubes as highly active bifunctional electrocatalysts for full water splitting, J. Catal. 353 (2017) 19-27.
https://doi.org/10.1021/acssuschemeng.8b01359

[86]   K. Qu, Y. Zheng, Y. Jiao, X. Zhang, S. Dai, S-Z. Qiao, Polydopamine-Inspired, Dual Heteroatom-Doped Carbon Nanotubes for Highly Efficient Overall Water Splitting, Adv. Energy Mater. 7 (2017)1602068.
https://doi.org/10.1002/aenm.201602068

[87]   A. Ali, D. Akyüz, M. A.Asghar, A. Koca, B. Keskin, Free-standing carbon nanotubes as non-metal electrocatalyst for oxygen evolution reaction in water splitting, Int. J. Hydrog. Energy 43(2018)1123-1128.
https://doi.org/10.1016/j.ijhydene.2017.11.060

[88]   C. Zhang, S. Bhoyate, M. Hyatt, B.L. Neria, K. Siam, P. K. Kahol, M. Ghimire, S. R. Mishra, F. Perez, R.K. Gupta, Nitrogen-doped flexible carbon cloth for durable metal free electrocatalyst for overall water splitting, Surf. Coat. Technol. 347 (2018) 407-413. https://doi.org/10.1016/j.ijhydene.2017.11.060

[89]   Y. Cheng, C. Xu, L. Jia, J.D. Gale, L. Zhang, C. Liu, P.K. Shen, S.P. Jiang, Pristine carbon nanotubes as non-metal electrocatalysts for oxygen evolution reaction of water splitting, Appl. Catal. B 163 (2015) 96-104.
https://doi.org/10.1016/j.apcatb.2014.07.049

[90]   A. Thomas, A. Fischer, F. Goettmann, M. Antonietti, J.O. Müller, R. Schlögl, J.M. Carlsson, Graphitic carbon nitride materials: variation of structure and morphology and their use as metal-free catalysts,J. Mater. Chem. 18 (2008) 4893-4908.
https://doi.org/10.1039/b800274f

[91]   Y. Zheng, J. Liu, J. Liang, M. Jaroniec, S.Z. Qiao, Graphitic carbon nitride materials: controllable synthesis and applications in fuel cells and photocatalysis, Energy Environ. Sci. 5 (2012) 6717-6731. https://doi.org/10.1039/c2ee03479d

[92]   H.X. Zhong, Q. Zhang, J. Wang, X.B. Zhang, X.L. Wei, Z.J. Wu, K. Li, F.L. Meng, D. Bao, J.M Yan, Engineering ultrathinC_3N_4 quantum dots on graphene as a metal-free water reduction electrocatalyst, ACS Catal. 8 (2018) 3965-3970.
https://doi.org/10.1021/acscatal.8b00467

[93]   Y. Zheng, Y. Jiao, Y. Zhu, L. H. Li, Y. Han, Y. Chen, A. Du, M. Jaroniec, S.Z. Qiao, Hydrogen evolution by a metal-free electrocatalyst, Nat. Commun. 5 (2014) 3783. https://doi.org/10.1038/ncomms4783

Electrochemical Water Splitting: Materials and Applications
Materials Research Foundations 59 (2019) 125-140

Materials Research Forum LLC
doi: https://doi.org/10.21741/9781644900451-5

Chapter 5

Ni-Based Electrocatalyst for Full Water Splitting

Atanu Roy[1], Samik Saha[1,2], Apurba Ray[1], Sachindranath Das[1]*

[1]Department of Instrumentation Science, Jadavpur University, Kolkata, India-700032

[2]Department of Physics, Jadavpur University, Kolkata, India-700032

sachindas15@gmail.com, sachindran.das@jadavpuruniversity.in

Abstract

Development of high-performance novel metal-free electrocatalysts operating for full water splitting (both oxygen evolution reaction (OER) and hydrogen evolution reaction (HER)) is desirable for energy storage and conversion process. However, it is thermodynamically unfavourable. Obtaining a feasible output for both HER and OER from the same catalyst and same electrolyte simultaneously is a very challenging task. Several Ni-composites or compounds based catalysts for full electrocatalysis of water. Several Ni-hydroxies, oxides, phosphides, nitrides, sulfides, selenides are capable of providing efficient output for full water splitting compared to the benchmark Pt and IrO_2 based electrocatalysts.

Keywords

HER, OER, Electrocatalysis, Water splitting, Ni-based Electrocatalyst

Contents

Electrochemical Water Splitting: Materials and Applications   Materials Research Forum LLC
Materials Research Foundations **59** (2019) 125-140   doi: https://doi.org/10.21741/9781644900451-5

1. Introduction

Electrolytic water splitting is the cleanest way to produce pure hydrogen and can act as large scale storage for renewable energies like solar, wind, etc. This H_2 generation process does not emit any pollutants, and it requires inexpensive water as a renewable source. Moreover, H_2 generated from this process is very pure and does not contain any carbon monoxide impurities which enable H_2 to be utilised in a fuel cell without any poisoning risk of anode catalysts. However, electrolytic water splitting process is thermodynamically unfavourable, and theoretically, it requires 1.23 V to start the reaction [1]. But a greater overpotential (η) above this thermodynamic potential is required to begin the unfavourable reaction at an acceptable rate.

Presently, ruthenium oxides (RuO_2) and iridium oxide (IrO_2) are the most favorable catalysts for oxygen evolution reaction (OER) [2,3]. On the other hand, platinum (Pt) is the best catalyst for hydrogen evolution reaction (HER) [4]. Both these oxides and Pt are very expensive as well as their limited availability makes the H_2 production from electrocatalysis an economically unfeasible process. Recently, several groups have attempted to develop the efficient OER and HER catalysts from earth-abundant transition metals such as Fe, Ni, Co, etc. The major challenge of these transition metal-oxides based catalysts is cost effectiveness and comparable performance to the conventional precious metals. Various metal oxides or hydroxides, perovskite oxides, and phosphides, polymeric carbon nitride, etc. have been used as the potential catalyst for OER.

On the other hand, transition metal, carbides, selenides, chalcogenides, phosphides, bi-metal alloys, nanocomposites, etc. have the potential to act as an efficient HER catalyst. However, the coupling of two different types of catalyst to form a water splitting cell for practical applications is challenging because of the mismatch of pH range of the electrolytes in which these catalysts provide optimum performance and stability. In

Electrochemical Water Splitting: Materials and Applications   Materials Research Forum LLC
Materials Research Foundations **59** (2019) 125-140   doi: https://doi.org/10.21741/9781644900451-5

addition to that, the production of different types of catalysts for HER and OER needs different processes and equipment which also enhance the cost. Thus, the development of a catalyst for both OER and HER in the same electrolyte has become utmost important yet challenging [4,5].

There are some composite materials based on Ni, Co, Mo, and Fe which show their activity for both HER and OER. These catalysts include metal hydroxides (e.g. $Ni(OH)_2$) [6], phosphides (e.g. Ni_5P_4) [7], nitrides ($TiN-Ni_3N$) [8], selenides ($NiSe$) [9] etc. However, when these materials are paired to form an electrolyser, most of them show much lower overall output compared to the benchmark Pt and IrO_2 based electrolyser. The most efficient $Pt//IrO_2$ based electrolyser generates water splitting current density of 10 mA cm^{-2} or more at a potential ~1.5 V [10]. None of these previously mentioned non-noble metal based materials can achieve such low potential. Moreover, the cyclic stability for long term performance of some of these materials also degrades during overall water splitting.

Some recent reports describe optimized overall water splitting performances of Ni based composites. Owing to rich redox reaction, good electrical/ionic conductivity, low cost, and excellent microstructure of Ni based composites encouraged researchers to study them as the active catalyst for water splitting. Several Ni-foam supported composites synthesized through hydrothermal deposition [11], electro deposition [12], and chemical vapour deposition [13] show promise as an efficient catalyst for water splitting. This is mainly due to the presence of extra-large macrospores (>200 μm) of Ni-foam which provides enormous open space to develop desirable nanostructures and microstructures with high porosity and enhanced and high surface area without hampering the gas bubble dissipation capability of Ni-foam. Growing Ni based composites on Ni foam enables binder free catalysts, which are important for rapid transport of electrons for catalysis.

2.   Water splitting

2.1   Brief history and basics of water splitting

Since the first report of water electrocatalysis in 1789 [14], it has been tremendously studied and applied in various applications. Different electrolysis techniques and electrolysers were developed between the 1920s and 1930s. During the Second World War, Canada and Norway developed several gigantic 100-MW size plants based on low-cost hydroelectric projects for the production of ammonia based fertilizers. After the Second World War, low-cost fossil fuels replaced the electrochemically generated H_2 and the researchers lost their interest in this field. This historic process again got popularity in the last two decades because of its cleanest pollutant free nature. In spite of such long

Electrochemical Water Splitting: Materials and Applications   Materials Research Forum LLC
Materials Research Foundations **59** (2019) 125-140   doi: https://doi.org/10.21741/9781644900451-5

efforts, water splitting still needs more advanced catalyst having long term stability and low cost applications. The electrolysis of water contains two half reactions, namely, HER, which occurs at the cathode, whereas the OER occurs at the anode [15,16]. The HER process includes two electrons transfer, whereas OER requires four electrons transfer.

2.2 Few parameters related to t oxygen evolution reaction, hydrogen evolution reaction and catalytic activity

A. Overpotential (η): The excess potential required to attain the current density of 10 mA cm^{-2} is known as overpotential. This current density is the expected current density for 12.3% efficient solar water splitting cell. The overpotential of the electrode is generally estimated by linear sweep voltammetry (LSV) and cyclic voltammetry (CV). Ideal material could generate high current density even with lower overpotential. The overpotential can be calculated by using these two equations,

$$\eta = 0 - V_{observed} \qquad \text{for HER} \tag{1}$$

$$\eta = V_{observed} - 1.23\,V \qquad \text{for OER} \tag{2}$$

B. Tafel slope (b): The information about reaction kinetics of OER and HER processes can be obtained from the knowledge of Tafel slope. This can be obtained from LSV data by fitting the linear region to the Tafel equation as given below,

$$\eta = b \log j/j_0$$

Here η is the previously mentioned overpotential, b is the Tafel slope, j is the observed current density and j_0 is known as exchange current density, which is defined as the current density when no electrolysis occurs at zero overpotential. The value of Tafel slope depends on different parameters such as reaction pathway, crystal categories, catalyst preparation, adsorption condition of active sites, etc. A good catalyst should have low Tafel slope.

C. Stability: Stability of a catalyst towards water splitting is an important parameter. Long-term stability of the catalyst can be ensured by CV and chronoamperometry (potential-time) or chronopotentiometry (current-time). For CV test, stability is usually evaluated by comparing the electrochemical performance (polarization curves) before and after the cyclic test of few thousands of cycles. On the other hand, chronoamperometry and chronopotentiometry are the measurements of potential or current with time at constant current density or constant potential, respectively. These variations are usually monitored for more than 10 hours. Any degradation in the stability of the electrocatalysis can be inferred from the observed departure from the constant potential (chronopotentiometry) or the constant current (chronoamperometry) value over the time duration.

D. Electrochemical active surface area (ECSA): The overall water splitting capability of an electrode material depends both on the intrinsic catalytic performance and electrochemical active surface area (ECSA) of that material itself. ECSA can be estimated from the double layer capacitance C_{dl} offered by the material using the following equation [13].

$$ECSA=C_{dl}/C_s$$

Here C_{dl} can be estimated from the slope of straight line fitting of scan rate vs current data at the higher scan rates (non-Faradic) and C_s is the electrochemical double layer capacitance of a standard catalyst [17], and its value is 40 μF cm^{-2} for 1M KOH electrolyte [18]. Since catalytic activity is a surface dependent phenomenon, so it will increase with the increase in ECSA. Thus catalyst with better ECSA ensures better material for catalytic activity.

E. Roughness factor (RF): Similar to the ECSA higher value of RF offers superior electrocatalytic performance. The atoms present on the surface can only contribute to OER and HER reaction. The RF can be obtained from the following equation [19].
RF= ECSA/geometric area of the catalyst

2.3 Mechanism of electrochemical water splitting

An electrolyser consists of three components: a cathode, an anode, and an electrolyte. To accelerate the water splitting process, active OER and HER catalysts are coated on anode and cathode, respectively. When an external potential is introduced to the electrodes, the water molecule disintegrated into oxygen and hydrogen. Highly pure hydrogen and oxygen can be collected from cathode and anode, respectively. The total water splitting reaction can be expressed in terms of two half reactions, which are generally different for different types of electrolytes.

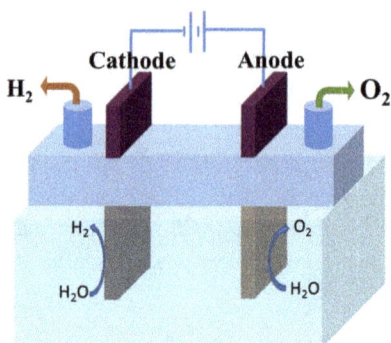

Fig. 1: Schematic diagram of water splitting.

In alkaline and neutral electrolyte:

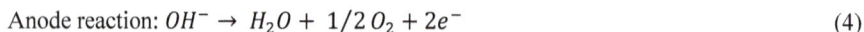

Cathode reaction: $2H_2O + 2e^- \rightarrow H_2 + 2OH^-$ $\qquad\qquad\qquad$ (3)

Anode reaction: $OH^- \rightarrow H_2O + 1/2\,O_2 + 2e^-$ $\qquad\qquad\qquad$ (4)

In acidic electrolyte:

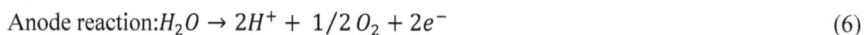

Cathode reaction: $2H^+ + 2e^- \rightarrow H_2$ $\qquad\qquad\qquad\qquad$ (5)

Anode reaction: $H_2O \rightarrow 2H^+ + 1/2\,O_2 + 2e^-$ $\qquad\qquad\qquad$ (6)

Although the required theoretical thermodynamic potential of water splitting is 1.23 V (at 25 °C and 1 atm) which is independent of the electrolyte in which the water splitting is taking place. However, in general, a higher potential is required to expedite water splitting. This higher potential results in the consumption of excess energy and decreases conversion efficiency. The excess potential, which is previously mentioned as overpotential (η), is required to overcome the intrinsic reaction activation barrier, solution resistance, contact resistance, etc. Many types of research are working on to develop an efficient catalyst with lower overpotential and low cost.

2.3.1 Hydrogen evolution reaction (HER)

HER is a multi-step procedure, and it occurs at the surface of the catalyst electrode. The HER generally follows the reaction:

$$2\,H^+ + 2\,e^- \rightarrow H_2 \tag{7}$$

The mechanism of hydrogen adsorption (H_{ad}) on the surface of the electrocatalyst consists of two steps. In the first step, known as Volmer step, the formation of H_{ad} occurs. Next, the formation of H_2 gas from H_{ad} occurs either by Tafel step or by Heyrovsky step. These steps can be represented by the chemical equations as follow

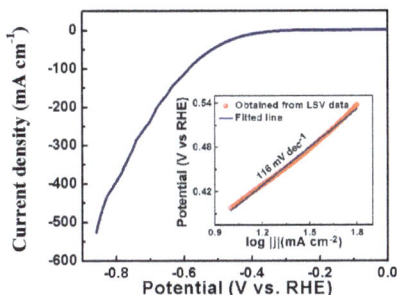

Fig. 2: Typical HER plot of a catalyst (inset) Tafel slope of the same data.

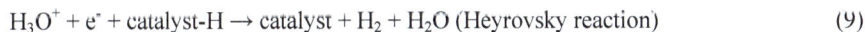

$$H_3O^+ + e^- + catalyst \rightarrow catalyst\text{-}H + H_2O \text{ (Volmer reaction)} \tag{8}$$

$$H_3O^+ + e^- + catalyst\text{-}H \rightarrow catalyst + H_2 + H_2O \text{ (Heyrovsky reaction)} \tag{9}$$

$$catalyst\text{-}H + catalyst\text{-}H \rightarrow 2\,catalyst + H_2 \text{ (Tafel reaction)}. \tag{10}$$

To acquire fast kinetics, a suitable electrocatalyst is needed for HER. Recently, novel electrocatalysts such as Pt group metals and Ru/Ir-based compound possess higher HER activities with very small Tafel slope, very low over potential and high exchange current density but their high cost and scarcity restrict their commercial usage. Therefore, the development of HER catalysts for water splitting with efficient, lower over potential, cheap, stable, and abundant is most important to achieve cost effective hydrogen generation.

2.3.2 Oxygen evolution reaction (OER)

The OER is another half of the water splitting reaction (2 $H_2O \rightarrow 2H_2 + O_2$). The overall water splitting reactions which occur at anode and cathode are different under acidic and alkaline conditions.

In acidic conditions

Cathode reaction:

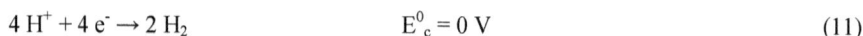

$$4\,H^+ + 4\,e^- \rightarrow 2\,H_2 \qquad\qquad E^0_c = 0\ V \qquad\qquad (11)$$

Anode reaction:

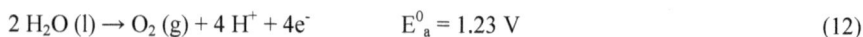

$$2\,H_2O\,(l) \rightarrow O_2\,(g) + 4\,H^+ + 4e^- \qquad E^0_a = 1.23\ V \qquad\qquad (12)$$

Possible OER mechanism under acidic conditions

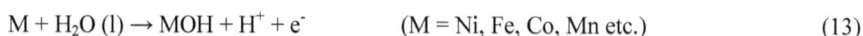

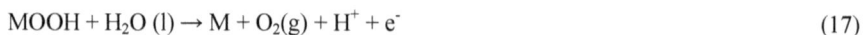

$$M + H_2O\,(l) \rightarrow MOH + H^+ + e^- \qquad (M = Ni,\ Fe,\ Co,\ Mn\ etc.) \qquad (13)$$

$$MOH + OH^- \rightarrow MO + H_2O\,(l) + e^- \qquad\qquad (14)$$

$$2MO \rightarrow 2M + O_2\,(g) \qquad\qquad (15)$$

$$MO + H_2O\,(l) \rightarrow MOOH + H^+ + e^- \qquad\qquad (16)$$

$$MOOH + H_2O\,(l) \rightarrow M + O_2(g) + H^+ + e^- \qquad\qquad (17)$$

Electrochemical Water Splitting: Materials and Applications   Materials Research Forum LLC
Materials Research Foundations **59** (2019) 125-140   doi: https://doi.org/10.21741/9781644900451-5

Alkaline conditions

Cathode reaction:

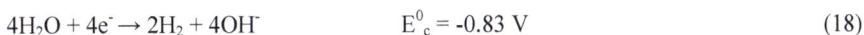

$$4H_2O + 4e^- \rightarrow 2H_2 + 4OH^- \qquad\qquad E^0_c = -0.83 \text{ V} \qquad\qquad (18)$$

Anode reaction:

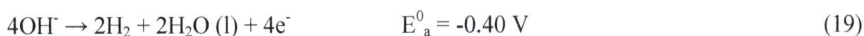

$$4OH^- \rightarrow 2H_2 + 2H_2O \text{ (l)} + 4e^- \qquad\qquad E^0_a = -0.40 \text{ V} \qquad\qquad (19)$$

Possible OER mechanism under alkaline conditions

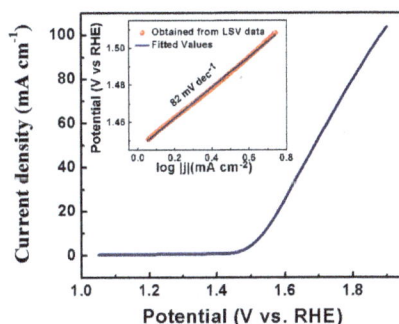

$$M + OH^- \rightarrow MOH \qquad\qquad\qquad\qquad (20)$$

$$MOH + OH^- \rightarrow MO + H_2O \text{ (l)} \qquad\qquad\qquad\qquad (21)$$

$$2MO \rightarrow 2M + O_2 \text{ (g)} \qquad\qquad\qquad\qquad (22)$$

$$MO + OH^- \rightarrow MOOH + e^- \qquad\qquad\qquad\qquad (23)$$

$$MOOH + OH^- \rightarrow M + O_2(g) + H_2O \text{ (l)} \qquad\qquad\qquad\qquad (24)$$

Fig. 3: Typical OER plot of a catalyst (inset) Tafel slope of the same data.

2.4 Recent advances on materials and performance of Ni based materials for overall water splitting

2.4.1 Ni- based oxides and hydroxides

Unlike the metallic states, Ni based oxide or hydroxide composites often demonstrate +2 oxidation states for Ni, which is adjacent to the oxidation states of the OER region. On the other hand, layered double hydroxides (LDH) contain layers of positively charge metal ions, and the anions are in intermediate layers. This type of layered structure possesses a large area for electrochemical interaction. Various NiFe oxides with well-defined structures provide excellent performance on overall water splitting. Haotian Wang et al. [10] used different metal oxides nanoparticles and converted them to ultra-small nanoparticles of 2-5 nm diameter using lithium induced conversion. They demonstrated the overall water-splitting capability of $NiFeO_x$ nanoparticles in 1M KOH solution which offered 10 mA cm^{-2} water-splitting current at 1.51 V. This composite exhibits long term stability for more than 200 h.

$Ni_{0.8}Co_{0.1}Fe_{0.1}O_xH_y$ has been developed by Q. Zhao and his group[20]. They have reported excellent OER, and HER activity and these catalysts have demonstrated overall electrical water splitting at a potential of 1.58 mV at 10 mA cm^{-2}. This doped hydroxide provides a pathway for the fabrication of overall water splitting activity. Sultana et al. [21] have demonstrated electrodeposited $Ni(OH)_2$ and $Co(OH)_2$, which contained well dispersed Au nanoparticles. The electrodeposition was first carried out on a glassy carbon electrode for a particular hydroxide then gold nanoparticles were deposited followed by electrodeposition of the second hydroxide layer. They showed that for a two electrode device, these electrodes exceeded the current density of 175 mA cm^{-2} at a potential of 1.9V for 6M aqueous NaOH electrolyte [21].

2.4.2 Ni-based phosphides

Several Ni-based phosphides such as Ni_2P, Ni_8P_3, Ni-P alloy, etc. have been used as the catalyst for HER activity in both acidic and base electrolyte. These composites offer high current densities at lower overpotential. Wang et al. [22] electrodeposited nickel phosphide electrocatalyst on carbon fibre paper. They demonstrated this self-supported electrode offers very small overpotential of 162 mV and 250 mV in acidic and alkaline medium, respectively towards hydrogen evolution. They also obtained a high OER current at an overpotential of 0.3V in 1MKOH. They had constructed an electrolyzer using the same electrode material for both cathode and anode that showed water splitting efficiency of 91% at 10 mA cm^{-2} for 100h [22].

Gao-Feng Chen et al. [23] have demonstrated overall water splitting using electrodes made of Ni_8P_3 and Ni_9S_8 deposited on commercial Ni-foam. They showed that Ni_8P_3 is the most stable catalyst with an onset potential of -0.07 V for HER and 1.49 V for OER in 1M KOH solution. Fang Yu et al. [24] synthesized dinickel phosphide on nickel foam, which was utilized to demonstrate the overall water splitting in aqueous alkaline solution. The electrode offered 10 mA cm^{-2} current density when 1.42 V potential was applied i.e. a very low overpotential which is as low as 0.19 V. They had also shown the durability test for 40h by applying 1.72 V across the electrode which then offered 500 mA cm^{-2} current density. Stern et al. [25] has demonstrated that nickel phosphide (Ni_2P) can be used as an efficient hydrogen evolving catalyst. They reported the overpotential of 290 mV in 1M KOH. They also assembled a two electrode device using the same material in both electrodes. This device demonstrated 10 mA cm^{-2} current density at 1.63 V onset potential. Ni-P alloy thin film which demonstrated by Liju Elias et al. [26] for the overall water splitting in alkaline medium. They optimized the mass loadings of the films and obtained onset potential of 0.43 V for O_2 evolution and – 1.3 V for H_2 evolution.

2.4.3 Ni-based nitrides

Several Ni based single phase nitrides, binary nitrides and nitrides based composites have been studied as the catalyst for both OER and HER activity over a wide pH range. Qiting Zhang et al. [8] reported an overall water splitting electrolyzer based on Sn and Ni_3N nanowire arrays. For hydrogen evolution reaction they demonstrated that these nanowire arrays offered overpotential of 15 mV where as an overpotential of 1.52 V (vs. RHE) was obtained during the OER. A device was assembled using the nanowire arrays in the form of two electrodes which could perform water splitting with an onset potential of 1.57 V and current retention of 63.8% after 16 h of operation. Ni-doped molybdenum nitride nanorods on Ni foam synthesized by Jun Ran Ji et al.[27]. They showed oxygen evolution reaction and hydrogen evolution reaction performance at a current density of 10 mA cm^{-2} in 1.0 M KOH with overpotentials of 218 mV and 15 mV, respectively.

Menny Shalom et al. [28] fabricated nickel nitride on nickel foam and demonstrated their electrocatalytic properties. These coated nickel foams demonstrated overpotential of 50 mV in alkaline solution for HER. They had also shown the OER performances of the bare nickel foam improved after the nickel nitride coating.

2.4.4 Ni-based sulfides

Ni based sulfides or binary sulfides have gained the attention of the researchers because of their high theoretical catalytic activity, wide applicability (pH of electrolyte), controllable microstructure, and low costs. Yuanyuan Wu et al. [29] synthesized Ni_3S_2

nanosheet decorated with $Ni_xCo_{3-x}S_4$arrays on Ni-foam. This self-standing, noble metal free catalyst was used as electrodes of an electrolyzer. This device reaches current densities of 10 mA cm^{-2}at very low overpotential of 1.53 V and 100 mA cm^{-2}at 1.80 V. Bin Dong *et al.* [30] reported the bifunctional electrocatalytic properties of NiCoS nanorods grown on Ni-foam. These foams were directly used as the electrodes to generate 100 mA cm^{-2} at an overpotential of 360 mV for OER and 50 mA cm^{-2} at an overpotential of 200 mV for HER in alkaline medium.

The overall electrocatalytic properties of nickel sulphide grown on nickel foam were reported by Jin-Tao Ren *et al.* [31]. For HER these electrodes required overpotential of -122 mV to achieve 10 mA cm^{-2} current density on the other hand current density of 20 mA cm^{-2} was achieved at a low overpotential of 320 mV for OER. They had also fabricated an electrolyzer using the coated nickel foam as electrodes. The developed electrolyzer operated at a cell voltage of 1.61 V to provide a current density of 10 mA cm^{-2}.

2.4.4 Ni-based selenides

Metal selenides are more attractive catalysts than metal oxides or sulphides because of their intrinsic metallic property which enables rapid transport of electrons during electrocatalysis. Moreover, atomic ratio and crystal structure of M_xSe (M=Ni, Co and Fe) can be tailored, which is important for electrocatalytic performance. Typically, nickel selenides can have a series nonstoichiometric form such as $NiSe_2$, Ni_3Se_2, $NiSe$, Ni_3Se_4, Ni_5Se_6 etc. Chun Tang *et al.* [9] hydrothermally synthesized NiSe nanowire on nickel foam. These coated foams were directly used the electrodes which exhibited overpotential of 270 mV for OER in 1M KOH. They also demonstrated alkaline water based electrolyzer with 10 mA cm^{-2} obtained at onset potential of 1.63 V.

Fe-doped NiSe was synthesized using solvothermal reaction and reported by Zexian Zou *et al.* [32]. They have reported catalytic activity of the composite. They optimized the doping percentage of Fe to optimize the performance. They had shown that $Fe_{7.4\%}$-NiSe required overpotential of 231 mV to obtain a current density of 50 mA cm^{-2} for OER, whereas during HER the electrode achieved 10 mA cm^{-2} at an over the potential of 163 mV. Xiao Li *et al.* [33] synthesized cobalt and nickel selenide based nanowalls on a conductive graphene coated nickel mesh substrate and demonstrated their electrocatalytic properties. A current density of 10 mA cm^{-2} at -78 mV was reached for HER, and 150 mA cm^{-2} was reached at 1.74 V for OER. They had also studied the electrode stability for 24 h, which showed very little degradation.

Conclusions

Several efforts have been going on for the improvement of new catalysts for OER, HER and water splitting to commercialize clean and green energy production. As a replacement of high expensive Pt and IrO_2 catalysts, Ni-based composites have been widely explored as the catalysts for the water splitting applications. Different binary or ternary alloys of Ni and other transition metal oxides have been established as one of the most capable catalysts for water splitting. Besides, other Ni-based compounds such as Ni nitrides, sulfides, and selenides are known to have OER, HER, and bifunctional activity. Despite the high catalytic performance, the thermal and electrochemical performances of these materials are limited. However, Ni based composites exhibit excellent catalytic performance with high current densities as well as low overpotentials which are compatible with the Pt/IrO_2 based electrolyser. Moreover, ease of synthesis, low cost, the electrochemically stable and excellent performance of Ni-based catalyst is the foremost important and the future outlook for the production of clean energy.

Acknowledgement

A. Roy (IF140920) wishes to thank the Department of Science and Technology (DST), INSPIRE, Government of India, for financial support through the 'INSPIRE Fellowship'. S. Saha (File No.–09/096(0898)/2017–EMR-I) and A. Ray (File No.–09/096(0927)/2018-EMR-I) are thankful CSIR, Government of India for fellowship. S. Das is thankful to the 'INSPIRE Faculty Award' (IFA13-PH-71), Department of Science and Technology (DST), Government of India, for providing research support.

References

[1]   Y. Yan, B.Y. Xia, B. Zhao, X. Wang, A review on noble-metal-free bifunctional heterogeneous catalysts for overall electrochemical water splitting, J. Mater. Chem. A. 4 (2016) 17587–17603. https://doi.org/10.1039/C6TA08075H

[2]   T.D. Nguyen, G.G. Scherer, Z.J. Xu, A facile synthesis of size-controllable iro2 and ruo2 nanoparticles for the oxygen evolution reaction, Electrocatalysis. 7 (2016) 420–427. https://doi.org/10.1007/s12678-016-0321-2

[3]   H.H. Pham, N.P. Nguyen, C.L. Do, B.T. Le, Nanosized IrxRu1-xO2 electrocatalysts for oxygen evolution reaction in proton exchange membrane water electrolyzer, Adv. Nat. Sci. Nanosci. Nanotechnol. 6 (2015) 2–7. https://doi.org/10.1088/2043-6262/6/2/025015

[4]   N.-T. Suen, S.-F. Hung, Q. Quan, N. Zhang, Y.-J. Xu, H.M. Chen, Electrocatalysis for the oxygen evolution reaction: recent development and future perspectives, Chem. Soc. Rev. 46 (2017) 337–365. https://doi.org/10.1039/C6CS00328A

[5]   S.Y. Tee, K.Y. Win, W.S. Teo, L.D. Koh, S. Liu, C.P. Teng, M.Y. Han, Recent Progress in Energy-Driven Water Splitting, Adv. Sci. 4 (2017). https://doi.org/10.1002/advs.201600337

[6]   S. Sun, Y.C. Zhang, G. Shen, Y. Wang, X. Liu, Z. Duan, L. Pan, X. Zhang, J.J. Zou, Photoinduced composite of Pt decorated Ni(OH)$_2$ as strongly synergetic cocatalyst to boost H2O activation for photocatalytic overall water splitting, Appl. Catal. B Environ. 243 (2019) 253–261. https://doi.org/10.1016/j.apcatb.2018.10.051

[7]   M. Antonietti, M. Shalom, H.-P. Steinrück, S. Krick Calderón, C. Papp, M. Ledendecker, The synthesis of nanostructured Ni$_5$P$_4$ films and their use as a non-noble bifunctional electrocatalyst for full water splitting , Angew. Chemie Int. Ed. 54 (2015) 12361–12365. https://doi.org/10.1002/anie.201502438

[8]   Q. Zhang, Y. Wang, Y. Wang, A.M. Al-Enizi, A.A. Elzatahry, G. Zheng, Myriophyllum -like hierarchical TiN@Ni$_3$N nanowire arrays for bifunctional water splitting catalysts, J. Mater. Chem. A. 4 (2016) 5713–5718. https://doi.org/10.1039/C6TA00356G

[9]   W. Xing, C. Tang, N. Cheng, X. Sun, Z. Pu, NiSe nanowire film supported on nickel foam: an efficient and stable 3d bifunctional electrode for full water splitting, Angew. Chemie Int. Ed. 54 (2015) 9351–9355. https://doi.org/10.1002/anie.201503407

[10]   H. Wang, H.W. Lee, Y. Deng, Z. Lu, P.C. Hsu, Y. Liu, D. Lin, Y. Cui, Bifunctional non-noble metal oxide nanoparticle electrocatalysts through lithium-induced conversion for overall water splitting, Nat. Commun. 6 (2015) 1–8. https://doi.org/10.1038/ncomms8261

[11]   W. Zhou, X.J. Wu, X. Cao, X. Huang, C. Tan, J. Tian, H. Liu, J. Wang, H. Zhang, Ni$_3$S$_2$ nanorods/Ni foam composite electrode with low overpotential for electrocatalytic oxygen evolution, Energy Environ. Sci. 6 (2013) 2921–2924. https://doi.org/10.1039/c3ee41572d

[12]   Z. Lu, W. Xu, W. Zhu, Q. Yang, X. Lei, J. Liu, Y. Li, X. Sun, X. Duan, Three-dimensional NiFe layered double hydroxide film for high-efficiency oxygen evolution reaction, Chem. Commun. 50 (2014) 6479–6482. https://doi.org/10.1039/C4CC01625D

[13]   I. Elizabeth, A.K. Nair, B.P. Singh, S. Gopukumar, Multifunctional Ni-NiO-CNT composite as high performing free standing anode for li ion batteries and advanced electro catalyst for oxygen evolution reaction, Electrochim. Acta. 230 (2017) 98–105. https://doi.org/10.1016/j.electacta.2017.01.189

[14] S. Trassati, Water electrolysis: who first?, J. Electroanal. Chem. 476 (1999) 90–91. https://doi.org/10.1016/S0022-0728(99)00364-2

[15] J. Nai, H. Yin, T. You, L. Zheng, J. Zhang, P. Wang, Z. Jin, Y. Tian, J. Liu, Z. Tang, L. Guo, Efficient electrocatalytic water oxidation by using amorphous Ni-Co double hydroxides nanocages, Adv. Energy Mater. 5 (2015) 1–7. https://doi.org/10.1002/aenm.201401880

[16] J. Wang, H.X. Zhong, Z.L. Wang, F.L. Meng, X.B. Zhang, Integrated three-dimensional carbon paper/carbon tubes/cobalt-sulfide sheets as an efficient electrode for overall water splitting, ACS Nano. 10 (2016) 2342–2348. https://doi.org/10.1021/acsnano.5b07126

[17] J. Wang, H.X. Zhong, Y.L. Qin, X.B. Zhang, An efficient three-dimensional oxygen evolution electrode, Angew. Chemie - Int. Ed. 52 (2013) 5248–5253. https://doi.org/10.1002/anie.201301066

[18] Z. Zhao, H. Wu, H. He, X. Xu, Y. Jin, A high-performance binary Ni-Co hydroxide-based water oxidation electrode with three-dimensional coaxial nanotube array structure, Adv. Funct. Mater. 24 (2014) 4698–4705. https://doi.org/10.1002/adfm.201400118

[19] C. Jin, F. Lu, X. Cao, Z. Yang, R. Yang, Facile synthesis and excellent electrochemical properties of NiCo2O4 spinel nanowire arrays as a bifunctional catalyst for the oxygen reduction and evolution reaction, J. Mater. Chem. A. 1 (2013) 12170. https://doi.org/10.1039/c3ta12118f

[20] Q. Zhao, J. Yang, M. Liu, R. Wang, G. Zhang, H. Wang, H. Tang, C. Liu, Z. Mei, H. Chen, F. Pan, Tuning electronic push/pull of ni-based hydroxides to enhance hydrogen and oxygen evolution reactions for water splitting, ACS Catal. 8 (2018) 5621–5629. https://doi.org/10.1021/acscatal.8b01567

[21] U.K. Sultana, J.D. Riches, A.P. O'Mullane, Gold doping in a layered Co-Ni hydroxide system via galvanic replacement for overall electrochemical water splitting, Adv. Funct. Mater. 28 (2018) 1–8. https://doi.org/10.1002/adfm.201804361

[22] X. Wang, W. Li, D. Xiong, D.Y. Petrovykh, L. Liu, Bifunctional nickel phosphide nanocatalysts supported on carbon fiber paper for highly efficient and stable overall water splitting, Adv. Funct. Mater. 26 (2016) 4067–4077. https://doi.org/10.1002/adfm.201505509

[23] Z.-Q. Liu, G.-F. Chen, S.-Z. Qiao, K. Davey, N. Li, Y.-Z. Su, T.Y. Ma, Efficient and stable bifunctional electrocatalysts Ni/Ni x M y (M = P, S) for overall water splitting, Adv. Funct. Mater. 26 (2016) 3314–3323. https://doi.org/10.1002/adfm.201505626

[24]   J. Bao, Z. Ren, S. Chen, H. Zhou, W.A. Goddard, F. Qin, F. Yu, J. Sun, Y. Huang, High-performance bifunctional porous non-noble metal phosphide catalyst for overall water splitting, Nat. Commun. 9 (2018) 1–9. https://doi.org/10.1038/s41467-017-02088-w

[25]   L.A. Stern, L. Feng, F. Song, X. Hu, Ni2P as a Janus catalyst for water splitting: The oxygen evolution activity of Ni2P nanoparticles, Energy Environ. Sci. 8 (2015) 2347–2351. https://doi.org/10.1039/C5EE01155H

[26]   L. Elias, V.H. Damle, A.C. Hegde, Electrodeposited Ni-P alloy thin films for alkaline water splitting reaction, IOP Conf. Ser. Mater. Sci. Eng. 149 (2016). https://doi.org/10.1088/1757-899X/149/1/012179

[27]   J. Jia, M. Zhai, J. Lv, B. Zhao, H. Du, J. Zhu, Nickel molybdenum nitride nanorods grown on ni foam as efficient and stable bifunctional electrocatalysts for overall water splitting, ACS Appl. Mater. Interfaces. 10 (2018) 30400–30408. https://doi.org/10.1021/acsami.8b09854

[28]   M. Shalom, D. Ressnig, X. Yang, G. Clavel, T.P. Fellinger, M. Antonietti, Nickel nitride as an efficient electrocatalyst for water splitting, J. Mater. Chem. A. 3 (2015) 8171–8177. https://doi.org/10.1039/C5TA00078E

[29]   X. Lian, X. Zou, D. Wang, Y. Liu, X. Zou, T. Asefa, L. Sun, Y. Wu, G.-D. Li, Efficient electrocatalysis of overall water splitting by ultrasmall NixCo3−xS4 coupled Ni3S2 nanosheet arrays, Nano Energy. 35 (2017) 161–170. https://doi.org/10.1016/j.nanoen.2017.03.024

[30]   C.-G. Liu, Z.-Z. Liu, W.-K. Gao, J.-Q. Chi, K.-L. Yan, Y.-M. Chai, B. Dong, J.-H. Lin, F.-N. Dai, Urchin-like nanorods of binary nicos supported on nickel foam for electrocatalytic overall water splitting, J. Electrochem. Soc. 165 (2018) H102–H108. https://doi.org/10.1149/2.0351803jes

[31]   J.T. Ren, Z.Y. Yuan, Hierarchical nickel sulfide nanosheets directly grown on ni foam: a stable and efficient electrocatalyst for water reduction and oxidation in alkaline medium, ACS Sustain. Chem. Eng. 5 (2017) 7203–7210. https://doi.org/10.1021/acssuschemeng.7b01419

[32]   Z. Zou, X. Wang, J. Huang, Z. Wu, F. Gao, An Fe-doped nickel selenide nanorod/nanosheet hierarchical array for efficient overall water splitting, J. Mater. Chem. A. 7 (2019) 2233–2241. https://doi.org/10.1039/C8TA11072G

[33]   X. Li, L. Zhang, M. Huang, S. Wang, X. Li, H. Zhu, Cobalt and nickel selenide nanowalls anchored on graphene as bifunctional electrocatalysts for overall water splitting, J. Mater. Chem. A. 4 (2016) 14789–14795. https://doi.org/10.1039/C6TA07009D

Electrochemical Water Splitting: Materials and Applications   Materials Research Forum LLC
Materials Research Foundations **59** (2019) 141-168   doi: https://doi.org/10.21741/9781644900451-6

Chapter 6

Transition-Metal Chalcogenides for Oxygen-Evolution Reaction

Kartick Chandra Majhi, Paramita Karfa, Rashmi Madhuri*

Department of Applied Chemistry, Indian Institute of Technology (Indian School of Mines), Dhanbad, Jharkhand 826 004, INDIA

Abstract

In order to search a green, sustainable, and highly effective energy creation to fulfill the energy demand of our modern society, different technologies have been developed for energy generation/conversion. But, after surveying different prospects, it was found that electrolysis of water is one of the best among them and can be stabilized as a very exciting/useful technology for energy generation. Oxygen evolution reaction (OER) is the half-cell reaction of water electrolysis and therefore has been greatly studied in the last few decades. Since four electrons are required for successful OER, and the result reaction kinetics is very slow, to accelerate the reaction rate, highly efficient catalysts are required. In recent times, based on the low cost, more feasible option, tunable properties, transition metal chalcogenides, i.e. sulfides, selenides, and tellurides based materials have come up to rescue the difficult kinetics of OER. They have been found as best candidates for OER in terms of activity, durability, earth abundance, and low cost, owing to their unique physical, optical, and chemical properties. In this chapter, we have briefly discussed the kinetics parameter, OER mechanism, and role of transition metal chalcogenides towards OER. As a concluding remark, their prospects have also been discussed in this chapter.

Keywords

Oxygen Evolution Reaction, Transition Metal Chalcogenides, Sulfides, Selenides, Tellurides

Contents

1.   Introduction

The continuous and speedy increase in the population, industry, and technology has led us to the border of an energy crisis, owing to the high-end consumption of fossil fuels. On the other hand, the by-product of these fossil fuels has increased the problem of environmentalists, who are trying to save the earth and wish to keep things "green". Many advanced concepts have been developed to discover more efficient energy conversion or storage systems with greener and cleaner way like water electrolysis, fuel cells, and metal-air batteries [1]. Production of oxygen and hydrogen from water electrolysis has gained a lot of importance in recent times, as an alternative of fossil fuels having large abundance, greenhouse, and other pollutant gases free emission as well as high efficiency. Electrolysis of water molecule to form hydrogen and oxygen gases is the suitable methods/approach to offer clean, green, and sustainable energy. The overall water splitting reaction can be represented as: $2H_2O \rightarrow 2H_2 + O_2$. The electrolysis of water molecule consists of two half-cell reactions, i.e. oxygen evolution reaction and hydrogen evolution reaction, which are performed at anode and cathode compartment, respectively (Fig. 1). The reactions are presented below depending on the different pH of the electrolyte:

Reaction in the anodic compartment i.e. OER, $E_a^0 = +1.23$ V vs. RHE

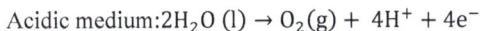

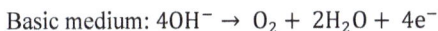

Acidic medium: $2H_2O\ (l) \rightarrow O_2(g) + 4H^+ + 4e^-$

Basic medium: $4OH^- \rightarrow O_2 + 2H_2O + 4e^-$

The reaction in the cathodic compartment, i.e. HER, $E_c^0 = 0$ V vs. RHE

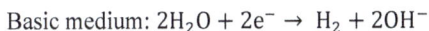

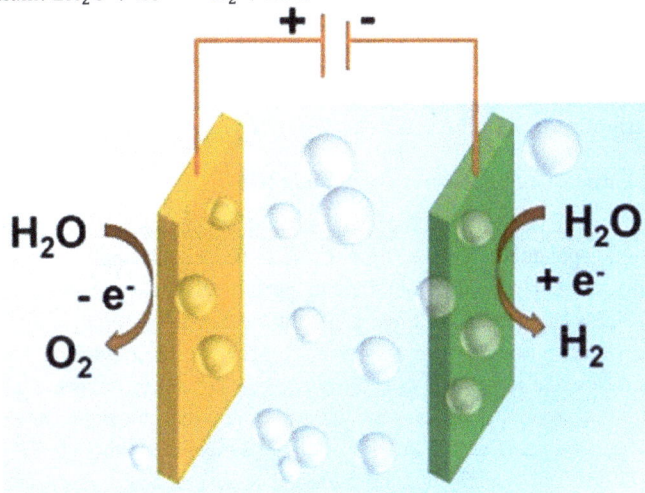

Acidic medium: $4H^+ + 4e^- \rightarrow 2H_2$

Basic medium: $2H_2O + 2e^- \rightarrow H_2 + 2OH^-$

Fig. 1: *Pictorial representation of overall water splitting, where OER and HER are take place at anode and cathode electrodes [1].*

Although the above reactions look simple in the first instance, in practical, they are very difficult to perform, owing to their sluggish kinetics. As we can see in the above reactions, OER requires four electrons and HER needs two electrons to form oxygen and hydrogen gases, respectively from water molecules. Therefore, it can be easily understood that OER required higher energy than the HER to overcome the kinetic energy barrier. Hence, especially OER needs the more efficient and effective catalyst to speed up the reaction with lesser energy requirement.

In the last few decades, many researchers/scientists are engaged to discover or fabricate suitable catalysts, which can enhance the kinetics of OER in different electrolytes and pH conditions. According to the literature, transition metal oxide like RuO_2 and IrO_2 are the best electrocatalysts reported so far for OER in alkaline as well as acidic medium [2-7]. But, owing to some disadvantages like their self-oxidation in the anodic potential region to covert in RuO_4 and IrO_3, respectively by dissolving in electrolyte have restricted their performance in industrial applications. As compared to RuO_2, IrO_2 is comparatively stable and able to tolerate the large anodic potential [8]. But both the catalysts have very high cost and therefore cannot be used for large scale production. Therefore, it is very necessary to find their replacement, which possesses their qualities but not their drawbacks. In this chapter, we have focused on the popular electrocatalysts used so far for OER, but before that, we have included here the fundamentals of OER also. Herein, we have incorporated the fundamentals of OER mechanism and essential parameters on which the performance of OER electrocatalyst can be judged.

1.1 Mechanism of oxygen evolution reaction (OER)

Oxygen evolution reaction is the half-cell reaction of electrochemical water splitting. Thermodynamic potential required for overall water splitting is +1.23 V, but experimentally more potential is required to perform the reaction or split the water molecule, called overpotential. As we know, the external potential is applied to the electrode for water splitting, which results in the generation of oxygen and hydrogen at the anode and cathode, respectively. The reaction mechanism of oxygen evolution from water splitting largely depends on the pH of the medium and therefore, mechanisms/reaction paths vary from acid to alkaline medium. Although the actual OER mechanism is still not completely understood, but several researchers/scientists have offered different possible mechanism in acid and alkaline mediums. But one thing common in all the mechanisms is the generation of reaction intermediates like EO and EOH (where E corresponds to the catalyst) [9]. Herein, we have discussed some of the mechanism reported in different kinds of literature for OER. Starting in acidic medium, firstly the water molecules adsorbed on the active surface of catalyst (equation 1), but contrary to this, in the alkaline medium, the hydroxyl ion gets bind on the active surface of catalyst (equation 6).

Possible mechanism for OER under aqueous acidic condition:

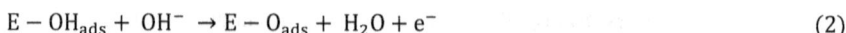

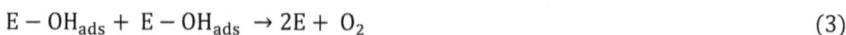

$$E + H_2O \rightarrow E - OH_{ads} + H^+ + e^- \tag{1}$$

$$E - OH_{ads} + OH^- \rightarrow E - O_{ads} + H_2O + e^- \tag{2}$$

$$E - OH_{ads} + E - OH_{ads} \rightarrow 2E + O_2 \tag{3}$$

Or

$$E - OH_{ads} + H_2O \rightarrow E - OOH_{ads} + H^+ + e^- \tag{4}$$

$$E - OOH_{ads} + H_2O \rightarrow E + O_2 + H^+ + e^- \tag{5}$$

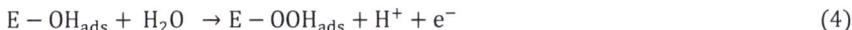

Possible mechanism for OER under aqueous basic condition:

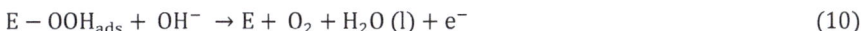

$$E + OH^- \rightarrow E - OH_{ads} + H^+ + e^- \tag{6}$$

$$E - OH_{ads} + OH^- \rightarrow E - O_{ads} + H_2O + e^- \tag{7}$$

$$E - OH_{ads} + E - OH_{ads} \rightarrow 2E + O_2 \tag{8}$$

$$E - OH_{ads} + OH^- \rightarrow E - OOH_{ads} + e^- \tag{9}$$

$$E - OOH_{ads} + OH^- \rightarrow E + O_2 + H_2O \ (l) + e^- \tag{10}$$

Although the same intermediates are formed in both the conditions, but the step in which they are generated is different based on the pH of the electrolytes. After adsorption of water or OH⁻ molecules on the catalyst surface, oxygen is produced by two pathways: the first one is the direct combination of 2 molecules of E-OH$_{ads}$ to produce $O_2(g)$, where the elementary reaction steps are shown in equations 2→3→4in acidic conditions and 7→8→9in alkaline condition. The second path is the formation of E-OOH$_{ads}$ intermediate by the combination of E-OH$_{ads}$ and H$_2$Oin acidic condition (equation 5)and E-OH$_{ads}$ and OH⁻ in basic condition (equation 9→10) and subsequently, produce $O_2(g)$. The pictorial representation for the mechanism of OER in acid and alkali conditions is shown in Fig. 2 [8].

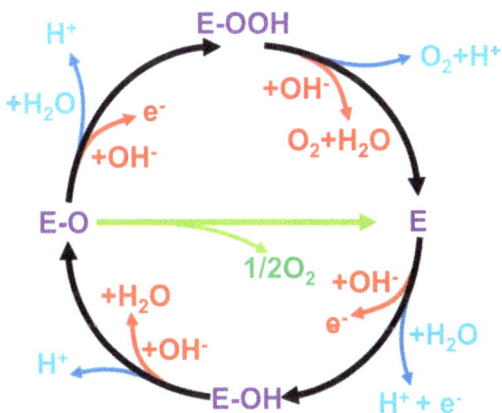

Fig. 2: *Schematic representation of OER mechanism in acid and alkali medium assign as blue and red lines, respectively [8].*

Electrochemical Water Splitting: Materials and Applications   Materials Research Forum LLC
Materials Research Foundations **59** (2019) 141-168   doi: https://doi.org/10.21741/9781644900451-6

1.2   Kinetic parameters used to find the suitable catalysts for OER

The material that is used to enhance the rate of electrochemical reactions is known as an electrocatalyst, basically charge transfer reactions, which is shown in the previous section. Actually, during the reaction, reactant gets adsorbed to the surface of electrocatalyst to form the intermediates, followed by charge transfer between the reactant and electrode. The electrocatalyst is used to enhance the reaction speed and ease the kinetics followed for OER. Therefore, the capability of electrocatalyst to enhance the OER depends on some kinetic parameters like overpotential, exchange current density, and Tafel slope value. The mentioned parameters not only suggest the potential of electrocatalyst towards OER but also offer significant information about the mechanism involved in OER. Therefore, before describing the role of electrocatalyst towards OER, we have discussed here some basic details about the mentioned kinetic parameters.

1.2.1   Overpotential

It is the key parameter for OER, and it is represented with the symbol 'η'. Ideally, the applied potential to initiate the electrochemical reactions should be equal to the equilibrium potential. But in practical, the required potential for OER, i.e. applied potential value is always higher than the equilibrium potential, needed to overcome the kinetic barriers. The applied potential (E_{app}) can be expressed following Nernst equation [10]:

$$E_{app} = E^0 + \frac{RT}{nF} \ln\left[\frac{O}{R}\right] \tag{11}$$

Where, E_{app} = applied potential, E^0 = Standard electrode potential at 25°C, T = Temperature, F = Faraday constant, O = Concentration of oxidant, R = Universal gas constant, R = Concentration of reductant, and n = Number of electron transfer. However, the overpotential can be measured by using following equation:

$$\eta = E_{app} - E_{eq} \tag{12}$$

In the electrochemical study, the overpotential is measured as the potential, at the current density of 10 mA cm^{-2}. An excellent electrocatalyst always exhibits lower overpotential value. Although, in the literature, sometime, different overpotential values were reported at different current density values. The typical polarization curves of HER and OER showing representation of different symbols and parameters are shown in Fig. 3 [10].

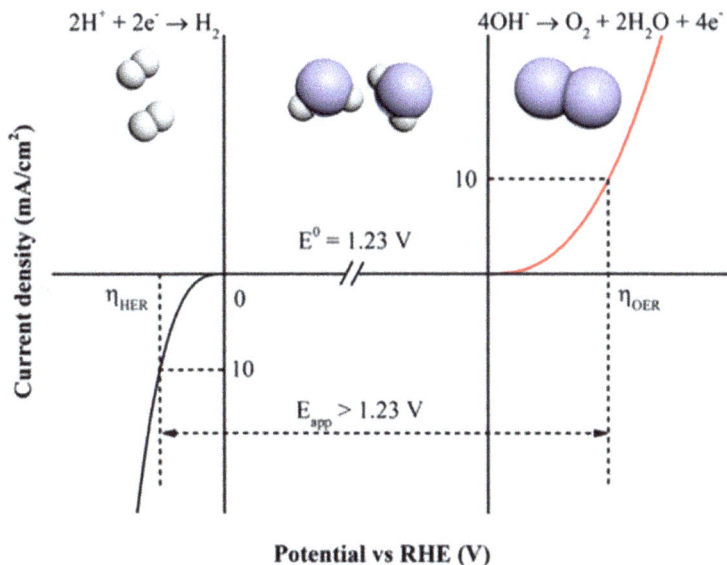

$2H^+ + 2e^- \rightarrow H_2$

$4OH^- \rightarrow O_2 + 2H_2O + 4e^-$

Current density (mA/cm^2)

$E^0 = 1.23$ V

η_{HER}

10

0

10

η_{OER}

$E_{app} > 1.23$ V

Potential vs RHE (V)

Fig. 3: *Polarization or LSV curves of HER (left side) and OER (right side). The thermodynamic equilibrium potential for water splitting is +1.23 V, but the practical applied potential is far greater than +1.23 V to achieve the water-splitting with current density (j) of 10 mA cm^{-2}. The η_{HER} and η_{OER} are the overpotentials to achieve current density 10 mA cm^{-2} for HER (at cathode) and OER (at anode), respectively [10].*

1.2.2. Exchange current density

Another important kinetic parameter is exchange current density and referred to as J_0. In order to find the exchange current density, the total current density is required and calculated by the equation shown as:

$$J = J_a + J_c \qquad (13)$$

The individual cathodic and anodic current density can be measured using the following equations:

$$J_a = nFK_a[R]exp^{\left(\frac{\alpha_a nFE}{RT}\right)} \qquad (14)$$

$$J_c = nFK_c[O]exp^{\left(-\frac{\alpha_c nFE}{RT}\right)} \qquad (15)$$

Where, J_a and J_c = Anodic and cathodic current density, respectively, n = Number of electrons transferred, K_a= Rate constant of anodic, and K_c = Rate constant of cathodic half-cell reaction, F = Faraday constant,α_a and α_c = Anodic and Cathodic trans coefficient, E = Applied potential, [R] = Concentration of reductant and [O] = Concentration of oxidant.

At equilibrium, applied potential equal to the equilibrium potential, it means overpotential become equal to zero (equation 12). At this condition, J_a and J_c have equal magnitude, but opposite in sign as a result total current density (J) also becomes zero. Therefore, the intercept at zero overpotential are known as exchange current. Dividing total current (J) by area of the working electrode (A) gives exchange current density (J_0):

$$J_0 = \frac{J}{A} \qquad (16)$$

The value of exchange current density directly indicates the charge transferring ability of electrocatalyst from proposed electrocatalyst and reactant molecule. Therefore, the higher the value of the exchange current density greater will be the charge transfer rate. As a conclusion, better electrocatalyst must have higher exchange current density. The state-of-art catalysts for OER, i.e. gold (Au), iridium (Ir) and rhodium (Rh) possesses high exchange current density than the others metal; therefore, popularly used as a best electrocatalyst.

1.2.3 Tafel equation and Tafel plot

Tafel equation and Tafel plot are also a very important parameter for OER. Tafel slope gives the information about the mechanism involved in the electrochemical reaction. Although, the smaller Tafel slope value always offer the best electrocatalyst, which also means that low overpotential signifies the faster reaction kinetics [12]. According to the literature, Tafel slope (b) can be measured by linear sweep voltammetry (LSV) study. According to this, from LSV runs, firstly, overpotential will be calculated and plotted against the logarithm of current density. The slope of the linear graph between overpotential versus log j is termed as Tafel slope (b) and can be obtained using the Tafel equation [13]:

$$\eta = b \times \log j + a \qquad (17)$$

Where η, b, j and a represent overpotential, Tafel slope, current density and reaction constant, respectively. This method is used to calculate the Tafel slope values for the low overpotential values only.

1.2.4 Electrochemical active surface area (ECSA)

Electrochemical active surface area is another important parameter to study the electrocatalytic activity of any catalyst towards electrochemical reactions. To determine the electrochemically active surface area, cyclic voltammetry (CV) study was performed in the particular potential range at different scan rates. Generally, five to ten CV runs are taken at different scan rates to select a rectangular type region called as a non-faradic region and used to calculate the difference in current density ($\Delta j = j_a - j_c$, where, j_c and j_a are the cathodic and anodic current density, respectively) at the particular potential. From the slope of current density (Δj) versus scan rate linear equation, electrochemical double-layer capacitance (C_{dl}) can be calculated. It has been found that greater the value of C_{dl} offers the larger number of active sites at the surface of the catalyst and therefore, results in higher performance as electrocatalyst towards OER [14].

1.2.5 Faraday efficiency (FE)

Faraday efficiency is the quantitate parameter, which describes the charge or electron transfer in the external circuit for conducting the electrochemical reaction. It has been greatly used for electrochemical reactions such as fuel cells and water splitting reactions (OER/HER). Faraday efficiency can be defined as the ratio of actual/experimental products (oxygen and hydrogen gas) and theoretical products. The amount of gas produced (hydrogen and oxygen) theoretically can be calculated based on the electrolysis time, current density, and electrode surface area, while the amount of gas produced by practical experiment can be measured using Gas chromatography or a water-gas displacement method [12].

1.3 Experimental methods used to study the OER behavior and stability of catalysts

While studying the OER behavior of any catalysts, it is essential that everyone will follow the same method and procedures, so that a comparative study can be performed on a similar kind of materials. For example, in the literature, two or more than two groups of researchers have studied the electrocatalytic activity of the same catalyst, but reported results from different groups are correct but quite different from each other [15]. This is because the different research groups have experimented with different conditions. That is why it is important to develop a standard protocol to measure the catalytic activity of proposed catalysts. For this, McCrory et al. [16] have developed a flow chart, where different techniques like CV and LSV can be used for catalytic activity measurement of the catalysts. A CV can be used for surface area and roughness factor measurement, while electrochemical active surface area (ECSA) and X-ray photoelectron microscopy

Electrochemical Water Splitting: Materials and Applications   Materials Research Forum LLC
Materials Research Foundations **59** (2019) 141-168   doi: https://doi.org/10.21741/9781644900451-6

(XPS) can be used for determination of the elemental composition of catalysts. Similarly, electrochemical impedance spectroscopy (EIS), faraday efficiency, stability, and extended stability are used to obtained catalytic activity [16]. Stability of the proposed catalyst is also an important part of the electrochemical study and their real time practical applications. In general, the stability of catalysts was tested by recording CV for multiple cycles or long-term stability by chronopotentiometry study (i.e., I-t curve). The electrocatalyst long period stability can be studied by recording constant current density for a long period [14].

2.   Transition metal chalcogenides as replacement of state-of-art catalyst for OER

According to the International Union of Pure and Applied Chemistry (IUPAC), an element having partially filled d-orbitals or the elements which form cation with partially filled d-orbital is known as transition metal element [17]. Therefore, the elements that belong to group 3 to 12 in the periodic table are called transition metals. The general electronic configuration for transition metal elements can be represented as $(n-1)d^{1-10}ns^2$.

Similarly, according to the IUPAC nomenclature, chalcogenides are the compounds which contain at least one chalcogen elements and the group 16 elements, i.e. sulfur, selenium, and telluriumare called chalcogen elements [18]. Although oxygen and polonium also belong to the same group but cannot classify and known as chalcogen. The valence shell electronic configuration of chalcogen elements can be represented as $ns^2\, np_x^1\, np_y^1\, np_z^2$. Here, two electrons are present in the s-orbital and two unpaired p-orbitals are present among the three orbitals. Generally, p-orbitals which contains one-one electron in each orbital were act as lone pair (not always true) but don't contribute in the formation of covalent bond and s-orbital are chemically inactive. During the formation of chemical compounds, the electronic configuration might get changed, which depends on the chemical elements present in the compound. But in general, chalcogen elements can form two covalent bonds with neighboring atoms. Naturally, sulfur and selenium can form chain or ring like structures whereas tellurium can form the chains structure only. But while all kind of structure formation, the paired s- and p- orbital electrons remain inert and act as lone pair [18].

In the last few decades, transition metal chalcogenides have originated as new exciting material for OER. This is mainly due to their high electrical conductivity, d-electron occupancy, unique electronic configuration, small size, large surface area to volume ratio, unique structural properties, and high resistance in alkali and acidic condition [19, 20]. In this chapter, we have discussed the electrocatalytic performance of sulfur, selenium, and tellurium based transition metal chalcogenides towards OER. The specific properties, which make transition metal chalcogenide a suitable candidate as electrocatalyst OER is summarized and portrayed in Fig. 4.

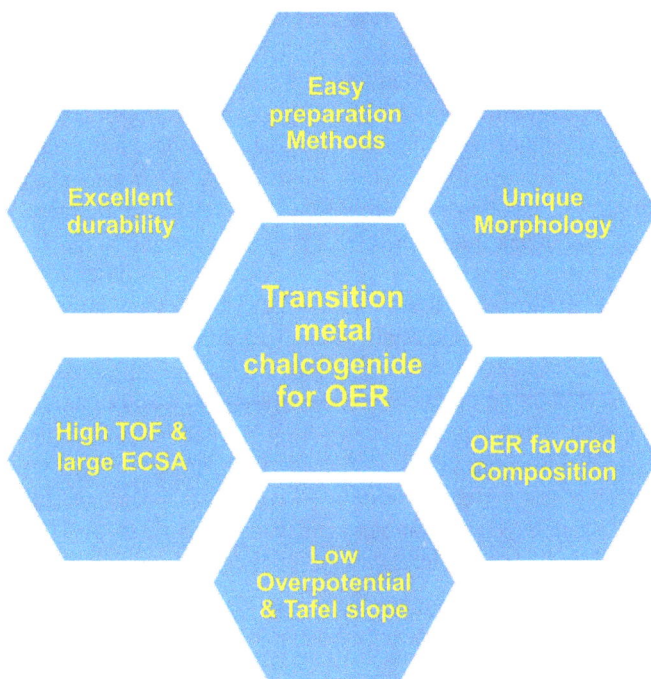

Fig. 4: Pictorial representation of specific properties transition metal chalcogenide, which them a suitable electrocatalyst for OER.

2.1 Transition metal sulphide for oxygen evolution reaction

Among the different class of transition metal chalcogenides, 'S' based electrocatalysts are the most studied one towards OER. Transition metal sulfides have some unique properties like crystal structure and accessible electronic states, which results in their superior performance towards OER. Strong bonding between the transition metal and chalcogen (M-S) and metalloid-metalloid covalent bond (S-S) present in the transition metal sulfides provides them high stability in harsh conditions and different electrolytes [21].

Some of the recent literature shows the efficiency of 'sulphur' based transition metal chalcogenides as OER catalyst are discussed below. Luo et al. [22] has prepared NiS_2 and NiS hollow microspheres by one step hydrothermal process followed by calcination and applied it for overall water splitting. They have studied the OER and HER performance of NiS and NiS_2, respectively in three electrode systems, where proposed NiS catalyst offered the low overpotential of 320 mV and 390 mV to a delivery current density of 10 and 20 mA cm^{-2}, respectively. Excellent OER activity of proposed NiS catalyst was strongly supported by the low Tafel slope value, i.e. 59 mV/dec and long-term stability, which suggested their practical applications. In the year 2017, Yang and coworkers [23] have synthesized Co_9S_8 hollow microsphere via simple hydrothermal route, which has a particle size of 1.0 μm and studied their OER performance in 1.0 M KOH. The designed catalyst has offered excellent OER performance because of low onset potential of +1.47 V versus RHE, overpotential of 420 mV vs. RHE and low Tafel slope of 113 mV/dec. In 2018, Feng et al. [24] were successfully synthesized differently shaped (hollow spheres and flower like structure) of Co_9S_8 by facile solvothermal route for OER in alkaline conditions (1 M KOH). But the hollow sphere structure of Co_9S_8 exhibited is a superior catalyst towards OER than a flower like structure. The hollow sphere structure of Co_9S_8 offered smaller overpotential 285 mV to attained 10 mA cm^{-2} current density and 58 mV/dec Tafel slope than flower like structure (overpotential 380 mV and Tafel slope 75.5 mV/dec). The obtained polarization curves of the hollow sphere and flower like the structure of Co_9S_8 are shown in Fig. 5.

Recently, according to one study, it was found that OER performance of a catalyst can be improved with increase in the number of transition metal present in the materials [25]. On this context, in the year 2017, Deka and coworkers [26] have synthesized Co_3S_4, $CuCo_2S_4$, and $Cu_{0.5}Co_{2.5}S_4$ by facile one step hydrothermal process. They have studied the OER activity of the prepared materials and found that $CuCo_2S_4$ showed best performance than the other two. The onset potentials were found to be +1.43, +1.54 and +1.61 V vs. RHE and Tafel slope of 86, 98, and 144 mV/dec for $CuCo_2S_4$, $Cu_{0.5}Co_{2.5}S_4$, and Co_3S_4, respectively. Huang et al. [27] have proved by density functional theory (DFT) study that the introduction of transition metal into transition metal sulfide lattice resulted in an increase in the Gibbs free energy, which directly affects the electrical conductivity of the electrocatalyst. This may be the reason, increase in the doping of Cu into the cobalt lattice lead to enhance the electrical conductivity and hence $CuCo_2S_4$ exhibited the best performance towards OER.

Another research group by Gao et al. [28] has prepared iron doped Co_9S_8 nano microsphere developed on nickel foam (Fe-Co_9S_8@NM@NF) via simple one step hydrothermal method and used it towards OER. They have investigated the OER study of

Fe-Co_9S_8@NM@NF, which requires the small overpotential of 0.270 V to reach the current density 10 mA cm^{-2}, showed low Tafel slope of 70 mV/dec and robust stability. Therefore, designed Fe-Co_9S_8@NM@NF offered remarkable OER activity attributed to the synergic effect between the doped Fe and Co_9S_8, which accelerate the charge transfer.

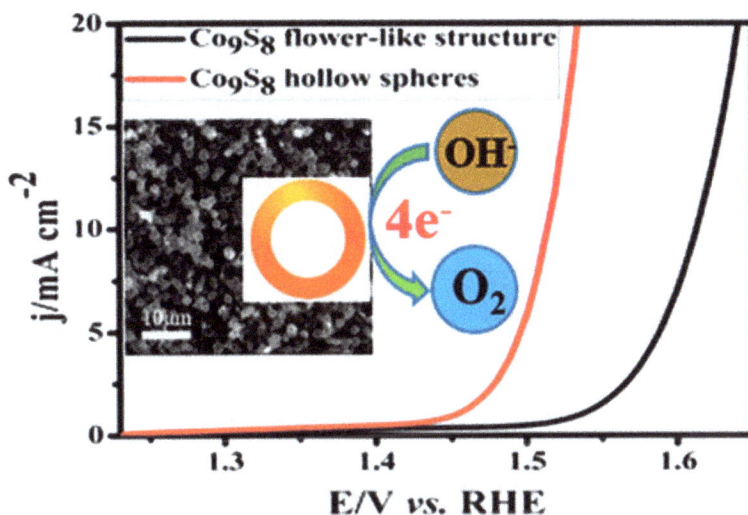

Fig. 5: LSV curves of hollow sphere and flower like structure of Co_9S_8 and inset is showing FE-SEM images of hollow sphere like structure of Co_9S_8 [24].

Recently, in the year 2019, Jadhav et al. [29] have prepared bimetallic transition metal oxide ($MnCo_2O_4$), and sulfide ($MnCo_2S_4$) nanoflakes on stainless steel (SS) mesh through electrodeposition process followed by ion-exchange method. They have compared the electrocatalytic performance of $MnCo_2O_4$ and $MnCo_2S_4$ towards OER in 1.0 M KOH $MnCo_2S_4$ exhibited better performance than the oxides analogue, with low overpotential 290 mV to attain current density of 10 mA cm^{-2}, small Tafel slope of 67 mV/dec. While $MnCo_2O_4$ needs overpotential of 310 mV to attain current density 10 mA cm^{-2} and showed small Tafel slope of 78 mV/dec. The better OER performance of $MnCo_2S_4$ could be attributed to their large C_{dl} value (52 mF cm^{-2}), i.e. high electrochemical surface area, electrical conductivity, and synergic effect between Mn, Co, and S.

Table 1: *Summary of some of the recently reported transition metal sulphides used as electrocatalyst for oxygen evolution reaction 1.0 M KOH as an electrolyte.*

S.N.	Catalyst	Method of preparation	Overpotential (mV)	Tafel Slope (mV/dec)	Ref.
1.	NiS_x flims	Vapor-phase atomic layer deposition	$408@10$ mA cm^{-2}	56	[32]
2.	Ni–Co–S NS film	Electrodeposition	$363@100$ mA cm^{-2}	109	[33]
	Ni-S		$383@100$ mA cm^{-2}	128	
	Co-S		$393@100$ mA cm^{-2}	132	
3.	NiS flim	Electrodeposition	$320@10$ mA cm^{-2}	-	[34]
4.	NiCoFe-S-Fe@NF	Solvothermal	$230@100$ mA cm^{-2}	94	[35]
	NiCoFe-S@NF		$330@100$ mA cm^{-2}	128	
5.	Ni-CoS$_2$/CC	Electrodeposition	$370@100$ mA cm^{-2}	119	[36]
	CoS$_2$/CC		$510@100$ mA cm^{-2}	127	
6.	NiCo$_2$S$_4$ NA/CC	Calcination	$340@100$ mA cm^{-2}	-	[37]
7.	CuCo$_2$S$_4$	Hydrothermal	$290@20$ mA cm^{-2}	-	[38]
8.	Fe(III) doped Ni$_3$S$_2$ NS	Electrodeposition	$213@10$ mA cm^{-2}	33.2	[39]
	Ni$_3$S$_2$ NS		$324@10$ mA cm^{-2}	72.1	
9.	NiS$_2$ NWs	Hydrothermal	$246@10$ mA cm^{-2}	94.5	[40]
10.	NiCo$_2$S$_4$ NGA	Hydrothermal	$310@10$ mA cm^{-2}	95	[41]
11.	NiS/NF	Hydrothermal	$320@20$ mA cm^{-2}	71	[42]

N.B.: N-MC = nitrogen doped mesoporous graphite, NS- nanosheets, NS- nickel foam, CC-carbon cloth, NA- nanowire array, NWs-nanowires, NGA- needle glass array.

Zhang et al. [30] have synthesized CoS_2, NiS_2, and $NiCoS_2$ by a simple hydrothermal route and compared their OER performance in 1.0 M KOH. Among them, $NiCoS_2$ nanosheets showed remarkable OER performance [in Fig. 6 (a)] with lower overpotential (270 mV) than CoS_2 (350 mV), and NiS_2 (310 mV) to attain the current density of 10 mA cm^{-2}. The Tafel slope of $NiCoS_2$ nanosheets is found to be 58 mV/dec, which is smaller than CoS_2 (107 mV/dec), and NiS_2 (123 mV/dec) [Fig. 6 (b)]. A small value of Tafel slope indicates the faster charge transfer rate [31]. The EIS data also support the higher catalytic activity of bimetallic analogue. As shown in the Fig. 6 (c), $NiCoS_2$ nanosheets have a small radius of the semi-circle, which means the small charge transfer resistance (R_{ct}) value than NiS_2 and CoS_2. Smaller R_{ct} value offered faster electronic transport/transmission. In addition, synthesized $NiCoS_2$ nanosheets also have large double-layer capacitance (C_{dl}) and robust stability with constant current density for 70000 s at +1.5 V than the NiS_2 and CoS_2 [shown in Fig. 6 (d and e)]. The OER mechanisms steps followed during the process are shown in Fig. 6(f). As shown in the Fig., third step i.e. formation of *OOH is found to be the potential limiting step (PLS) by DFT calculation. The overpotential values were found to be 0.51 V for $NiCoS_2$ nanosheets, which is much smaller than the CoS_2 ($\eta = 0.54$ V) and NiS_2 ($\eta = 2.00$ V). So, from the data, it is very clear that $NiCoS_2$nanosheets have shown remarkable electrocatalytic behavior towards OER in the alkaline medium than monometallic sulfide materials. The pictorial representation showing OER is shown in Fig. 6(g), where the reaction takes place on (100) plane of $NiCoS_2$ nanosheets. In addition to the discussed materials, some recently reported transition metal chalcogenide electrocatalyst towards oxygen evolution reaction (OER) are summarized in and depicted in Table 1 [32-42].

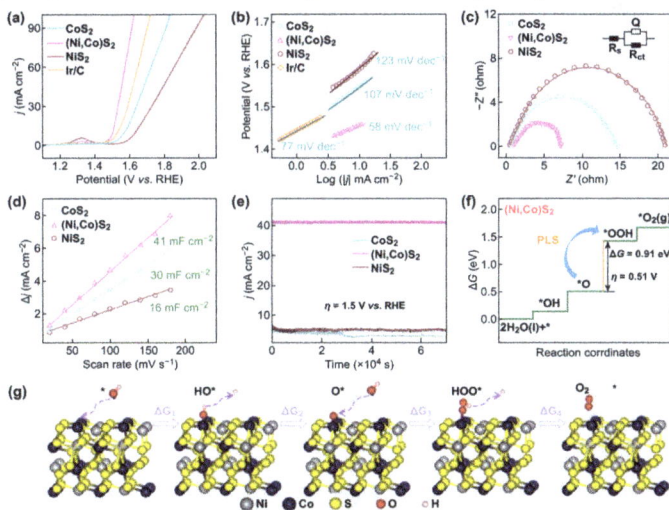

Fig. 6: (a) Polarization plots of CoS_2, (Ni, Co)S_2, NiS_2, and Ir/C at optimized scan rate 5 mV s^{-1} in 0.1 M KOH. (b) Tafel slopes of mentioned materials. (c) EIS curves of CoS_2, (Ni, Co)S_2, and NiS_2 and inset is circuit diagram. (d) ECSA and C_{dl} of CoS_2, (Ni, Co)S_2, and NiS_2. (e) Controlled electrolysis for 7×10^4 s. (f) Representation of Gibbs free energy of each four elementary steps during OER on (100) surface/plane of (Ni,Co)S_2. (g) Pictorial representation for the formation oxygen from water on (100) surface/plane (Ni,Co)S_2 [30].

2.2   Transition metal selenide for oxygen evolution reaction

Electrocatalysts towards oxygen evolution reaction are the key for green and sustainable energy. A remarkable electrocatalyst for OER essentially consist of a high number of active sites for adsorption or desorption, have fast charge transfer kinetics, long period of stability, and robust structure/morphology [43]. Last few years transition metal selenide has been extensively used as an electrocatalyst for overall water splitting, due to their unique d-electron configuration, high electrical conductivity, optical properties, stability in both alkali and acid electrolyte and high earth abundance [44]. If we follow the Horn's prediction, with increasing the metallic character or covalence of transition metal chalcogenide the electrocatalytic activity also get increased [45]. Therefore, transition metal selenide may exhibit excellent electrocatalytic performance than transition metal sulfide and/or oxide.

Based on these theories and assumptions, several kinds of literature have been reported showing the outstanding capability of 'selenium' based transition metal chalcogenides for OER. Among different types of transition metal selenides, nickel based selenides have been greatly used as electrocatalyst for OER in comparison to other transition metal selenides, owing to the high electrical conductivity of nickel selenide. For example, a NiSe nanowire with particle size 20-80 nm was synthesized by the hydrothermal route, and their OER performance was studied by Tang et al. [46]. The remarkable OER performance of NiSe nanowire can be depicted by their low overpotential of 270 mV needed to achieve a current density of 20 mA cm^{-2} and long period stability. The obtained remarkable results could be attributed to the formation of Ni-OOH on the surface of the catalyst during the OER process. Wu et al. [47] have synthesized hexagonal $Ni_{0.85}Se$ by electrodeposition process. The prepared hexagonal $Ni_{0.85}Se$ had excellent electrocatalytic activity for OER in alkaline medium. It has overpotential of 302 mV to achieve 10 mA cm^{-2} current density and small Tafel slope of 62 mV/dec. The remarkable OER activity of hexagonal $Ni_{0.85}Se$ can be attributed to the presence of vacancies in the hexagonal structure of $Ni_{0.85}Se$.

In 2013, Du et al. [48] has successfully synthesized $NiSe_2$, $FeSe_2$ and $Ni_xFe_{1-x}Se_2$ (where, x= 0.25, 0.5 and 0.75) nanomaterials by thermal decomposition method. It was found that OER performance of prepared nanomaterials gets increased by the introduction of transition metals (monometal transition to bimetallic transition chalcogenide). As a result, the overpotential of $NiSe_2$, $FeSe_2$, $Ni_{0.5}Fe_{0.5}Se_2$, $Ni_{0.25}Fe_{0.75}Se_2$ and $Ni_{0.75}Fe_{0.25}Se_2$ were found to be 305, 332, 235, 271 and 250, respectively with Tafel slope of 64.2, 51.3, 34.7, 53.8, and 45.8 mV/dec, respectively. The obtained results can be attributed to the synergic effect and structural composition of the prepared nanomaterial. Ganesan et al. [49] have successfully synthesized $CoSe_2$ nanoboxes using Co-Prussian blue analogue

Electrochemical Water Splitting: Materials and Applications   Materials Research Forum LLC
Materials Research Foundations **59** (2019) 141-168   doi: https://doi.org/10.21741/9781644900451-6

(PBA) using a metal organic framework (MOF) as a template by etching and selenization steps. Prepared $CoSe_2$ nanobox has exhibited remarkable electrocatalytic activity towards OER with low over potential 335 mV (for 10 mA cm^{-2}) and small Tafel slope 54.2 mV/dec with long term stability. This remarkable result of $CoSe_2$ nanobox could be attributed to their hollow morphology and high conductivity.

In 2016, Liao et al. [50] have synthesized coral like $CoSe_2$ by hydrothermal method and studied their OER performance in basic medium. The overpotential was found to be 295 mV at a current density of 10 mA cm^{-2} and Tafel slope of 40 mV/dec with long term stability.

Flake like surface morphology of Co_7Se_8 was prepared by electrodeposition method and studied their OER performance in alkaline medium by Masud et al. [51]. It had low overpotential 260 mV to achieve the current density 10 mA cm^{-2}, small Tafel slope 32.6 mV/dec and long-term stability (12 h). Masud et al. [51] have synthesized copper selenides (Cu_2Se) by different techniques like direct electrodeposition, hydrothermal, and chemical vapor deposition and studied their performances towards OER in 1 M KOH Authors observed that based on the difference in the synthesis process, different overpotential values were obtained for Cu_2Se, i.e. for electrodeposition, hydrothermal and chemical vapor deposition method they needed 270, 290 and 300 mV, respectively to reach the current density 10 mA cm^{-2} and with different Tafel slope values of 107.6, 136.6 and 90.9 mV/dec, respectively.

Kwak et al. [52] have successfully synthesized $NiSe_2$ and $CoSe_2$ nanocrystals with an average particle size of 20 nm and compared their electrocatalytic activity towards OER. The prepared nanocrystals exhibited superior OER performance with onset potential +1.43 and +1.55 V vs. RHE for $NiSe_2$ and $CoSe_2$, respectively. They offered overpotential 430 mV@$CoSe_2$ and 250 mV@$NiSe_2$to achieve current density 10 mA cm^{-2} with Tafel slope 38 and 53 mV/dec for $NiSe_2$ and $CoSe_2$, respectively. Remarkably little overpotential and small Tafel slope proved that $NiSe_2$ nanocrystals exhibited better catalytic activity than $CoSe_2$ nanocrystals. The greater electrocatalytic activity of $NiSe_2$ can be again attributed to the formation of oxide layers (Ni-OOH layer), which are found to be more active than those of $CoSe_2$nanocrystals oxide layer (Co-OOH layer).

To improve the catalytic efficacy, doping, or introduction of extra transition metals are also done by some of the researchers. For instance, Zhang et al. [53] have prepared $NiSe_2$, Fe, and Co doped $NiSe_2$ by one step solvothermal route. The OER performance of prepared catalysts was studied in alkaline medium. The overpotential for $NiSe_2$, $Fe_{0.09}Ni_{0.91}Se_2$ and $Fe_{0.08}Co_{0.09}Ni_{0.83}Se_2$ were obtained 375, 325 and 268 mV, respectively to reach the current density 10 mA cm^{-2} and Tafel slope were 89.36, 57.53 and 54.78

mV/dec, respectively. The excellent OER performance with robust stability of the prepared catalyst could be attributed by high electrical conductivity and extended synergic effect in $Fe_{0.08}Co_{0.09}Ni_{0.83}Se_2$, due to the incorporation of extra transition metal in the chalcogenides.

Table 2: *Summary of some of the recently reported transition metal selenides based electrocatalyst for oxygen evolution reaction 1.0 M KOH as an electrolyte.*

S.N.	Catalyst	Preparation Method	Overpotential (mV)	Tafel Slope (mV/dec)	Ref.
1.	$Co_{0.85}Se$/NiFe LDH NS	Hydrothermal	1.51 V@250 mA cm^{-2}	57	[46]
	$Co_{0.85}Se$		-	73	
	NiFe-LDH		-	86	
2.	NiSe-NF	Hydrothermal	270@20 mA cm^{-2}	64	[55]
3.	NiSe@Ti mesh	Electrodeposition	295@20 mA cm^{-2}	172	[56]
4.	$Ni_{0.5}Co_{0.5}SeO_3$/NF	Electrodeposition	380@100 mA cm^{-2}	44	[57]
	$NiSeO_3$/NF		500@157.08 mA cm^{-2}	69	
	$CoSeO_3$/NF		500@138.16 mA cm^{-2}	118	
5.	$NiSe_2$/Ni	Immersion–selenization	235@10 mA cm^{-2}	63.1	[58]
6.	Sn–Ni_3S_2 NS	Hydrothermal	270@100 mA cm^{-2}	52.7	[59]
7.	NiSe/Ni_3Se_2/NF	Hydrothermal	260@20 mA cm^{-2}	69.2	[60]
8.	NiSe	Solvothermal	271@10 mA cm^{-2}	80	[61]
9.	Hierarchical $(Ni,Fe)_3Se_4$	Solvothermal	225@10 mA cm^{-2}	41	[62]
10.	Textured $NiSe_2$ Film	Hydrothermal	140@10 mA cm^{-2}	56.6	[63]
		Electrodeposition	200@100 mA cm^{-2}	48.7	

N.B.: LDH- layer double hydroxides, NS-nanosheets, NF-nanofiber.

Recently, Wang et al. [54] have prepared ternary Ni-Co-Se hexagonal nanoalloy and quaternary Ni-Co-S-Se hexagonal nanoalloy by simple one step solvothermal method. They had excellent OER performance in alkaline medium, but quaternary $Ni_{0.25}Co_{0.65}S_{0.4}Se_{0.6}$ nanocrystals exhibited 7-times greater OER activity than binary CoSe in term of turn over frequency. Quaternary $Ni_{0.25}Co_{0.65}S_{0.4}Se_{0.6}$ and $Ni_{0.3}Co_{0.73}S_{0.2}Se_{0.8}$ nanocrystals offered low overpotential of 262 and 272 mV, respectively with corresponding Tafel slope of 64 and 64 mV/dec, respectively. Whereas, ternary $Ni_{0.3}Co_{0.74}Se$ nanocrystals offered higher overpotential as well as Tafel slope (76 mV/dec) than quaternary nanocrystals, which indicates the quaternary nanocrystals had a faster reaction rate than ternarynanocrystals. Additionally, some of the recently reported transition metal selenide used as electrocatlyst for oxygen evolution reaction are portrayed in Table 2 [46,55-63].

2.3 Transition metal telluride for oxygen evolution reaction

In the previous sections, we have briefly discussed the electrocatalytic activity of transition metal sulfides and selenides towards OER. Here, we will discuss the electrocatalytic activity of transition metal tellurides for OER. Since the electronegativity of chalcogenide, i.e. sulfur to tellurium decreases and resulted covalence of transition metal chalcogenide increased. According to the previous literature, the OER activity large depend on the covalence of the metal anion by means higher covalence higher is the OER activity [64]. Moreover, from the qualitative band structure study has proved that tellurides have more band alignment in water splitting than selenides [65, 66].

Based on the theory, some work has been reported for the electrocatalytic activity of transition metal tellurides towards OER. For instance, Silva et al. [67] has successfully synthesized Ni_3Te_2 by simple hydrothermal and electrodeposition route and studied its OER performance. It had excellent electrocatlytic activity towards OER than oxide (NiO_x), sulfides (Ni_3S_2) and selenides (Ni_3Se_2). The telluride analogue offered low overpotential than others as: 360, 330, 280, and 180 mV for NiO_x, Ni_3S_2, Ni_3Se_2, and Ni_3Te_2, respectively to achieve current density of 10 mA cm^{-2}, however, the electronegativity of oxygen, sulfur, selenide, and tellurium is 3.44, 2.58, 2.55 and 2.10, respectively. The Ni_3Te_2 was synthesized here by two different synthesis method, i.e. hydrothermal and electrodeposition routes. It is found that lower Tafel slope and higher electrocatalytic activity was obtained for Ni_3Te_2 synthesized via hydrothermal (61.5mV/dec) than electrodeposition route (64.3 mV/dec).

In 2018, Wang et al. [68] has prepared NiTe nanosheets grown on nickel foam via tellurization to fabricate effective electrocatalyst for OER. The prepared catalysts exhibited low overpotential 262 mV to reach current density 10 mA cm^{-2} and showed robust behavior in alkaline medium. Similarly, nickel ditelluride nanosheets ($NiTe_2$) revealed remarkable catalytic activity towards OER with small overpotential 315 mV to achieve the current density 10 mA cm^{-2} synthesized by Wang et al. in 2018. Bhat et al. [69] has successfully synthesized porous hollow nickel telluride nanosheets via ion exchange reaction followed by the hydrothermal route and studied their role as effective OER catalyst in alkali medium. It had overpotential 659 mV to reach the current density 10 mA cm^{-2} and small Tafel slope 151 mV/dec.

Electrochemical Water Splitting: Materials and Applications   Materials Research Forum LLC
Materials Research Foundations **59** (2019) 141-168   doi: https://doi.org/10.21741/9781644900451-6

In 2016, McKendry et al. [70] prepared cobalt telluride ($CoTe_2$) and studied their electrocatalytic activity towards water oxidation. It had remarkable catalytic activity with a low overpotential of 340 mV to attain current density of 10 mA cm^{-2} with Tafel slope 58 mV/dec. Other than these, Patil et al. [71] have successfully prepared cobalt telluride (CoTe) nanotube and examined their elctrocatlytic activity towards OER in 1.0 M KOH. The prepared CoTe nanotube offered low overpotential of 370 mV. Their better electrocatalytic activity could be attributed to the large BET surface area (14.33 m^2 g^{-1}) and low charge transfer resistance.

Gao et al. [72] first time successfully synthesized the phase selective uniform hierarchical $CoTe_2$ and CoTe nanofleeces by chemical transformation method. The nanostructured $CoTe_2$ exhibited smaller overpotential (357 mV) than nanostructured CoTe to attain current density 10 mA cm^{-2}. The Tafel slope of the nanostructured $CoTe_2$ and CoTe were found to be 32 and 73 mV/dec, respectively, which are lower than the commercially available state-of-art catalysts, i.e. Pt/C (123 mV/dec) and RuO_2 (79 mV/dec).

Recently in the year 2019, Wu and coworkers [73] have successfully synthesized ternary PbNiTe, binary PbTe and NiTe nanostructure using a different weight of nickel precursor (1.0 and 0.3 g) by two steps solvothermal route. The ternary nanostructures exhibited higher catalytic activity towards OER than the binary NiTe and PbTe.They optimized NiTe-PbNiTe (1.0 g) offered low overpotential (387 mV) than NiTe-PbNiTe (0.3 g), i.e. 516 mV and pure NiTe (398 mV). The obtained Tafel slopes of prepared nanostructures catalysts were found to be 96, 144, 155 and 270 mV/dec for NiTe-PbNiTe (1.0 g), NiTe-PbNiTe (0.3 g), NiTe and PbTe, respectively. The higher catalytic activity of NiTe-PbNiTe (1.0 g) could be attributed to heterostructures nanorods like morphology of nanomaterials, which have strong synergic effect between NiTe and PbTe. Combination of these two led to the formation of more interfaces and active sites, which directly lowers the activation energy and simultaneously enhance the reaction rate of OER.

Conclusion and Future prospective

To improve the reaction rate of sluggish OER, transition metal sulfide/selenide/telluride has been proven as a best catalyst with high activity, durability, and low cost. Moreover, it was also found that catalytic activity of transition metal chalcogenides towards OER increased from transition metal sulfides to transition metal tellurides. This is mainly due to the increase in covalence of transition metal chalcogenide with a decrease in electronegativity of chalcogens from sulfur to tellurium. Here, we have summarized the efficient and recently reported transition metal based sulfide/selenide/telluride towards OER/water splitting. Different strategies have been tried to enhance the catalytic activity of transition metal sulfides, selenides, and tellurides towards OER. For instance, the most commonly followed strategy is making a composite of transition metal chalcogenides and the other transition metal chalcogenides or others nanomaterials. Doping of foreign element/candidate including transition metal and/or chalcogenide onto the transition metal chalcogenides also worked well and successfully as it increases the surface area, electrical conductivity, synergic effect, active site, and surface reaction. Fast, inexpensive, low cost synthetic methods (i.e., solvothermal, electrodeposition, hydrothermal and chemical vapor deposition methods) have been used to get different but uniform surface morphology of the transition metal chalcogenides, which resulted in their higher catalytic activity towards OER in alkaline medium.

Although the transition metal chalcogenides have exhibited good catalytic activity towards OER, to further improve the catalytic activity, more study in this area is still needed. For example, the formation of nanocomposite by means transition metal chalcogenide composite with graphene oxide, reduced graphene oxide, or nickel foam or other nanomaterials may be formed to improve their existing properties.

Author declaration
Mr. Majhi has given the major contribution in writing this book chapter along with drawing the Figures and Tables, taking the copyright permission, etc.

References

[1] X. Luo, S. Qi, P. Yecan, H. Xiaoqing, Trimetallic molybdate nanobelts as active and stable electrocatalysts for the oxygen evolution reaction, ACS Catal. 2 (2018) 1013-1018. https://doi.org/10.1021/acscatal.8b04521

[2] R. Frydendal, E. A. Paoli, B. P. Knudsen, B. Wickman, P. Malacrida, I. E. Stephens, I. Chorkendorff, Benchmarking the stability of oxygen evolution reaction catalysts: the importance of monitoring mass losses, Chem. Electro. Chem. 12 (2014) 2075-81. https://doi.org/10.1002/celc.201402262

[3] Y. Lee, J. Suntivich, K. J. May, E. E. Perry, Y. Shao-Horn, Synthesis and activities of rutile IrO_2 and RuO_2 nanoparticles for oxygen evolution in acid and alkaline solutions, J. Phy. Chem. Lett, 3 (2012) 399-404. https://doi.org/10.1021/jz2016507

[4] T. Reier, O. Mehtap, S. Peter, Electrocatalytic oxygen evolution reaction (OER) on Ru, Ir, and Pt catalysts: a comparative study of nanoparticles and bulk materials, ACS Catal. 8 (2012) 1765-1772. https://doi.org/10.1021/cs3003098

[5] M. Vukovic, Oxygen evolution reaction on thermally treated iridium oxide films, J. Appl. Electrochem. 4 (1987) 737-745. https://doi.org/10.1007/BF01007809

[6] E. Antolini, Iridium as catalyst and cocatalyst for oxygen evolution/reduction in acidic polymer electrolyte membrane electrolyzers and fuel cells, ACS Catal. 5 (2014) 1426-1440. https://doi.org/10.1021/cs4011875

[7] R. Kotz, H. Neff, S. Stucki, Anodic iridium oxide films XPS-studies of oxidation state changes, J. Electrochem. Soc. 1 (1984) 72-77. https://doi.org/10.1149/1.2115548

[8] N.T. Suen, F.H. Sung, Q. Quan, Z. Nan, X. J. Yi, C. M. Hao, Electrocatalysis for the oxygen evolution reaction: recent development and future perspectives, Chem. Soc. Rev. 2 (2017) 337-365. https://doi.org/10.1039/C6CS00328A

[9] S. Farid, R. Suzhen, H. Ce, MOF-derived metal/carbon materials as oxygen evolution reaction catalysts, Inor. Chem. Comm. 94 (2018) 57-74. https://doi.org/10.1016/j.inoche.2018.06.008

[10] F. Lu, Z. Min Z. Yuxue, Z. Xianghua, First-row transition metal based catalysts for the oxygen evolution reaction under alkaline conditions: Basic principles and recent advances, Small 45 (2017) 1701931. https://doi.org/10.1002/smll.201701931

[11] J.A. Bard, F.R. Larry, L. Johna, Z.G. Cynthia, Electrochemical methods: Fundamentals and applications. Vol. 2. New York: wiley, 1980.

[12] F. Yu, L. Yu, I.K. Mishra, Y. Yu, Z.F. Ren, H.Q. Zhou, Recent developments in earth-abundant and non-noble electrocatalysts for water electrolysis, Mater. Today Phys. 7 (2018) 121-138. https://doi.org/10.1016/j.mtphys.2018.11.007

[13] K. Xiao, Z. Lei, S. Mingfei, W. Min, Fabrication of (Ni, Co)$_{0.85}$ Se nanosheet arrays derived from layered double hydroxides toward largely enhanced overall water splitting, J. Mater. Chem. A 17 (2018) 7585-7591. https://doi.org/10.1039/C8TA01067F

[14] W. Zhang, L. Daohao, Z. Longzhou, S. Xilin, Y. Dongjiang, NiFe-based nanostructures on nickel foam as highly efficiently electrocatalysts for oxygen and hydrogen evolution reactions, J. Energy. Chem. (2019). https://doi.org/10.1016/j.jechem.2019.01.017

[15] R. Frydendal, P.A. Elisa, K.P. Brian, W. Bjorn, M. Paolo, S. Ifan EL, C. Ib, Benchmarking the stability of oxygen evolution reaction catalysts: the importance of monitoring mass losses, Chem. Electro. Chem. 12 (2014) 2075-2081. https://doi.org/10.1002/celc.201402262

[16] C.L.C. McCrory, J. Suho, F.M. Ivonne, C.M. Shawn, P.C. Jonas, J.F. Thomas, Benchmarking hydrogen evolving reaction and oxygen evolving reaction electrocatalysts for solar water splitting devices, J. Am. Chem. Soc. 13 (2015) 4347-4357. https://doi.org/10.1021/ja510442p

[17]   M. Nic, H. Ladislav, J. Jiri, K. Bedrich, Z. Jiri, IUPAC compendium of chemical terminology-the gold book, IUPAC, 2005.

[18]   V.A. Kolobov, T. Junji, Two-dimensional transition-metal dichalcogenides, Vol. 239. Springer, 2016. https://doi.org/10.1007/978-3-319-31450-1

[19]   T.A. Swesi, M. Jahangir, N. Manashi, Transition metal chalcogenide based elelctrocatalysts for facile water oxidation/reduction, In Meeting Abstracts, The Electrochem. Soc. 29 (2016) 1438-1438.

[20]   H. Yuan, K. Long, L. Tao, Z. Qiang, A review of transition metal chalcogenide/graphene nanocomposites for energy storage and conversion, Chinese Chem. Lett. 12 (2017) 2180-2194. https://doi.org/10.1016/j.cclet.2017.11.038

[21]   S. Anantharaj, E.R. Sivasankara, S. Kuppan, K. Kannimuthu, M. Soumyaranjan, K. Subrata, Recent trends and perspectives in electrochemical water splitting with an emphasis on sulfide, selenide, and phosphide catalysts of Fe, Co, and Ni: A review, ACS Catal. 12 (2016) 8069-8097. https://doi.org/10.1021/acscatal.6b02479

[22]   P. Luo, Z. Huijuan, L. Li, Z. Yan,D. Ju, X. Chaohe, H. Ning, W. Yu, Targeted synthesis of unique nickel sulfide (NiS, NiS$_2$) microarchitectures and the applications for the enhanced water splitting system, ACS Appl. Mater. Interfaces 3 (2017) 2500-2508. https://doi.org/10.1021/acsami.6b13984

[23]   Y. Zhang, C. Shujun, W. Xiaobing, H. Huijuan, B. Zhengyu, Y. Lin, Hierarchical Co$_9$S$_8$ hollow microspheres as multifunctional electrocatalysts for oxygen reduction, oxygen evolution and hydrogen evolution reactions, Electrochim. Acta 246 (2017) 380-390. https://doi.org/10.1016/j.electacta.2017.06.058

[24]   X. Feng,J. Qingze,L. Tong, L. Qun, Y. Mengmeng, Z. Yun, L. Hansheng, F. Caihong,Z. Wei, Facile synthesis of Co$_9$S$_8$ hollow spheres as a high-performance electrocatalyst for the oxygen evolution reaction, ACS Sustain. Chem. Engineer. 2 (2017): 1863-1871. https://doi.org/10.1021/acssuschemeng.7b03236

[25]   Z. Zeng, X. Rui, Z. Huaping, Z. Huanming, X. Shipu, L. Yong, exploration of nanowire-and nanotube-based electrocatalysts for oxygen reduction and oxygen evolution reaction, Mater. Today Nano (2018). https://doi.org/10.1016/j.mtnano.2018.11.005

[26]   M. Chauhan, P.R. Kasala, G.S. Chinnakonda, D. Sasanka, Copper cobalt sulfide nanosheets realizing a promising electrocatalytic oxygen evolution reaction, ACS Catalysis 9 (2017) 5871-5879. https://doi.org/10.1021/acscatal.7b01831

[27]   Z.F. Huang, S. Jiajia, L. Ke, T. Muhammad, T.W. Yu, P. Lun, W. Li, Z. Xiangwen, J.Z. Ji, Hollow cobalt-based bimetallic sulfide polyhedra for efficient all-

pH-value electrochemical and photocatalytic hydrogen evolution, J. Am. Chem. Soc. 4 (2016): 1359-1365. https://doi.org/10.1021/jacs.5b11986

[28]   W.K. Gao, F.Q. Jun, W. Kai, L.Y. Kai, Z.L. Zi, H.L. Jia, M.C. Yong, G.L. Chen, D. Bin, Facile synthesis of Fe-doped Co_9S_8 nano-microspheres grown on nickel foam for efficient oxygen evolution reaction, Appl. Surface Sci. 454 (2018) 46-53. https://doi.org/10.1016/j.apsusc.2018.05.099

[29]   H.S. Jadhav, R. Animesh, M.T. Gaurav, J.C. Wook, G.S. Jeong, Hierarchical free-standing networks of $MnCo_2S_4$ as efficient Electrocatalyst for oxygen evolution reaction, J. Indus. Engineer. Chem. 71 (2019) 452-459. https://doi.org/10.1016/j.jiec.2018.12.002

[30]   J. Zhang, B. Xiaowan, W. Tongtong, X. Wen, X. Pinxian, W. Jinlan, G. Daqiang, W. John, Bimetallic nickel cobalt sulfide as efficient electrocatalyst for Zn–air battery and water splitting, Nano-micro lett. 1 (2019): 2. https://doi.org/10.1007/s40820-018-0232-2

[31]   Z. Cai, X. Bu, P. Wang, Y. Gao, X. Wang, Recent advances on layered double hydroxide electrocatalysts for oxygen evolution reaction, J. Mater. Chem. A 1 (2013) 1-22

[32]   H. Li, S. Youdong, S. Yantao, G. Yuanhong, W. Xinwei, Vapor-phase atomic layer deposition of nickel sulfide and its application for efficient oxygen-evolution electrocatalysis, Chem. Mater. 4 (2016) 1155-1164. https://doi.org/10.1021/acs.chemmater.5b04645

[33]   T. Liu, S. Xuping, M.A. Abdullah, H. Yuquan, One-step electrodeposition of Ni–Co–S nanosheets film as a bifunctional electrocatalyst for efficient water splitting, Int.J. Hydrogen Energy 18 (2016) 7264-7269. https://doi.org/10.1016/j.ijhydene.2016.03.111

[34]   O. Mabayoje, S. Ahmed, R.W. Bryan, C.B. Mullins, The role of anions in metal chalcogenide oxygen evolution catalysis: Electrodeposited thin films of nickel sulfide as "pre-catalysts", ACS Energy Lett. 1 (2016) 195-201. https://doi.org/10.1021/acsenergylett.6b00084

[35]   X. Zhao, S. Xiao Q. Yun, D. Bin, Q.H. Guan, L. Xiao, R.L. Yan, C. Qi, C.M. Yong, L. G. Chen, Electrodeposition-solvothermal access to ternary mixed metal Ni-Co-Fe sulfides for highly efficient electrocatalytic water oxidation in alkaline media, Electrochim. Acta 230 (2017) 151-159. https://doi.org/10.1016/j.electacta.2017.01.178

[36]   W. Fang, L. Danni, L. Qun, S. Xuping, M.A. Abdullah, Nickel promoted cobalt disulfide nanowire array supported on carbon cloth: an efficient and stable

bifunctional electrocatalyst for full water splitting, Electrochem. Commun. 63 (2016) 60-64. https://doi.org/10.1016/j.elecom.2015.10.010

[37]   D. Liu, L. Qun, L. Yonglan, S. Xuping, M.A. Abdullah, $NiCo_2S_4$ nanowires array as an efficient bifunctional electrocatalyst for full water splitting with superior activity, Nanoscale 37 (2015) 15122-15126. https://doi.org/10.1039/C5NR04064G

[38]   C. Zequine, B. Sanket, W. Fangzhou, L. Xianglin, S. Khamis, P.K. Kahol, K.G. Ram, Effect of solvent for tailoring the nanomorphology of multinary $CuCo_2S_4$ for overall water splitting and energy storage, J. Alloys Compounds 784 (2019) 1-7. https://doi.org/10.1016/j.jallcom.2019.01.012

[39]   L. Wang, L. Yibin, S. Qiangqiang, Q. Qi, S. Yuqian, M. Yi, W. Zenglin, Z. Chuan, Ultralow Fe(III) ion doping triggered generation of Ni_3S_2 ultrathin nanosheet for enhanced oxygen evolution reaction, Chem. Cat. Chem. 11 (2019) 2011-2016. https://doi.org/10.1002/cctc.201801959

[40]   Y. Guo, D. Guo, Y. Feng, K. Wang, S. Zhongqi. Synthesis of lawn-like NiS_2 nanowires on carbon fiber paper as bifunctional electrode for water splitting, Int. J. Hydrogen Energy 27 (2017) 17038-17048. https://doi.org/10.1016/j.ijhydene.2017.05.195

[41]   C. Zequine, B. Sanket, S. Khamis, K. K. Pawan,K. Nikolaos, M. Christian, J. H. Steven, Needle grass array of nanostructured nickel cobalt sulfide electrode for clean energy generation, Surf. Coatings Technol. 354 (2018) 306-312. https://doi.org/10.1016/j.surfcoat.2018.09.045

[42]   J.T. Ren, Y.Y. Zhong, Hierarchical nickel sulfide nanosheets directly grown on Ni foam: a stable and efficient electrocatalyst for water reduction and oxidation in alkaline medium, ACS Sustain. Chem. Engineer. 8 (2017) 7203-7210. https://doi.org/10.1021/acssuschemeng.7b01419

[43]   Z. Fang, P. Lele, L. Haifeng, Z. Yue, Y. Chunshuang, W. Shengqi, K. Pranav, W. Xiaojun, Y. Guihua, Metallic transition metal selenide holey nanosheets for efficient oxygen evolution electrocatalysis, ACS nano 9 (2017) 9550-9557. https://doi.org/10.1021/acsnano.7b05481

[44]   A. T. Swesi, M. Jahangir,N. Manashi, Nickel selenide as a high-efficiency catalyst for oxygen evolution reaction, Energy Environ. Sci. 5 (2016) 1771-1782. https://doi.org/10.1039/C5EE02463C

[45]   J. Suntivich, J.M. Kevin, A. G. Hubert, B. G. John B. S. H. Yang, A perovskite oxide optimized for oxygen evolution catalysis from molecular orbital principles, Science 6061 (2011) 1383-1385. https://doi.org/10.1126/science.1212858

[46]   C. Tang, C. Ningyan, P. Zonghua, X. Wei, S. Xuping, NiSe nanowire film supported on nickel foam: an efficient and stable 3D bifunctional electrode for full water splitting, Angewandte Chemie International Edition 32 (2015) 9351-9355. https://doi.org/10.1002/anie.201503407

[47]   X. Wu, H. Denghong, Z. Hongxiu, L. Hao, L. Zhongjian, Y. Bin, L. Zhan, L. Lecheng, Z. Xingwang Zhang, $Ni_{0.85}Se$ as an efficient non-noble bifunctional electrocatalyst for full water splitting, Int. J. Hydrogen Energy 25 (2016) 10688-10694. https://doi.org/10.1016/j.ijhydene.2016.05.010

[48]   Y. Du, C. Gongzhen, L. Wei, $NiSe_2/FeSe_2$ nanodendrites: a highly efficient electrocatalyst for oxygen evolution reaction, Catal. Sci. Technol. 20 (2017) 4604-4608. https://doi.org/10.1039/C7CY01496A

[49]   V. Ganesan, K. Jinkwon, Prussian blue analogue metal organic framework-derived $CoSe_2$ nanoboxes for highly efficient oxygen evolution reaction, Mater. Lett. 223 (2018) 49-52. https://doi.org/10.1016/j.matlet.2018.03.125

[50]   M. Liao, Z. Guangfeng, L. Tingting, J. Zhaoyu, W. Yujue, K. Xingming, X. Dan. Three-dimensional coral-like cobalt selenide as an advanced electrocatalyst for highly efficient oxygen evolution reaction, Electrochim.Acta194 (2016) 59-66. https://doi.org/10.1016/j.electacta.2016.02.046

[51]   J. Masud, A. T. Swesi, P. R. W. Liyanage, N. Manashi, Cobalt selenide nanostructures: an efficient bifunctional catalyst with high current density at low coverage, ACS Appl. Mater. Interfaces 27 (2016) 17292-17302. https://doi.org/10.1021/acsami.6b04862

[52]   I.H. Kwak, S.I. Hyung, M.J. Dong, W.K. Young, P. Kidong, R.L.Young, H.C. Eun, P. Jeunghee, $CoSe_2$ and $NiSe_2$ nanocrystals as superior bifunctional catalysts for electrochemical and photoelectrochemical water splitting, ACS Appl. Mater. Interfaces 8 (2016) 5327-5334. https://doi.org/10.1021/acsami.5b12093

[53]   J. Zhang, J. Bei, Z. Jingru, L. Ruguang, Z. Nana, L. Ruixin, L. Jiakai, Z. Daojun, Z. Renchun, Facile synthesis of $NiSe_2$ particles with highly efficient electrocatalytic oxygen evolution reaction, Mater. Lett. 235 (2019) 53-56. https://doi.org/10.1016/j.matlet.2018.09.163

[54]   M. Wang, D. Zhiya, P. Mirko, V.S. Dipak, D.T. Luca, M. Liberato Manna, Ni–Co–S–Se Alloy nanocrystals: Influence of the composition on their in situ transformation and electrocatalytic activity for the oxygen evolution reaction, ACS Appl. Nano Mater.10 (2018) 5753-5762. https://doi.org/10.1021/acsanm.8b01418

[55]   Y. Hou, R.L. Martin, Z. Jian, L. Shaohua, Z. Xiaodong, F. Xinliang, Vertically oriented cobalt selenide/NiFe layered-double-hydroxide nanosheets supported on

exfoliated graphene foil: an efficient 3D electrode for overall water splitting, Energy Environ. Sci. 2 (2016) 478-483. https://doi.org/10.1039/C5EE03440J

[56]   Z. Pu, L. Yonglan, A.M. Abdullah, S. Xuping, Efficient electrochemical water splitting catalyzed by electrodeposited nickel diselenide nanoparticles based film, ACS Appl. Mater. Interfaces 7 (2016) 4718-4723. https://doi.org/10.1021/acsami.5b12143

[57]   J. Xin, T. Hua, L. Zhihe, Z. Lili, X. Junfeng, S. Yuanhua, Z. Weijia, W. Aizhu, L. Hong, J. W. Jian, Facile synthesis of hierarchical porous $Ni_xCo_{1-x}SeO_3$ networks with controllable composition as a new and efficient water oxidation catalyst, Nanoscale 7 (2019) 3268-3274. https://doi.org/10.1039/C8NR09218D

[58]   J. Zhang, W. Ying, Z. Chi, G. Hui, L. Lanfen, H. Lulu, Z. Zhonghua, Self-supported porous $NiSe_2$ nanowrinkles as efficient bifunctional electrocatalysts for overall water splitting, ACS Sustain. Chem. Engineer. 2 (2017) 2231-2239. https://doi.org/10.1021/acssuschemeng.7b03657

[59]   J. Jian, Y. Long, Q. Hui, S. Xuejiao, Z. Le, L. He, Y. Hongming, F. Shouhua, Sn–Ni_3S_2 Ultrathin Nanosheets as Efficient Bifunctional Water-Splitting Catalysts with a Large Current Density and Low Overpotential, ACS Appl. Mater. Interfaces 47 (2018) 40568-40576. https://doi.org/10.1021/acsami.8b14603

[60]   F. Zhang, P. Yu, G. Yuancai, C. Hang, C. Steven, D. Pei, C. Jun, M.A. Pulickel, Y. Mingxin, S. Jianfeng, Controlled synthesis of eutectic $NiSe/Ni_3Se_2$ self-supported on Ni foam: An excellent bifunctional electrocatalyst for overall water splitting, Adv. Mater. Interfaces 8 (2018) 1701507. https://doi.org/10.1002/admi.201701507

[61]   H. Ren, H. H. Zheng, Y. Zhiyu, T. Shujun, K. Feiyu, L. Ruitao, Facile synthesis of free-standing nickel chalcogenide electrodes for overall water splitting, J. energy chem. 6 (2017) 1217-1222. https://doi.org/10.1016/j.jechem.2017.10.004

[62]   J. Du, Z.Z. Jing, L. Chen, X. Cailing, Hierarchical Fe-doped Ni_3Se_4 ultrathin nanosheets as an efficient electrocatalyst for oxygen evolution reaction, Nanoscale 11 (2018) 5163-5170. https://doi.org/10.1039/C8NR00426A

[63]   A.T. Swesi, M. Jahangir, L.P.R. Wipula, U. Siddesh, B. Eric, M. Julia, N. Manashi, Textured $NiSe_2$ film: Bifunctional electrocatalyst for full water splitting at remarkably low overpotential with high energy efficiency, Sci. Reports 1 (2017) 2401. https://doi.org/10.1038/s41598-017-02285-z

[64]   J. Masud, I. Polydoros-Chrysovalantis, L. Nikolaos, K. Panayotis, N. Manashi, A Molecular Ni-complex containing tetrahedral nickel selenide core as highly efficient electrocatalyst for water oxidation, Chem. Sus. Chem. 22 (2016) 3128-3132. https://doi.org/10.1002/cssc.201601054

[65]   F.A. Rasmussen, S.T. Kristian Computational 2D materials database: electronic structure of transition-metal dichalcogenides and oxides, J. Phy. Chem. C 23 (2015) 13169-13183. https://doi.org/10.1021/acs.jpcc.5b02950

[66]   Y. Xu, A.A.S. Martin, The absolute energy positions of conduction and valence bands of selected semiconducting minerals, American Mineralogist 3-4 (2000) 543-556. https://doi.org/10.2138/am-2000-0416

[67]   D.U. Silva, J. Masud, N. Zhang, Y. Hong, L.P.R. Wipula, A.Z. Mohsen, N. Manashi, Nickel telluride as a bifunctional electrocatalyst for efficient water splitting in alkaline medium, J. Mater. Chem. A 17 (2018) 7608-7622. https://doi.org/10.1039/C8TA01760C

[68]   Z. Wang, Z. Lixue In situ growth of NiTe nanosheet film on nickel foam as electrocatalyst for oxygen evolution reaction, Electrochem. Commun. 88 (2018) 29-33. https://doi.org/10.1016/j.elecom.2018.01.014

[69]   K.S. Bhat, C.B. Harish, H.S. Nagaraja, Porous nickel telluride nanostructures as bifunctional electrocatalyst towards hydrogen and oxygen evolution reaction, Int. J. hydrogen energy 39 (2017) 24645-24655. https://doi.org/10.1016/j.ijhydene.2017.08.098

[70]   I.G. McKendry, C.T. Akila, S. Jianwei, P. Haowei, P.P. John, R.S. Daniel, J.Z. Michael, Water oxidation catalyzed by cobalt oxide supported on the mattagamite phase of $CoTe_2$, ACS Catal. 11 (2016) 7393-7397. https://doi.org/10.1021/acscatal.6b01878

[71]   S. Patil, A. Supriya, K. Eun-Kyung, K.S. Nabeen, C. Jinho, K.J. Joong, H.H. Sung, Formation of semimetallic cobalt telluride nanotube film via anion exchange tellurization strategy in aqueous solution for electrocatalytic applications, ACS Appl. Mater. Interfaces 46 (2015) 25914-25922. https://doi.org/10.1021/acsami.5b08501

[72]   Q. Gao, C.Q.H. Qiang, M.J.Yi, M.R. Gao, J.W. Liu, A. Duo, C. Chun-Hua, R.Z. Ya, L. Wei-Xue,H. Y. Shu-Hong, Phase-selective syntheses of cobalt telluride nanofleeces for efficient oxygen evolution catalysts, Angewandte Chemie 27 (2017) 7877-7881. https://doi.org/10.1002/ange.201701998

[73]   Q. Wang, Z. Junyong, W. Huanhuan, Y. Sichao, W. Xiaohong, Anchoring NiTe domains with unusual composition on $Pb_{0.95}Ni_{0.05}Te$ nanorod as superior lithium-ion battery anodes and oxygen evolution catalysts, Mater. Today Energy 11 (2019) 199-210. https://doi.org/10.1016/j.mtener.2018.12.001

Electrochemical Water Splitting: Materials and Applications   Materials Research Forum LLC
Materials Research Foundations **59** (2019) 169-178   doi: https://doi.org/10.21741/9781644900451-7

Chapter 7

Interface-Engineered Electrocatalysts for Water Splitting

Nur Azimah Abd Samad, Chin Wei Lai *, Mohd Rafie Johan

Nanotechnology & Catalysis Research Centre (NANOCAT), 3rd Floor, Block A, Institute for Advanced Studies (IAS), University of Malaya, 50603 Kuala Lumpur, Malaysia

* cwlai@um.edu.my

Abstract

Three critical points in tailoring the surface/interface of electrocatalyst are (a) light harvest, (b) charge kinetics, and (c) active sites. In improving light harvesting capability, we need always to remember that we want to harvest all three regions of the solar spectrum; ultraviolet rays (10^{-5} to 400 nm), visible light (400 to 800 nm), and infrared (800 to 1060 nm) so as to completely utilise the solar energy. Enhanced photoelectrochemical (PEC) water splitting performance by interface-engineered electrocatalysts can best be improved by implementing impurity doped-electrocatalyst, creating the surface plasmon resonance (SPR) effect at the surface/interface of electrocatalyst, and building Z-scheme water splitting system.

Keywords

Surface/Interface, Electrocatalyst, Photocatalyst, Photoelectrochemical (PEC), Water Splitting

Contents

1.   The surface/interface mechanism in photoelectrochemical water splitting

Research and development (R&D) around the world is facing the low performance of electrocatalyst. Thus, it impedes the industry-scale of solar-driven water splitting system. However, the re-engineering of the surface/interface of materials may escalate the electrocatalyst performance. Three critical points in tailoring the surface/interface of electrocatalyst are (a) light harvest, (b) charge kinetics, and (c) active sites [1]. These three points are closely bound in escalating the electrocatalyst performance as explained in Fig. 1.

In improving the light harvesting capability, it should be always remembered that all three regions of the solar spectrum need to be harvested; ultraviolet rays (10^{-5} to 400 nm), visible light (400 to 800 nm), and infrared (800 to 10^6 nm) so as to completely utilise the solar energy. Photon absorbed by the surface atoms in a material generates the electron-hole pairs. Consequently, electron excites from the valence band to conduction band (photoexcitation). Upon photoexcitation, the electron movement from the valence band to conduction band can be steered by manipulating the surface/interface (between photon harvest centres and active sites) in order to reduce the electron-hole recombination [1].

2.   Enhanced photoelectrochemical water splitting performance by interface-engineered electrocatalysts

2.1   Impurity doping

Implementing impurity doped-electrocatalyst, creating the surface plasmon resonance (SPR) effect at the surface/interface of electrocatalyst, and building Z-scheme water splitting system are the strategies to efficiently utilise the solar energy [1-4]. Cation/anion doping is a fine method for extending light absorption from UV-Vis-NIR light absorption. There are two impurity dopings; interstitial impurity doping and substitutional impurity doping. Theoretically, interstitial impurity doping acquires a relatively small ionic radius in comparison to the anion. Meanwhile, substitutional impurity doping, either cation or anion will substitute the host ion, which normally contains most similar electrical sense [5]. To achieve substitutional impurity doping to occur, the substitutional impurity must have nearly similar in ionic size and charge as those in host ions to maintain the electroneutrality within the solid. However, there is event where the substitutional impurity doping has a different charge, and this can be accomplished by the occurrence of lattice defects [5]. Yan, H., et al. [6] studied the 3d transition metals: vanadium and chromium, 4d transition metals: zirconium, niobium and molybdenum, 5d transition metals: hafnium, tantalum and tungsten to replace the Ti atom, it was found that Nb, W, Zr, Ta, and Hf were effective dopants as these elements caused a little change at the conduction band maximum (CBM) but not Cr, V, and Mo, since these caused deepest defect level inside the band gap (E_g) away from the CBM in the host atom because Cr, V, and Mo have lower 3d orbital energy as compared to Ti. Therefore, rendering these ineffective in Ti substitution. All the above analysis have been done by density functional theory (DFT) calculations that is performed by project augmented wave (PAW) pseudo potentials [6].

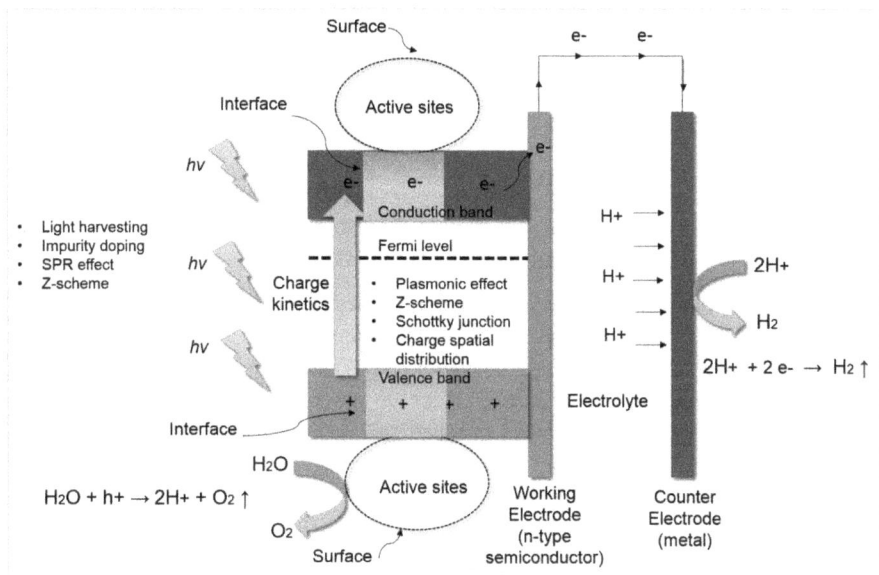

Fig. 1: *Schematic diagram surface/interface design in PEC water splitting and its electron-hole transportation (charge kinetics) [1].*

2.2 Surface plasmon resonance effect

Surface plasmon resonance (SPR) effect towards photoelectrochemical water splitting is improved-catalysis and charge separation from the combination of materials nanocomposites (metal/semiconductor). Specifically, SPR is oscillation collection of conduction band electrons in metal nanoparticles that have been excited by the incident light electromagnetic [1, 7].

Electrochemical Water Splitting: Materials and Applications   Materials Research Forum LLC
Materials Research Foundations **59** (2019) 169-178   doi: https://doi.org/10.21741/9781644900451-7

Referring to Warren, S.C., and Thimsen, E. [9] there are motivations of using SPR as improved-interface for photoelectrochemical water splitting. Most semiconductors have short minority carrier distance, L. Bulk recombination of electron-hole is expected when the semiconductor structure is having L smaller than its physical size. The recombination can be reduced by confining the light absorption at the semiconductor surface. Thus, minority carriers can travel less than L. Plasmonic metal nanoparticle as an antenna can change the charge carrier location by localizing the optical energy and transfer to semiconductor within 10nm of the metal.

Additionally, less amount of plasmonic metal nanoparticles contribute in addressing the low absorption coefficient of the semiconductor. By efficiently scattering the light, it increases the length of the light path through the semiconductor with a large size of plasmonic metal nanoparticles. Moreover, either light can be coupled into wave guiding influenced by semiconductor geometry, or scattered light can be trapped for semiconductor less than half the wavelength of light. By reflecting light between the top and bottom of the surface, this process can be achieved [8].

Besides, plasmonic nanoparticles can also inject electrons into the conduction band. By efficiently scattered the light, plasmonic metal nanoparticles also help in decreasing the reflection losses owned by most semiconductors. Thus, plasmonic metal nanoparticles reduce the semiconductor refractive index to a nearly negligible level. Besides, implementation of plasmonic metal nanoparticles onto semiconductor provides a high surface area. Hence, improving the charge separation at the near-surface region by altering minority carrier (Fermi level) and therefore, increases the photovoltage in PEC water splitting system [9].

As discussed by Nie, J. et al., [10] $Au-TiO_2$ is better employed for the visible light illumination (500nm cut off filter) as compared to bare TiO_2. The trapped electron-hole can be seen from the electron paramagnetic resonance (EPR) signals, which indicated the $Au-TiO_2$ at 500 nm visible light illumination. In fact, at 500 nm visible light illumination, the H_2 production can be observed for $Au-TiO_2$ but none for bare TiO_2. Experiments proved that Au-SPR could induce hot-electron to TiO_2 conduction band and simultaneously, trigger the electron-hole generation at a wavelength above than 500 nm and electron at the valence band of TiO_2 can excite to Au-related surface before been excited to TiO_2 conduction band. Hence, the formation of $Au-TiO_2$ provides two pathways as indicated above. This reasonable analysis also explained the good photocatalytic activitiy for SiO_2 homogeneously covered by plasmonic metal nanoparticles [10].

Electrochemical Water Splitting: Materials and Applications   Materials Research Forum LLC
Materials Research Foundations **59** (2019) 169-178   doi: https://doi.org/10.21741/9781644900451-7

2.3   Z-scheme system

30 years ago, the Z-scheme system has been triggered, but only about a decade ago, the visible-driven Z-scheme water splitting was found [11]. Z-scheme water splitting system is a two-step photoexcitation from heterogeneous semiconductor to split water into H_2 and O_2 under wide range of UV-Vis light (400nm- 700nm wavelength) by using redox mediator in order to create active sites, hence, improving surface/interface chemical reaction and simultaneously prevent backward reaction (electron-hole recombination) [11-14].

Theoretically, highly efficient Z-scheme PEC water splitting can be achieved by considering these two main factors: (1) even in the presence of an electron donor will direct to the outstanding selectively metal oxide photocatalyst, (2) nanoparticulate cocatalyst – modified photocatalyst is also an effective way of controlling the selectivity characteristic for forwarding reaction [11]. A few researchers demonstrated that H_2 and O_2 could be produced from Z - scheme water splitting under visible light illumination (up to 670 nm) by combining in the presence of a redox mediator two different narrow – band gap semiconductors [13, 15, 16]. Fe^{3+}/Fe^{2+}, IO_3^-/I^-, and $[Co \, (bpy)_3]^{3+/2+}$ $[Co \, (phen)_3]^{3+/2+}$ co-catalyst and solid electron mediator such as reduced graphene oxide and Au as the redox mediator for Z-scheme water splitting [12, 13, 17-22].

Although, few types of research proved excellent Z-scheme water splitting could be done without redox mediator as demonstrated by Wang et al. [11]. The production of H_2 and O_2 has occurred simultaneously on $SrTiO_3$: Rh, La and $BiVO_4$ photoelectrocatalyst. The combined materials immobilised on top of Au layer that deposited on a substrate. The charge kinetics are transported between these two materials and subsequently to the thin Au layer [23, 24]. This result attributed to the well-controlled particle sizes, hence, increases the hetero-interfaces density. Nevertheless, the porous structure on top of $SrTiO_3$:Rh and $BiVO_4$ photoelectrocatalyst contributed in the transportation of substrate and product within photoelectrocatalyst itself [13].

Two main hetero structures play a role in charge kinetics (type I and type II hetero-junction) (Fig. 2). For type, I (straddling gap hetero-junction) where electron-hole will accumulate at semiconductor B as conduction band (valence band) for semiconductor B is more positive (negative) than semiconductor A [25, 26]. Whilst, type II (staggered gap hetero-junction) is almost similar to the Z-scheme system yet it has inverted electron movement. In type II hetero-junction, the electron in conduction band of semiconductor B will transport to conduction band in semiconductor A and vice versa for the transportation of holes. In contrast for Z-scheme system, electron at semiconductor A conduction band will transport directly to semiconductor B valence band **(Fig. 3)** and p-n junction is a typical type II hetero-junction system [25, 27, 28].

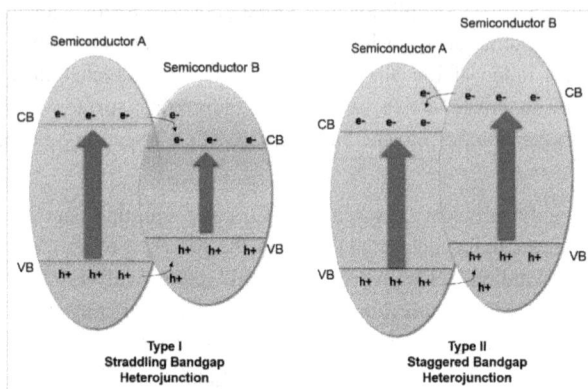

Fig. 2: *Schematic diagram for Type I straddling band gap hetero-junction and type II staggered band gap hetero-junction [25].*

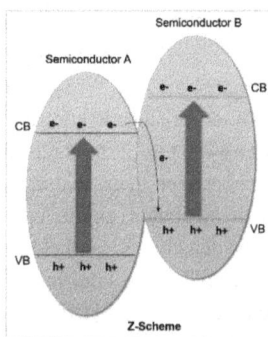

Fig. 3: *Schematic diagram of simplified Z-scheme system [25].*

The scheme also comprises Schottky junction, and it is one of the ways to suppress the recombination of electron-hole (Fig. 4) [25, 29, 30]. Schottky junction allows the transportation of electron from semiconductor A to semiconductor B through an interface so the Fermi energy level alignment will occur [25, 31]. Hence, semiconductor B has excess electrons, and semiconductor A possess more holes at the interface, thus, this contributes to the formation of space charge layer and conduction band, and valence band of semiconductor A is bent upward (Schottky junction). Finally, the Schottky junction serves as an electron trap and efficiently holds the photo-induced electrons, hence provides better photocatalytic activity [29, 31, 32].

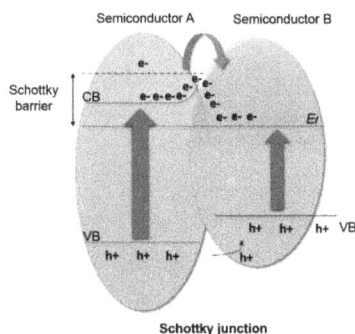

Fig. 4: Schematic diagram of simplified Schottky junction [25].

Acknowledgements

This work is financially supported by University of Malaya Research Grant (No. RP045A&B-17AET).

References

[1]   Y. Hu, C. Gao, Y. Xiong, Surface and interface design for photocatalytic water splitting, Dalton Trans. 47 (2018) 12035-12040. https://doi.org/10.1039/C8DT02885K

[2]   X. Li, J. Wen, J. Low, Y. Fang, J. Yu, Design and fabrication of semiconductor photocatalyst for photocatalytic reduction of CO2 to solar fuel, Science China Materials 57 (2014) 70-100. https://doi.org/10.1007/s40843-014-0003-1

[3]   Z. Wang, Y. Liu, B. Huang, Y. Dai, Z. Lou, G. Wang, X. Zhang, X. Qin, Progress on extending the light absorption spectra of photocatalysts, Phys. Chem. Chem. Phys. 16 (2014) 2758-2774. https://doi.org/10.1039/c3cp53817f

[4]   S. Dong, J. Feng, M. Fan, Y. Pi, L. Hu, X. Han, M. Liu, J. Sun, J. Sun, RSC Adv. 2015, 5, 14610–14630. https://doi.org/10.1039/C4RA13734E

[5]   W.D. Callister, D.G. Rethwisch, Materials science and engineering: an introduction, John Wiley & Sons New York2007.

[6]   H. Yan, X. Wang, M. Yao, X. Yao, Band structure design of semiconductors for enhanced photocatalytic activity: the case of TiO2, Prog. Nat. Sci. Mater. 23 (2013) 402-407. https://doi.org/10.1016/j.pnsc.2013.06.002

[7]   H. Dong, Z. Wu, Y. Gao, A. El-Shafei, B. Jiao, Y. Dai, X. Hou, A nanostructure-based counter electrode for dye-sensitized solar cells by assembly of silver nanoparticles, Org. Electron. 15 (2014) 1641-1649. https://doi.org/10.1016/j.orgel.2014.03.004

[8]   B.E. Saleh, M.C. Teich, Fundamentals of Photonics, John Wiley & Sons, Inc., Canada (1991). https://doi.org/10.1002/0471213748

[9]   S.C. Warren, E. Thimsen, Plasmonic solar water splitting, Energy Environ. Sci. 5 (2012) 5133-5146. https://doi.org/10.1039/C1EE02875H

[10]   J. Nie, J. Schneider, F. Sieland, L. Zhou, S. Xia, D.W. Bahnemann, New insights into the surface plasmon resonance (SPR) driven photocatalytic H 2 production of Au–TiO 2, RSC Adv. 8 (2018) 25881-25887. https://doi.org/10.1039/C8RA05450A

[11]   K. Maeda, Z-scheme water splitting using two different semiconductor photocatalysts, ACS Catalysis 3 (2013) 1486-1503. https://doi.org/10.1021/cs4002089

[12]   Q. Wang, Y. Li, T. Hisatomi, M. Nakabayashi, N. Shibata, J. Kubota, K. Domen, Z- scheme water splitting using particulate semiconductors immobilized onto metal layers for efficient electron relay, J. Catal. 328 (2015) 308-315. https://doi.org/10.1016/j.jcat.2014.12.006

[13]   S. Okunaka, H. Tokudome, R. Abe, Z-scheme water splitting into H2 and O2 under   visible light over photocatalyst panels consisting of Rh-doped SrTiO3 and BiVO4 fine particles, Chem. Lett. 45 (2015) 57-59. https://doi.org/10.1246/cl.150968

[14]   Y. Chang, Y. Xuan, C. Zhang, H. Hao, K. Yu, S. Liu, Z-Scheme Pt@ CdS/3DOM-SrTiO3 composite with enhanced photocatalytic hydrogen evolution from water splitting, Catal.Today 327 (2019) 315-322. https://doi.org/10.1016/j.cattod.2018.04.033

[15]   R. Abe, Development of a new system for photocatalytic water splitting into H2 and O2 under visible light irradiation, Bull. Chem. Soc. Jpn 84 (2011) 1000-1030. https://doi.org/10.1246/bcsj.20110132

[16]   AA Ismail, W. Bahnemann, Sol. Energy Mater. Sol. C 128 (2014) 85. https://doi.org/10.1016/j.solmat.2014.04.037

[17]   H. Kato, M. Hori, R. Konta, Y. Shimodaira, A. Kudo, Construction of Z-scheme type heterogeneous photocatalysis systems for water splitting into H2 and O2 under

visible light irradiation, Chem. Lett. 33 (2004) 1348-1349.
https://doi.org/10.1246/cl.2004.1348

[18]   H. Kato, Y. Sasaki, A. Iwase, A. Kudo, Role of iron ion electron mediator on photocatalytic overall water splitting under visible light irradiation using Z-scheme systems, Bull. Chem. Soc. Jpn. 80 (2007) 2457-2464.
https://doi.org/10.1246/bcsj.80.2457

[19]   Y. Sasaki, A. Iwase, H. Kato, A. Kudo, The effect of co-catalyst for Z-scheme photocatalysis systems with an Fe3+/Fe2+ electron mediator on overall water splitting under visible light irradiation, J. Catal. 259 (2008) 133-137.
https://doi.org/10.1016/j.jcat.2008.07.017

[20]   H. Kato, Y. Sasaki, N. Shirakura, A. Kudo, Synthesis of highly active rhodium-doped SrTiO3 powders in Z-scheme systems for visible-light-driven photocatalytic overall water splitting, J. Mater. Chem. A 1 (2013) 12327-12333.
https://doi.org/10.1039/c3ta12803b

[21]   Y. Sasaki, H. Kato, A. Kudo, [Co (bpy) 3] 3+/2+ and [Co (phen) 3] 3+/2+ electron mediators for overall water splitting under sunlight irradiation using Z-scheme photocatalyst system, J. Am. Chem. Soc. 135 (2013) 5441-5449.
https://doi.org/10.1021/ja400238r

[22]   A. Iwase, Y.H. Ng, Y. Ishiguro, A. Kudo, R. Amal, Reduced graphene oxide as a solid-state electron mediator in Z-scheme photocatalytic water splitting under visible light, J. Am. Chem. Soc. 133 (2011) 11054-11057. https://doi.org/10.1021/ja203296z

[23]   Y. Sasaki, H. Nemoto, K. Saito, A. Kudo, Solar water splitting using powdered photocatalysts driven by Z-schematic interparticle electron transfer without an electron mediator, J. Phys. Chem. C 113 (2009) 17536-17542.
https://doi.org/10.1021/jp907128k

[24]   Q. Jia, A. Iwase, A. Kudo, BiVO4–Ru/SrTiO3: Rh composite Z-scheme photocatalyst for solar water splitting, Chem. Sci. 5 (2014) 1513-1519.
https://doi.org/10.1039/c3sc52810c

[25]   T. Su, Q. Shao, Z. Qin, Z. Guo, Z. Wu, Role of interfaces in two-dimensional photocatalyst for water splitting, ACS Catalysis 8 (2018) 2253-2276.
https://doi.org/10.1021/acscatal.7b03437

[26]   D. Jiang, J. Li, C. Xing, Z. Zhang, S. Meng, M. Chen, Two-dimensional CaIn2S4/g-C3N4 heterojunction nanocomposite with enhanced visible-light

photocatalytic activities: interfacial engineering and mechanism insight, ACS appl. mater.interfaces 7 (2015) 19234-19242. https://doi.org/10.1021/acsami.5b05118

[27]   L. Xu, W.Q. Huang, L.L. Wang, Z.A. Tian, W. Hu, Y. Ma, X. Wang, A. Pan, G.F. Huang, Insights into enhanced visible-light photocatalytic hydrogen evolution of g-C3N4 and highly reduced graphene oxide composite: the role of oxygen, Chem. Mat. 27 (2015) 1612-1621. https://doi.org/10.1021/cm504265w

[28]   S.J. Moniz, S.A. Shevlin, D.J. Martin, Z.X. Guo, J. Tang, Visible-light driven heterojunction photocatalysts for water splitting–a critical review, Energy Environ. Sci. 8 (2015) 731-759. https://doi.org/10.1039/C4EE03271C

[29]   R. Peng, L. Liang, Z.D. Hood, A. Boulesbaa, A. Puretzky, A.V. Ievlev, J. Come, O.S. Ovchinnikova, H. Wang, C. Ma, In-plane heterojunctions enable multiphasic two-dimensional (2D) MoS2 nanosheets as efficient photocatalysts for hydrogen evolution from water reduction, ACS Catalysis 6 (2016) 6723-6729. https://doi.org/10.1021/acscatal.6b02076

[30]   J. Ran, J. Zhang, J. Yu, M. Jaroniec, S.Z. Qiao, Earth-abundant cocatalysts for semiconductor-based photocatalytic water splitting, Chem. Soc. Rev. 43 (2014) 7787-7812. https://doi.org/10.1039/C3CS60425J

[31]   J. Ran, G. Gao, F.T. Li, T.Y. Ma, A. Du, S.Z. Qiao, Ti3C2 MXene co-catalyst on metal sulfide photo-absorbers for enhanced visible-light photocatalytic hydrogen production, Nature commun. 8 (2017) 13907. https://doi.org/10.1038/ncomms13907

[32]   H. Wang, R. Peng, Z.D. Hood, M. Naguib, S.P. Adhikari, Z. Wu, Titania composites with 2 D transition metal carbides as photocatalysts for hydrogen production under visible-light irradiation, Chem. Sus .Chem. 9 (2016) 1490-1497. https://doi.org/10.1002/cssc.201600165

Electrochemical Water Splitting: Materials and Applications   Materials Research Forum LLC
Materials Research Foundations **59** (2019) 179-214   doi: https://doi.org/10.21741/9781644900451-8

Chapter 8

Application of Prussian Blue Analogues and Related Compounds for Water Splitting

Próspero Acevedo-Peña[1] and Edilso Reguera[2,*]

[1]CONACyT-Instituto Politécnico Nacional, Centro de Investigación en Ciencia Aplicada y Tecnología Avanzada, Unidad Legaria, CDMX, Mexico

[2] Instituto Politécnico Nacional, Centro de Investigación en Ciencia Aplicada y Tecnología Avanzada, Unidad Legaria, CDMX, Mexico

edilso.reguera@gmail.com

Abstract

This chapter summarizes state-of-the-art regarding the reported application of Prussian blue analogues and related coordination polymers for water splitting. This family of materials has been evaluated as redox mediator coupled with metal sulfides and/or metal oxides based semiconductors, under neutral conditions (pH 7), with promising results for the oxygen evolution reaction (oxidation of the water molecule). The hydrogen evolution reaction (water reduction) has received less attention. Moreover, according to their electronic and crystal structures, these materials have promising potentialities for applications on chemical energy production from sunlight. The discussion emphasizes supporting such a hypothesis.

Keywords

Prussian Blue Analogues, Porous Cyanometallates, Water Splitting, Oxygen Evolution Reaction, Hydrogen Evolution Reaction, Artificial Photosynthesis, Coordination Polymers

Contents

1.   Introduction

The current global energy consumption is about 17 TW, and it is projected to reach 40 TW by 2050 [1]. This is an extremely low fraction (<0.4 %) of the energy received by the earth as sunlight. Every day 1,65,000 TW of sunlight hits the planet earth and is partially converted to biomass through the natural photosynthesis process as well as diverse types of energy like thermal, hydraulic, eolic, oceanic (waves), and photovoltaic energy. The latter based on a technology fully developed by man. The fossil fuels, whose industrial exploitation has made possible the current technological and social development, have their origin in the natural photosynthesis process where the sunlight (energy) is transformed into chemical energy through the use of water and carbon dioxide as raw materials [2,3]. The formed biomass that has been accumulated for millions of years are now available as carbon, oil, natural gas, and other forms of fossil fuels. These are non-renewable energy sources undergoing progressive depletion and producing carbon dioxide emission during combustion as one of the main sources of climate change. Nowadays, the problems above are triggering a shift toward alternative environmentally friendly sources of energy, thus requiring new technologies, for instance, in efficient solar energy harvesting. This explains the interest for implementation of artificial photosynthesis processes to produce chemical-based energy vectors directly from sunlight [2,3].

In the natural photosynthesis process, a water molecule is split into $O_2 + 2H^+ + 2e^-$, through a photocatalytic reaction assisted by the Mn_4CaO_5 cluster in the photosystem II [4]. The formed molecular oxygen evolves to the atmosphere while two protons and electrons are separated and then used for carbon dioxide reduction to form glucose as well as other chemical products through complex biochemical reactions [3,4]. The simplest energy vector from water splitting is the production of hydrogen (H_2) as depicted in Fig. 1. According to this scheme, the incident photons would be able to promote the extraction of electrons from a donor center (D), which are transferred to an acceptor center (A). The resulting holes in the donor center are now available for the water molecule splitting into $O_2 + 2H^+$. The energy level for the generated electron holes in the donor center should be low enough in order to allow the oxidation of the water molecule. This partial step is usually known as the oxygen evolution reaction (OER). At the acceptor center, protons and electrons are recombined to form H_2 in a step known as hydrogen evolution reaction (HER). The practical implementation of these two partial reactions to produce chemical energy (H_2) from sunlight has been the focus of many researchers involved in the area of artificial photosynthesis [5]. Their research has been driven by the design of donor and acceptor centers that enable the water molecule splitting, while at the same time hinder the electron-hole recombination (a charge separation process). This last process requires high electron mobility towards acceptor centers, which must be characterized by a high electron affinity.

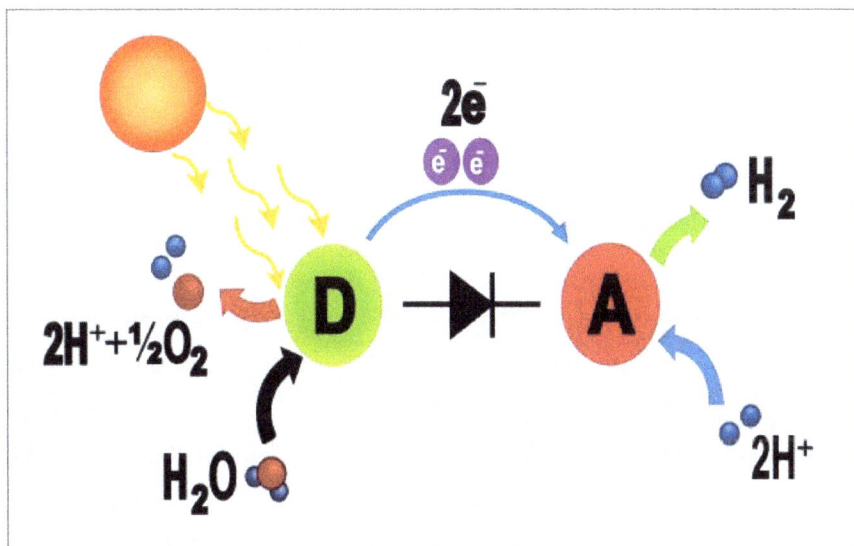

Fig. 1: *Illustrative scheme for the water splitting into $H_2 + O_2$ using sunlight.*

The ideal overall reaction, H_2O + energy $\rightarrow O_2 + H_2$, involves an energy consumption of 1.23 eV. In this ideal process, the energy is provided by the photons of incident sunlight. In practice, only a quite limited number of materials can simultaneously produce O_2 and H_2 by using sunlight as an energy source and for a limited time [5–7]. Usually, the reaction must be assisted by an imposed potential in a combined photo-electrochemical process. Additional complexity for the reaction emerges from thermodynamic and kinetic effects related to solid/liquid/gas interfaces [8]. The limiting step of the overall reaction concerns to the OER side, which requires high energy consumption. And even more relevant is its high kinetic energy barrier accounting for a great efficiency loss. The nature of such high kinetic energy barrier is briefly discussed below.

As donor centers, the most studied materials are semiconductors with an upper edge of the valence band positive enough to allow the water oxidation reaction, and with a band gap (E_g) that enable the absorption of photons from incident light causing the promotion of electrons from the valence band to the conduction band. In this sense, the first and most studied semiconductor has been titanium dioxide (TiO_2 anatase phase), with a band gap of 3.2 eV [9]. This material is capable of absorbing solar radiation in the ultraviolet region, which represents a marginal fraction of the sunlight that hits the earth surface. A similar behavior has been reported for other nd^0 metal oxides, e. g. ZrO_2, $KTaO_3$, $SrTO_3$ and WO_3[10,11]. In order to tune their band gap to have access to the visible and near infrared spectral region, many of these metal oxides have been doped but, practically without exception, their efficiency is still low because the doping process creates defects in the solid that may limit the electron-hole separation process. In a semiconductor, the most electronegative atoms determine the character of the valence band, while the conduction band is more influenced by the metal. This may explain why solids containing less electronegative atoms such as N, S, and P have been considered in the design of semiconductors for water splitting. Consequently, this may induce an energy increase for the upper edge of the valence band, contributing to tune the material band gap to the visible light spectral range. Likewise, in the material design, the metal selection is also relevant because it determines the conduction band features.

Within metal sulfide based semiconductors, cadmium sulfide (CdS), with a band gap of 2.3 eV, is probably the most attractive compound for water splitting. It combines upper and lower band gap edge positions, at appropriate energy levels to allow both the OER and HER reactions [11]. An attractive feature of metal sulfides as semiconductors for water splitting application is found in the possibility to implement band gap engineering to prepare solid solutions or alloys with fine control of defects, e. g. $(CuGa)_{1-x}Zn_{2x}S_2$ [12]. However, such potentiality for metal sulfide faces a limiting drawback, their instability related to the occurrence of photocorrosion under the operating conditions for the water

Electrochemical Water Splitting: Materials and Applications   Materials Research Forum LLC
Materials Research Foundations **59** (2019) 179-214   doi: https://doi.org/10.21741/9781644900451-8

splitting reaction [11,12]. The introduction of N atoms in metal sulfides and metal oxides based semiconductors is a route that has been evaluated to raise the upper edge of the valence band, with attractive results in terms of materials design but missing for a promising route leading to optimum materials for water splitting. In transition metal oxides the bonding properties of the oxygen atom and the metal, together with structural features within the formed solid, could be combined to produce semiconductor solids with strong light absorption in the visible spectral region. Such are the cases of Co_3O_4, α-Fe_2O_3 and Cu_2O [11,13]. However, the main drawback for their application in the water splitting reaction is related with their poor charge transport properties and defect concentration. In these materials, always an applied BIAS potential is required to promote the electrons transport from the semiconductor to the counter electrode (or vice versa) to perform the water splitting in the overall cell. For these reasons, several groups around the globe are currently conducting their research towards modifying the surface of these semiconductors with electrocatalysts, that decrease the energy barrier for oxygen evolution reaction (OER) or hydrogen evolution reaction (HER), thus minimizing or vanishing the BIAS potential needed for water splitting.

Another approach related to artificial photosynthesis systems concerns to the development of biomimetic structures, emulating the Mn_4CaO_5 cluster found in the natural photosystem II. This approach has been intensively studied by Nocera and coworkers [6] for years, through the preparation of a cobalt, oxygen, and phosphate nanostructure of still unknown structure, but with attractive performance. It seems that the cobalt cation is mainly responsible for the water oxidation reaction while the phosphate ion is contributing to maintain the catalyst stability.

For Prussian blue analogues (PBAs) and related cyanide-based coordination polymers band gaps in the range of 2.0 to 3.5 eV have been reported [14], and this value senses the energy difference or gap between highest occupied molecular orbital (HOMO) and lowest unoccupied molecular orbital (LUMO) of the involved metals. This characteristic depends on the corresponding 10Dq parameters and on the metal electronic configuration. To the best of our knowledge, the only exception corresponds to silver nitroprusside, $Ag_2[Fe(CN)_5NO]$, with a reported band gap of 1.65 eV, which is able to split the water molecule assisted by the application of a low potential [15]. In the following, we will summarize the reported applications of PBAs and other cyanide-based materials, with a previous discussion on their crystal structure, coordination chemistry, and related properties.

Electrochemical Water Splitting: Materials and Applications   Materials Research Forum LLC
Materials Research Foundations **59** (2019) 179-214   doi: https://doi.org/10.21741/9781644900451-8

2.   The coordination chemistry of Prussian blue analogues and other metal cyanides

The electronic structure of the cyanide ligand is relatively simple. Fig. 2 shows its molecular orbital energy level diagram [16]. The highest occupied molecular orbital (HOMO) has a σ character, while the lowest unoccupied molecular orbital (LUMO), which determines the CN acceptor features, has a π^* character. Since this orbital has an antibonding nature, the charge retro-donation process weakens the $CN^- \pi$ bond. The σ orbital where the unpaired electron resides is located at the N side, the most electronegative atom of the CN molecule. This is a strong ligand, which forms low spin complexes with transition metals at its C end, while the bond at the N atom has a more ionic character. The complexes have octahedral geometry for a maximum of six electrons in the metal nd orbitals, have square planar geometry for metals with eight electrons in these orbitals (Ni^{2+}, Pd^{2+} and Pt^{2+}), are tetrahedral for nd^{10} divalent metals (Zn^{2+}, Cd^{2+} and Hg^{2+}), and linear for monovalent nd^{10} metals (Cu^+, Ag^+ and Au^+) [17]. Other coordination geometries are also possible.

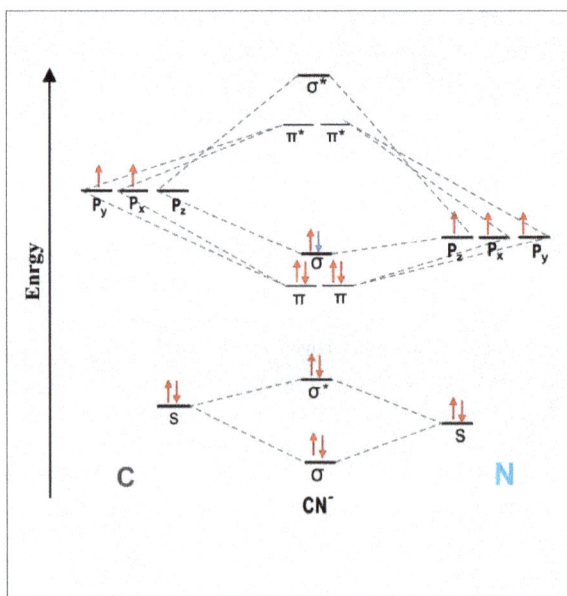

Fig. 2: *Diagram of energy levels for the molecular orbitals of the cyanide. Blue color signals the extra electron responsible for the anionic feature of the CN ligand.*

Fig. 3 illustrates the coordination of the CN ligand to form the octahedral complex that characterizes the PBAs. The ligand forms a strong covalent bond with the metal. During the complex formation, the π^* orbitals of the ligands and the t_{2g} orbitals of the metal result partially overlapped. This allows certain electron density retro-donation from the metal to the ligand. The retro-donated charge is found accumulated on the N end, the most electronegative atom of the CN group. When a second transition metal is linked at the N atom, that charge is partially donated to that metal. The net effect is an overlap of the electron clouds of the two metals (Fig. 3). This explains the attention received by PBAs as prototypes of molecular magnets [18].

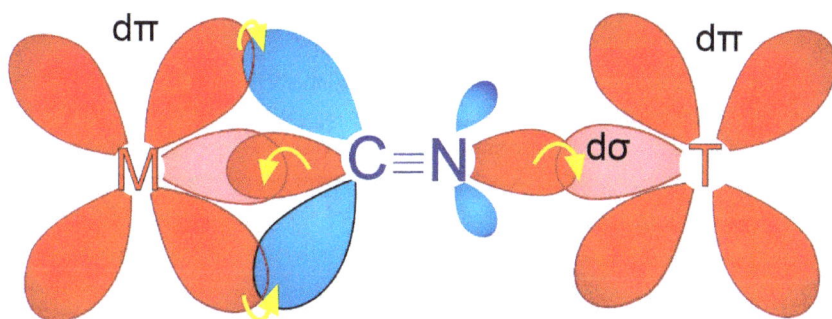

Fig. 3: Illustration for the charge donation processes detected for the hexacyanometallate ion formation and then when it is bonded, through the N atom of the CN ligand, to a second transition metal to form a coordination polymer.

In the classical Prussian blue, $Fe_4[Fe(CN)_6]_3 \cdot xH_2O$, the two metal centers are iron atoms, low spin Fe(II) for the inner metal, which is found coordinated to the C end, and high spin (3+) when the metal is bonded at the N end. In order to satisfy the charge neutrality, 25 % of the structural sites for the octahedral building block, $[Fe(CN)_6]$ remains vacant, resulting in a coordination polymer with a porous framework. The resulting free volume (cavity) has a diameter of ~9 Å and this volume is occupied by coordinated and hydrogen bonded water molecules. In these materials, the two metal centers are found with octahedral coordination. As PBAs are known those coordination polymers formed by the assembling of the hexacyanometallate anionic building block, $[M^{m+}(CN)_6]^{m-6}$, through a second transition metal cation (T^{n+}), tend to form an extended 3D network of ....$\equiv$N-T-N$\equiv$C-M-C$\equiv$N-.... chains. The formula unit for PBAs is $T_xA_z[M^m(CN)_6]_y$ where $nx + z = y(6-m)$, and A is an alkaline charge-compensating cation, that could be found occupying

Electrochemical Water Splitting: Materials and Applications   Materials Research Forum LLC
Materials Research Foundations **59** (2019) 179-214   doi: https://doi.org/10.21741/9781644900451-8

the interstitial structural voids related with the finite length for the CN ligand. The best sensor for the charge transfer involved in the coordination polymer formation is the frequency for the $\nu(CN)$ stretching vibration in the IR spectra [19]. As already mentioned, the charge removed from inner metal via π-back bonding is accumulated on the N end, particularly on the partially occupied σ (Fig. 2), which has a certain antibonding character for the C≡N bridge. On the coordination polymer formation, that charge is partially donated to the metal (T). And it is detected as an increase in the frequency for the $\nu(CN)$ band by about 40-80 cm^{-1}, depending on the metal polarizing power, which senses its ability to subtract electron density. The frequency shift is about ±14 cm^{-1} for a valence change in the outer metal (T) [18]. The frequency of the $\nu(CN)$ band also serves as a sensor for the coordination geometry for the T metal [18].

In the square-planar complexes, the ligand bonding interaction to the metal takes place through its x^2-y^2 orbital, the remaining nd orbitals (z^2, xy, xz, yz) are occupied by eight electrons. Similar to octahedral complexes, in this case, the inner metal is found in a diamagnetic state. Whereas for the tetrahedral and linear complexes, which are observed for nd^{10} metals, such coordination geometries suppose a sp hybridization of the metal.

An interesting series of CN-based coordination polymers, in the context of their application for artificial photosynthesis, is the one formed by transition metal salts of the pentacyanonitrosylferrate anion, $[Fe(CN)_5NO]^{2-}$. To form this building block, a CN ligand in the hexacyanoferrate (II) anion is replaced by a nitrosyl group. This last ligand remains bonded to the iron atom through the N atom but it is unable to form a coordination bond through the O end. From this fact, transition metal nitroprussides have a microporous framework where the microporosity depends on the orientation of the NO group in the resulting framework. This ligand has a large ability to subtract charge from the iron atom, through a low energy π^*-orbital via the metal xz and yz orbitals. The charge subtracted from the iron atom is located on the unbonded O end [20]. The large π^*-back donation toward the NO group reduces the electron density on the iron atom which is available to participate in an analogue effect via the CN ligands, at least compared with the observed effect for hexacyanometallates.

3.   Crystal structure of Prussian blue analogues and related coordination polymers

When in PBAs the two metals (T and M) are found with octahedral coordination, the solid crystallizes in a cubic unit cell, for which the cell edge corresponds to the T-N≡C-M-C≡N-T chain length (Fig. 4). The solid framework has an extended porosity when T:M > 1, related with the appearance of systematic vacancies for the building block, $[M(CN)6]$. For T:M = 3:2, a 33.3 % of the structural sites for the octahedral block remain vacant. In this case, the free space available in the solid depends on the accommodation mode for the vacancies, and it reaches 50 % of the framework volume when the vacancies remain ordered. It is also 50 % for T:M = 2:1. At the surface of a cavity generated by a vacancy, six partially naked metal (T) centers are found. In the as-synthesized material, these available coordination positions are occupied by water molecules, which in turn, stabilize weakly bonded water molecules inside the cavity through the formation of a network of hydrogen bonds (Fig. 4). This could be relevant for the application of this family of materials for the photo-activated water splitting, or as electrocatalyst in OER and HER.

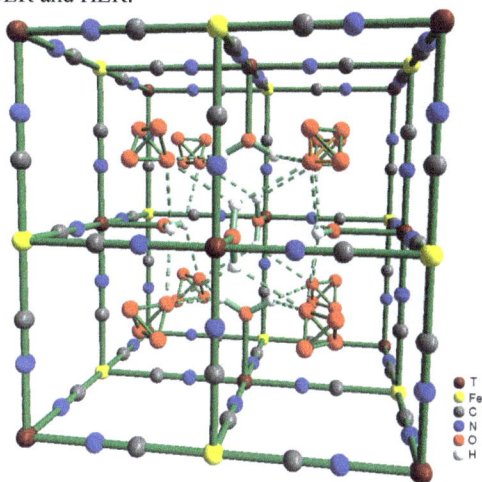

Fig. 4: Framework of porous PBAs. In the as-synthesized material, the free space resulting from vacancies for the building block, [M(CN)$_6$], are occupied by coordinated and hydrogen bonded water molecules.

When T = Co, Zn, the tetrahedral coordination for this metal is also possible, and the material crystallizes with a hexagonal unit cell. This structure also has an extended porosity but without water molecules coordinated to the metal, and probably it does not play a role as redox mediator coupled to semiconductor solids. For T$_3$[M(CN)$_6$]$_2$ with M = Fe, Co, Rh, Ir, anhydrous solids are obtained [21]. In metal dicyanides, T(CN)$_2$, with T = Zn, Cd and Hg, the metal also has saturated its coordination sphere with CN ligands and it is congruent with their anhydrous character.

Transition metal tetracyanometallates, $T[M(CN)_4] \cdot xH_2O$, with M = Ni, Pd, Pt, have a layered structure, of atomic thickness. In the as-synthesized materials, the axial coordination sites for the outer metal (T) are occupied by water molecules which, in turn, stabilize additional water molecules in the interlayer region by the formation of a network of hydrogen bonds (Fig. 5). Such a network of hydrogen bonding interaction supports the 3D framework for these materials [22]. Monovalent nd^{10} metal cyanides, MCN, with M = Cu, Ag, Au, have a linear structure, with the local disorder for the CN in the chain [17]. These metal cyanides form anhydrous solids, although the metal has four available coordination sites. It seems that the polarizing power for the metal in the chain is quite low to make possible the stabilization of water molecules in its equatorial coordination sites.

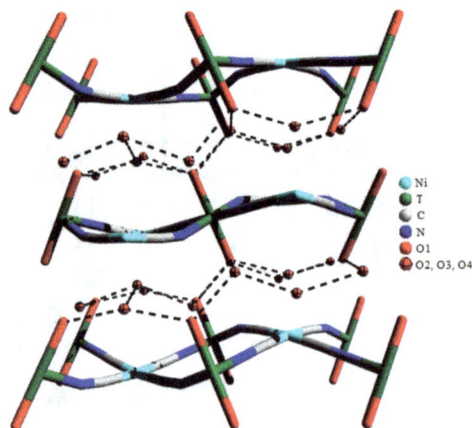

Fig. 5: *Layered structure for transition metal tetracyanometallates, $T[M(CN)_4] \cdot xH_2O$. Their 3D structure results from a network of hydrogen bonds interaction in the interlayer region.*

The other series considered herein, transition metal nitroprussides, $T[Fe(CN)_5NO] \cdot xH_2O$, with T = Mn, Fe, Co, Ni, Cu, are characterized by a wide structural diversity related with the orientation or accommodation mode for the NO group in the framework and the number of water molecules found in their porous framework [20,23]. For T = Fe, Co and Ni, the solid obtained in the precipitation reaction crystallize with the cubic unit cell typical of PBAs. In this structure, the NO group of six neighboring building blocks are oriented to a structural position corresponding to the metal (T), which remains vacant. In order to satisfy the charge neutrality, a site for the building block must also be unoccupied, generating a vacancy similar to that found in PBAs (Fig. 6). At the surface of such vacancy, six metals (T) with a partially naked coordination environment. In the as synthesized material, coordinated water molecules occupy these sites. The filling of the available free volume is completed by additional water molecules stabilized within the cavity through hydrogen bonding interaction with the coordinated ones. For Mn, Zn and Cd, the most stable phase crystallize as dehydrates with an orthorhombic unit cell, where a water molecule is found coordinated to the metal. For Cu, the as-synthesized powder is also a dehydrate but with a layered structure and two water molecules occupying the axial coordination sites for the copper atom [24]. The solid formed for silver, $Ag_2[Fe(CN)_5NO]$, results anhydrous, crystallizes with a monoclinic unit cell, and has two structural sites for the silver atom [25].

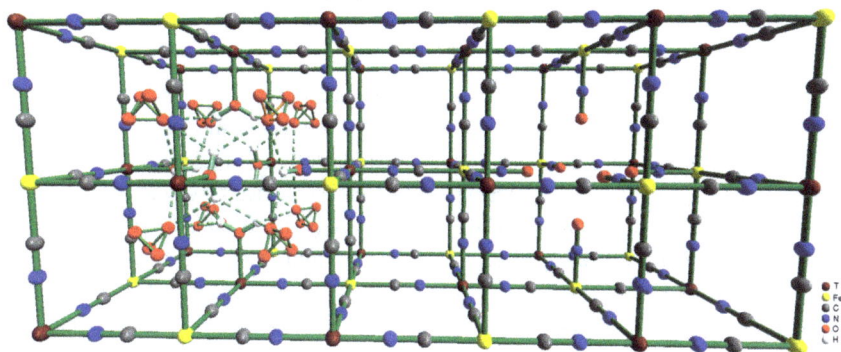

Fig. 6: *Atomic packing for cubic nitroprussides, $T[Fe(CN)_5NO] \cdot xH_2O$ with T = Fe, Co, Ni. Similar to PBAs, the largest cavities remain occupied by coordinated and hydrogen bonded water molecules.*

4. Photo-induced charge transfer in Prussian blue analogues and related solids

As already discussed, the materials considered herein are characterized by a strong overlapping of the electron cloud for the involved metals through the CN bridge. This reduces the height of the energy barrier for the charge transfer between M and T metals, and for some of them it is close to the photon energy (hv) in the UV-vis spectral region. In terms of semiconductors, this is equivalent to a low band gap energy. Such features for the electronic structure of PBAs and related solids have received a large attention from researchers involved in the study of photo-magnetism [25,26]. The possibility to induce magnetic order and magnetic transition using light opens attractive technological

Electrochemical Water Splitting: Materials and Applications   Materials Research Forum LLC
Materials Research Foundations **59** (2019) 179-214   doi: https://doi.org/10.21741/9781644900451-8

opportunities for fast information storage devices, in spintronic, and for advanced sensors and actuators. The potential applications of these photo-active properties of PBAs for sunlight energy harvesting remain practically ignored.

The deep blue color of PB results from the photo-induced charge transfer between Fe^{3+} and Fe^{II} metal centers through the CN bridge in $Fe_4[Fe(CN)_6]_3 \cdot xH_2O$, according to:

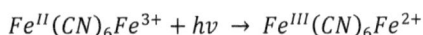

$$Fe^{II}(CN)_6Fe^{3+} + hv \rightarrow Fe^{III}(CN)_6Fe^{2+}$$

The charge transfer induced by the photon interaction with the ground state produces a ferrous hexacyanoferrate (III) species. Such an intense blue color of PB indicates that this transition is strongly allowed, with a life time typical of electronic transition, below 10^{-10} s. The excited state returns to the ground state in that short time through a combination of relaxation channels, including fluorescence.

The intense color and its change on illumination for many PBAs and related cyanide-based coordination polymers are visual evidence for the occurrence of photo-induced charge transfer effects in these materials. This is the case, for instance, of cobalt(2+) hexacyanoferrate (III). When this material is obtained by precipitation method, from aqueous solutions of a soluble Co(2+) salt and $K_3[Fe(CN)_6]$, the formed powder has a purple color, which on aging or after soft heating turns into intense grey-blue. This color change results from a spontaneous charge transfer from Co(2+) to Fe(III) to form a Co(III)Fe(II) species. This charge transfer process is accompanied by a spin transition from high spin Co(2+) to low spin Co(III) [18,26]. The resulting –Co(III)-N≡C-Fe(II)-C≡N-Co(III)-chain has diamagnetic character. The inverse transition, as a photo-induced process, has been intensively studied in the past two decades as a prototype of photo-induced magnets [26]. This transition is observed when the sample is irradiated with visible light (500 – 700 nm) while it is maintained at low temperature (< 15 K, the critical temperature, Tc). When the temperature increases, the thermal energy (kT) favors the reverse transition, and the magnetic order disappears. An analogue behavior has been reported for $(Fe_{0.40}Mn_{0.60})_3[Cr(CN)_6]_2 \cdot xH_2O$ [26] and $(Fe_{0.2}Cr_{0.8})_3[Cr(CN)_6]_2 \cdot xH_2O$ [18,26]. A common feature for all these PBAs is that they have a porous framework due to the presence of systematic vacancies for the octahedral building block, $[M(CN)_6]$. The metal valence changes on the photo-induced charge transfer to involve expansion and contraction for the interatomic metal-ligand distance, and such bond length variations must be locally absorbed by the material framework without requiring large structural change as a whole. In addition, the free volume generated by these vacancies remains occupied by coordinated and hydrogen bonded water molecules, and this could be

Electrochemical Water Splitting: Materials and Applications   Materials Research Forum LLC
Materials Research Foundations **59** (2019) 179-214   doi: https://doi.org/10.21741/9781644900451-8

relevant for a potential application of these materials for light induced water splitting process. The interstitial free space in the material structure, of about 4 Å diameter, is enough to facilitate the water molecule transport through the solid framework.

The photo-induced charge transfer has also been reported for other cyanide-based coordination polymers, among them, metal octacyanides: $Cu_2[Mo(CN)_8] \cdot 8H_2O$, $Co_3(pyrimidine)_4[W(CN)_8]_2 \cdot 6H_2O$, $Co_3(pyrimidine)_2(4\text{-}$ methylpyridine$)_2[W(CN)_8]_2 \cdot 6H_2O$, $Fe_2(4\text{-pyridinealdoxime})_8[Nb(CN)_8] \cdot 2H_2O$, and $Fe_2(4\text{-}$ bromopyridine$)_8[Nb(CN)_8](4\text{-bromopyridine}).2H_2O[27]$, and transition metal nitroprussides.

Photo-induced charge transfer in metal nitroprussides is related to the existence of photo-accessible metastable states on the NO ligand. For $Ni[Fe(CN)_5NO] \cdot xH_2O$, the photo- irradiation at 475 nm causes a charge transfer from the iron atom to the NO ligand, which results in the appearance of two paramagnetic centers on Fe and NO, with an antiferromagnetic coupling. The interaction between the spins on Ni and Fe atoms is ferromagnetic, resulting in a photo-induced ferrimagnetic order in the material. The photo-excited state persists on heating up to close 200 K [28]. Similar photo-induced transition is probably possible for other metals, but, to the best of our knowledge, it has not been reported. Within 3d metals, to Ni(2+) corresponds the higher polarizing power and the strongest coupling for the metal electron cloud through the CN ligand. This results in a relatively strong magnetic interaction between metal centers, which is observed at a relatively high temperature. This suggests the convenience of exploring the potentialities of this series of porous materials in photo-induced water splitting reactions.

5.   Electrochemical behavior of PBAs in aqueous solutions

PB analogues have fascinated the scientific community since the first report of electrochemical processes in iron hexacyanoferrate [29], due to its capacity to reversibly insert cations from solution, during the reduction of metal ions in the material, and extract them during oxidation of the same metal ions [30–33]. Charge-transfer of inner and external metals in PBAs are closely related to the above discussed features for the CN ligand. The redox processes in PBAs can be represented according to:

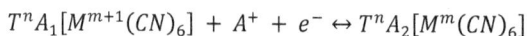

$$T^{n+1}[M^{m+1}(CN)_6] + A^+ + e^- \leftrightarrow T^n A_1[M^{m+1}(CN)_6]$$

$$T^n A_1[M^{m+1}(CN)_6] + A^+ + e^- \leftrightarrow T^n A_2[M^m(CN)_6]$$

Fig. 7 shows the expected cyclic voltammetry behavior for a generic PBA in which T and M metals, both have a reversible redox process. Normally, the redox processes of external transition metal (T) are observed at a lower potential than the one corresponding to the inner metal (M). The potential at which this process takes place and the extension of the electrochemical process depend on: the electrolyte conditions like nature of cation to be inserted and concentration, nature of both T and M metals, type of substrate in which PBA is supported and its processing procedure used to form the film. Actually, there are some reports that have observed the certain impact of the anions in solution on the obtained electrochemical results [34,35]. However, this topic remains controversial, since no influence of anions on these processes has been reported [36].

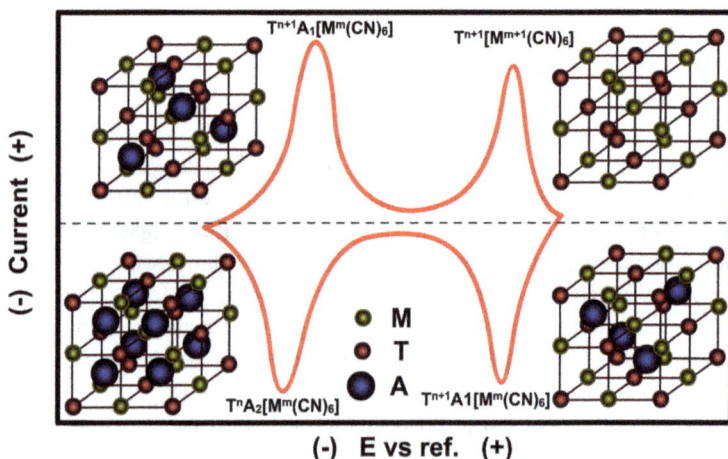

Fig. 7: *Cyclic voltammetry behavior expected for a PBA with two redox process related to outer (T) and inner (M) metals, inpresence of a cation (A) that could be inserted in (or removed from) tetrahedral interstitial structural spaces.*

The above-mentioned experimental parameters must be taken into consideration when employing PBAs and related solids for OER or HER since a proper control of these reactions allows tuning the generation of active catalytic sites in the material. Related to the electronic configuration of inner and external metal ions, some PBAs do not exhibit redox processes before the OER and HER reactions, however, intermediary species of high or low oxidation estate, respectively, of these metals are assumed to be responsible for catalytic behavior of PBAs, as will be discussed below.

6.   The water splitting reaction using transition metal cyanides

The mitigation of fossil fuel dependence by moving towards cleaner hydrogen based technologies is closely dependent on the progress of advantageous, green hydrogen generation technologies such as water splitting via electrolysis, photoelectrolysis or photocatalysis. In all these methodologies, there is a key piece in which PBAs and related solids could impact, and it is the development of catalytic materials made of earth abundant elements. Certain attention has been paid to cyanometallates due to the possibility of having a fine control of the amount and ratio of transition metals in the material, including two or more transition metal cations in the same compound. In addition, resulting coordination polymer could be designed to have an extended surface with a system of interconnected cavities available for the water molecules (discussed above). Particularly, CN is a very attractive ligand since it has a π-conjugated system that can effectively mobilize electrons to promote a steady charge separation through π-back donation.

The reports on the use of PBAs and related solids for HER and OER date since 1997. However, this was not until the past 5 years that more attention has been devoted to these materials for reactions involved in water splitting. Until now, the main applications of these materials in that area concern to: electrolysis cells, photo-assisted oxygen production and, as co-catalyst coupled with inorganic n-type semiconductor for OER.

A reliable comparison on the performance of PBAs reported in the literature, for water splitting reaction, appears to be complicated since several electrolytes have been employed, even those in which PBAs are not stable. A trivial change in electrolyte conditions may affect outcome in a substantial variation for the interface catalysis, particularly because inner redox processes in PBAs are highly dependent on the nature and concentration of ions in solution [34,35], affecting the generation of active sites (high oxidized metal centers) for the water oxidation reaction. Additionally, there have been established proper benchmarking to compare the behavior of the catalysts to OER reaction in acidic and basic media[37], but there are no widely accepted testing conditions for neutral electrolytes, in which PBAs are normally evaluated. Finally, several strategies and supports have been employed to form PBA films, which determine their final performance, and not all the authors report the geometric mass loading of the catalyst, for instance. Considering all these factors, a summary on the reported researches in the literature on PBA to perform both reactions, OER and HER, is complicated, but it is attempted below.

Electrochemical Water Splitting: Materials and Applications   Materials Research Forum LLC
Materials Research Foundations **59** (2019) 179-214   doi: https://doi.org/10.21741/9781644900451-8

6.1   Oxygen evolution reaction (OER)

The generation of molecular oxygen at the anode is a challenging process from both thermodynamic and kinetic points of view because this reaction needs the concerted transference of four electrons and four protons to occur. Thus, comprehending and improving the water oxidation process represents the most challenging step to overcome in order to develop an efficient sunlight harvesting process to produce chemical energy. PBAs appear to be viable water oxidation catalysts, as an alternative to the use of metal oxides or sulfides for promoting water oxidation reaction. These materials, which exhibit superior or equal kinetics, great stability in neutral and acidic media, are formed by abundant elements in the earth crust and can be simply obtained as films or powders in a wide range of sizes and forms [38,39].

PBAs were firstly reported as electrocatalysts for OER by Mascarós' research group [40]. These authors obtained a $K_{2x}Co_{(2-x)}[Fe(CN)_6]$ (0.85 $<x<$ 0.95) (CoFePBA) film through a two-step process: first, metallic cobalt is deposited onto FTO through electrodeposition, and then, the film is anodically dissolved in a $[Fe(CN)_6]^{3-}$ solution to obtain the CoFePBA. The final film was tested in a neutral pH towards OER, exhibiting long term stability and comparable performance to those reported for cobalt oxides. Authors also showed that CoFePBA film does not transform during OER, discarding the formation of other catalysts on the electrode surface. After the catalytic test, Co in PBA change its oxidation state from 2+ to 3+, however, no changes in the catalytic performance for the oxidized or reduced film was observed.

Another approach to obtain CoFePBA films onto FTO was reported by Bui *et al.* [41], and Han *et al.* [42]. For this approach, a film of 1D needles of cobalt hydroxycarbonate was firstly formed onto FTO by hydrothermal treatment; followed by an ion-exchange step in a solution of $[Fe(CN_6)]^{3-}$ at 60 °C. The obtained film exhibited exceptional high catalytic performance towards OER, and it was reported to have better characteristics than the one previously documented by Mascarós' group [40], with high stability in acidic solutions, at least compared the observed behavior for CoO_x. According to the authors, the material did not exhibit any structural changes up to a pH value of 13, where the formation of oxi-hydroxides is detected. Nonetheless, a change in the water oxidation mechanism was reported at pH values above 9. The increased performance of the film was ascribed to the presence of a larger amount of active sites (cobalt species) compared to the PBA film obtained from metallic cobalt electrodeposit [40]. It is worth pointing out that formed films by ion-exchange method exhibited considerably better performance for OER compared to films formed from powders. This result shows that the material processing to grow the film is a highly determinant factor for the overall performance of the obtained catalyst.

Karadas' group [43] has reported the possibility of improving the performance of CoFePBA by coordinating poly(4-vinylpyridine) to the iron atom in the molecular block, eliminating one axial cyanide group in the hexacyanoferrate. This was accomplished by synthesizing sodium pentacyanoaminoferrate, $Na_3[Fe(CN)_5NH_3] \cdot 3H_2O$, and then replacing labile NH_3 group by PVP in aqueous solutions. $CoFe(CN)_6$ and $Fe(CN)_5$-PVP films were directly deposited onto FTO by spreading an aliquot of $[Fe(CN)_6]^{4-}$ or $[Fe(CN)_5$-PVP], respectively, using a spin coater, followed by an immersion of the electrode in a $Co(NO_3)$ solution. The incorporation of PVP affected the redox process of $Co^{III/II}$ and $Fe^{III/II}$ centers in the coordination compound, shifting the potential for the redox reaction towards less positive values. Furthermore, catalytic currents for water oxidation were increased by the incorporation of PVP, which was mainly attributed to a 7-fold increment in the number of active Co sites with respect to $CoFe(CN)_6$@FTO film. The formation of composites of PBAs with other polymers has been reported by Zhang *et al.* [44]. This study evaluates the possibility of improving the electrochemical performance of $KCo[Fe(CN)_6] \cdot 3H_2O$ by growing a shell of PANI, via *in situ* oxidative chemical polymerization on PBA surface. In this case, the polymer is not directly coordinated to the iron atom in the molecular block. Instead it was used to improve the film conductivity to favor the charge transport towards the current collector. The electrocatalytic performance of powders was evaluated in a 1 M KOH electrolyte, using a film formed over a glassy carbon by drop casting. The formed composite exhibited much better performance than bare PBA or PANI, with a comparable behavior of that measured for IrO_2/C electrocatalyst. Formation of composites made of $Co_3[Fe(CN)_6]_2$ with Sb-doped SnO_2 (ATO) has also been recently reported [45]. In this composite, it is expected that (ATO) provides electronic conductivity, while $Co[Fe(CN)_6]$ would be responsible for catalytic activity ($Co^{III/II}$)towards OER. The interaction between the two materials was evidenced by a change in vibration signal of CN group, v(CN), towards lower frequency values (FTIR); and a displacement of the $Co^{III/II}$ and $Fe^{III/II}$ redox processes in the composite towards less positive potentials. Composite exhibited better performance than that observed for raw materials (ATO and PBA). It is worth mentioning that authors reported a negligent activity of PBA as water oxidation catalysts when it was supported on a Ti disk, evidencing the impact of substrate employed to characterize the performance of catalysts.

Alsaç et al [46] studied the electrocatalytic performance of several cobalt hexacyanometallates ($[Fe(CN)_6]^{4-}$; $[Fe(CN)_6]^{3-}$; $[Cr(CN)_6]^{3-}$; $[Co(CN)_6]^{3-}$) obtained by precipitation, aiming to understand the impact of the molecular block in the PBAs performance in the electrocatalytic reaction. The obtained materials were evaluated using a film of PBAs formed onto FTO by drop-casting. The activity of synthesized PBAs towards OER changed in the decreasing order as: Co^{II}-Co^{III}> Co^{II}-Cr^{III}> Co^{II}-Fe^{III}> Co^{II}-Fe^{II}. A direct correlation was observed between electrochemical performance and electronic density on the external cobalt (Co^{II}) atom in the PBAs, estimated from change in ν(CN) frequency, and in its binding energy. Such changes in the electronic properties, observed experimentally, were supported by DFT calculations. These results shed light on the impact of the chemical environment of Co-active sites over its catalytic performance in PBAs and related solids, and on the possibility of tuning the performance of electrocatalyst by altering the molecular block. As already discussed, cyanometallates offer a wide variety of molecular blocks, with different coordination features, that could be used to confer diverse characteristics to the external metal to perform the OER.

Indra *et al.* [47] employed the template method to obtain cobalt hexacyanocobaltate on a carbon cloth substrate. Authors showed that using cobalt hydroxycarbonate fiber as a template to obtain PBA by a chemical route, in a solution with the cobalt hexacyanide $[Co(CN)6]^{3-}$, results in the formation of films with better electrocatalytic performance than the one observed for PBAs without a template, or by classic precipitation method. Furthermore, spectroscopic characterization showed that Co^{II}-Co^{III} PBA is completely transformed into a double-layered hydroxide, after electrochemical test in 1 M KOH for OER. This hydroxide conserves the morphology of PBA (cubes) and maintains the mixed valence of cobalt. Actually, the *in-situ* formed hydroxide is more active than hydroxides obtained from other methods without using the template method.

Additionally, authors showed the possibility of obtaining hydroxides of Co and Ni by using as template a Co hydroxicarbonate, or a Ni hydroxicarbonate, and forming the corresponding cyanometallates from a solution of $[Ni(CN)_4]^{2-}$ or $[Co(CN)_6]^{3-}$, respectively. This combination ended up in a more active electrocatalyst, due to synergetic interaction between these two transition metals during OER. Several groups have used PBAs and related solids as templates to obtain new material to be employed as electrocatalyst in OER. There are numerous examples of several compounds (oxides, hydroxides, sulphides, selenides) obtained from PBAs particles as precursors [48–51].

While hexacyanometallates have received the most attention as water oxidation catalysts, the evaluation of other related materials (dicyanoamides, nitroprussides or tetracyanonickellates) used for this application is very limited or inexistent. All these molecular blocks would change the chemical environment of cobalt active sites, modulating its electronic properties and, in consequence, its electrocatalytic behavior. Nune *et al.* [52], reported the behavior of transition metal (Co, Ni, and Fe) dicyanomides formed by precipitation and supported onto FTO by drop casting. Synthesized materials exhibited better performance than previously reported PBAs by the same research group [46], and opens the possibility to increase the performance of the complex by incorporating other transition metal cation. Recently, Rahut *et al.* [53] have synthesized nanosheets of copper nitroprusside, $Cu[Fe(CN)_5NO]$, by conventional precipitation, using polyvinylpyrrolidone as shape-directing agent. The electrocatalytic performance of the formed powders was evaluated in a 0.1-2.0 M KOH electrolyte, using a film formed by spin coating over an FTO substrate. $Cu[Fe(CN)NO]$ nanosheets exhibited better performance than highly active RuO_2 electrocatalyst, displaying considerably large currents (100 mAcm^{-2}) at overpotentials lower than 600 mV. Contrary to the previous report for cyanometallates, where catalytic centers are attributed to external metal, commonly Co^{II}, in this study authors ascribed the high electrocatalytic performance of nitroprusside to the electrophilic character of nitrosyl group in the complex.

The electrocatalytic characteristics reported in the literature for the materials herein are summarized in Table 1. A drastic change in overpotentials for the indicated current density, as well as Tafel slopes, are observed when pH of electrolyte employed to evaluate the electrocatalyst moves towards higher values; mostly when highly concentrated KOH is employed. This phenomenon might be associated with the formation of highly active hydroxides on the material surface, or a total transformation of coordination compound, as have already been reported by some authors [42,47]. Furthermore, materials evaluated in neutral or acidic electrolytes exhibit very similar behavior, with some changes that could be attributed to the processing route (composition, a method for synthesis, substrate, etc.). Particularly, the very close Tafel slopes reported for all PBAs allow us to assume that all PBAs and related solids perform water oxidation following the same mechanism. In conventional oxide electrocatalyst, the Tafel value obtained for most of reported PBAs is associated with the first electron charge transfer as the rate determining step [54]. Nevertheless, little have been discussed regarding the mechanism for water oxidation using PBAs and related solids. Mascarós' group [42] proposed a five steps mechanism integrated by four simultaneous electron and proton transfers (Fig. 8). However, Karadas' group [13] found experimental evidence indicating that in cobalt hexacyanocobaltate, second electron transfer during OER does not involve a proton transfer, at least in a pH range from 6 to 11. It is proposed that proton transfer takes place in a succeeding step; steps II' and II'' in Fig. 8, respectively. Furthermore, this group proposes the nucleophilic attack of water in third electron-proton transfer, step III in Fig. 8, as the rate-determining step, since materials with lower electronic density in the Cu active sites in PBAs exhibited larger catalytic currents for OER. Computational chemistry might be very helpful to propose which are the most probable species formed on the catalytic center of PBAs, and probably could contribute in clearing up this controversy, however, there are no reports on this regard.

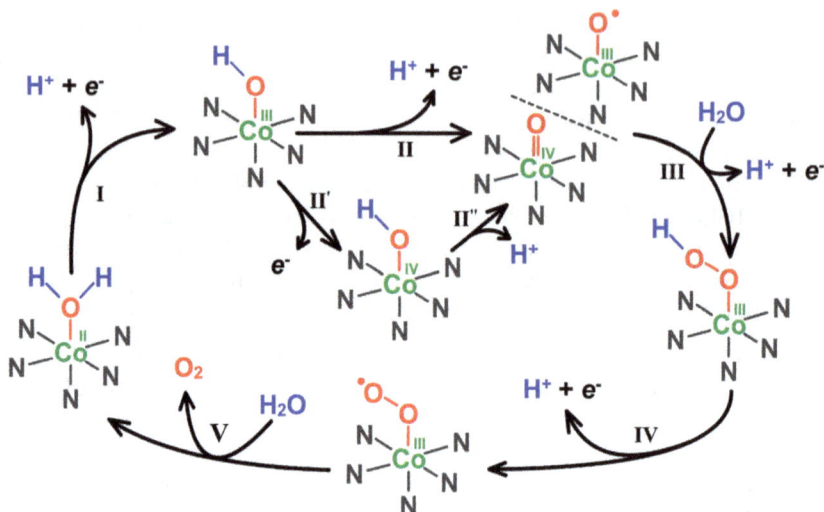

Fig. 8: *Summary of the steps involved in the water oxidation mechanism onto PBAs proposed in literature [42,46].*

Beyond the electrochemical evaluation of PBAs as a water oxidation catalyst, there are some reports that directly measure the catalytic performance of these materials using a photo assisted OER [55–57]. For this reaction, ruthenium bypiridine is employed to absorb visible photons ($E \geq 2.45$ eV), and persulfate is used as electron scavenger, leaving the hole in the HOMO of photosensitizer to be regenerated by catalysts, through water oxidation process Fig. 9. This methodology allows measuring the amount of oxygen generated in the time, which is an indirect measure of the catalytic performance of the PBA for OER.

Goberna-Ferrón *et al.* [55] used this procedure to evaluate the performance of Co, Cu, Mn, Ni, and Fe hexacyanoferrates and Fe, Co and Mn hexacyanocobaltates in that reaction. Catalytic performance was mainly detected when cobalt was present as external metal in PBA, being Co^{II}-Co^{III} coordination polymer the one with the highest activity. This result highlights the role that external Co^{II} plays during water oxidation. Fukuzumi's group [56,57] employed this methodology to evaluate the impact of partially substitution of $[Co(CN)_6]^{3-}$ by $[Pt(CN)_6]^{2-}$ to form $Co_n[Co_{1-x}Pt_x(CN)_6]$ [56], and the incorporation of Ca^{2+} in the interstices of $Co_{1.5}[Co(CN)_6]$[57]. In both cases, an improvement in catalytic performance was observed, which was attributed to electronic and structural changes induced by the heteropolynuclear cyanide complex, and the incorporation of redox-inactive metal ions as Lewis acids, respectively, which can tune the interaction of water with catalytically active sites during water oxidation. Additionally, as mentioned earlier in the text, changes in the molecular block and cations inserted in the interstitials could modulate the oxidation process of the external cobalt atom in the PBA.

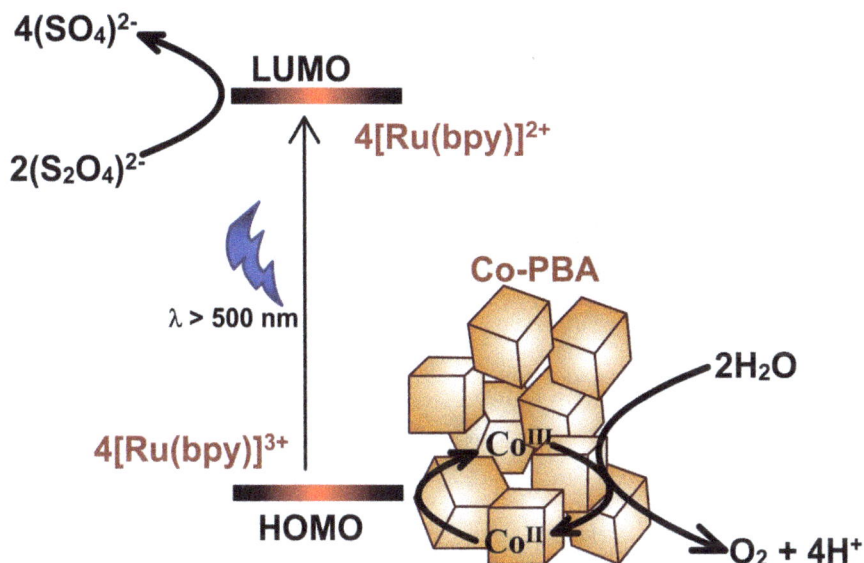

Fig. 9: *Schematic representation of the photoassisted oxygen generation employed to evaluate the catalytic performance of PBAs for OER. A PBA with Co*^{II} *as external metal is shown as an example, since this is the most employed in literature [55–57].*

Table 1: *Electrocatalytic characteristics reported for Prussian Blue Analogues and related solids, during OER in different electrolytes.*

Material	Electrolyte (pH)	Tafel slope (mV dec^{-1})	Overpotential (V)	Ref.
K$_{2x}$Co$_{(2-x)}$[Fe(CN)$_6$] (0.85 <x< 0.95)) / FTO	50 mM KPi (pH = 7.0)	85-95	--	[40]
Co$_3$[Fe(CN)$_6$]$_2$ / FTO	phosphate buffer, pH 7	--	0.57 @ 1 mAcm^{-2} 0.88 @ 10 mAcm^{-2}	[41]
	1 M KOH	--	0.42 @ 10 mAcm^{-2}	
Co$_4$(Fe(CN)$_6$)$_{2.67}$(H$_2$O)$_{15.33}$ / FTO	(pH = 1.0)	146		[42]
	(pH = 2.0)	108		
	(pH = 3.0)	95		
	(pH = 6.0)	85		
	0.1 M KPi (pH = 7.0) + 1M KNO$_3$	85	0.50 @ 1 mAcm^{-2} 0.80 @ 10 mAcm^{-2}	
	(pH = 8.0)	87		
	(pH = 9.0)	95		
	(pH = 10.0)	78		
	(pH = 11.0)	59		
	(pH = 12.0)	53		
Co[Fe(CN)$_6$] /FTO	50 mM KPi (pH = 7.0)	111	--	[43]
Co[Fe(CN)$_5$PVP] /FTO		121	0.51 @ 1 mAcm^{-2}	
KCo[Fe(CN)6] / GC†	1 M KOH	148	0.43 @ 10 mAcm^{-2}	[44]
KCo[Fe(CN)6]@PANI / GC		70	0.33 @ 10 mAcm^{-2}	
Co[Fe(CN)$_6$] / FTO	50 mM KPi (pH = 7.0)	120	1.08 @ 1 mAcm^{-2}	[46]
Co$_3$[Fe(CN)$_6$]$_2$ / FTO		130	0.72 @ 1 mAcm^{-2}	
Co$_3$[Cr(CN)$_6$]$_2$ / FTO		90	0.60 @ 1 mAcm^{-2}	
Co$_3$[Co(CN)$_6$]$_2$ / FTO		100	0.57 @ 1 mAcm^{-2}	
Co$_3$[Co(CN)$_6$]$_2$/ CC‡	1 M KOH	79	0.24 @ 10 mAcm^{-2}	[47]
Co[N(CN)$_2$]$_2$DMF / FTO	0.1 M KPi (pH = 7.0) + 1M KNO$_3$	94	0.58 @ 1 mAcm^{-2}	[52]
Co$_{0.9}$Ni$_{0.1}$[N(CN)$_2$]$_2$DMF /FTO		81	0.51 @ 1 mAcm^{-2}	
Cu[Fe(CN)$_5$NO] / FTO	1 M KOH	47	--	[53]

† Glassy Carbon; ‡ Carbon Cloth

6.2   Hydrogen evolution reaction (HER)

There are considerably lower efforts devoted to understand water reduction process over PBAs and related solids compared to OER, even though HER is a less exigent process, and that cyanometallates offer the possibility of synthesize solids made of abundant transition metals with well recognized catalytic behavior for this reaction. Additionally, this family of materials exhibits large stability in acidic electrolytes, facilitating the production of hydrogen by direct protons reduction.

One of the pioneers works in this field was carried out by Kaneko *et al.* [58]. Here, the behavior of Prussian White (PW) formed by electrodeposition at constant voltage (0.5 V vs Ag/AgCl) onto Pt towards HER in an acid electrolyte (pH 1.35) is reported. Results showed an increased performance compared with bare Pt, leading to hydrogen gas generation about 12 times as high as that measured for bare Pt electrode, at zero overpotential (-0.274 V vs Ag/AgCl in pH 1.35). The performance of PW modified Pt electrode was directly proportional to the PW amount deposited onto the electrode. Furthermore, the Faradaic efficiency to produce H_2 was larger than 80 %, and the material exhibited a decrease lower than 10% in its behavior, after one hour of constant electrolysis, proving its stability as a catalyst for this reaction. The same group also reported the behavior of ferric ruthenocyanide in proton reduction, evaluating the impact of the counter-ions (anions and cations) over electrochemical processes taking place in PBA [35]. Surprisingly, authors observed that catalytic hydrogen reduction does not take place in the presence of K^+, which was ascribed to larger mobility of this ion in the PBA compared to Na^+. The latter has to compete with H^+ to compensate the charge during the redox process of the iron ions in the PBA. This phenomenon was detected by cyclic voltammetry behavior of films in the presence of different cations; while in the presence of $K^{+,}$ only one process was detected. When Na^+ was added into the solution, it provoked the formation of two close current peaks observed during voltammetry characterization, related to the redox process of the external iron ($Fe^{II/III}$) in the PBA. This phenomenon is ascribed to the formation of a phase deficient in sodium ions and another phase rich in these ions found in the interstitial positions. The charges in redox reactions for deficient and rich in sodium forms are expected to be compensated by H^+ (probably partially) and Na^+, respectively. Anion employed for evaluation of ferric ruthenocyanide also affected the reduction of protons. When dihydrogen phosphate ions were employed, a high initial catalytic current and hydrogen evolution were registered. However, catalytic current and current peaks associated to deficient and rich in sodium forms decreased with time (cycles). Meanwhile, in the presence of chloride ions, the catalytic current started at low values and increased with time, together with a detriment mainly in currents associated to deficient in sodium form up to a constant current was reached for all the processes. From both studies, the authors suggested that cyanide ions behave as active sites for protons reduction presumably by coordinating H^+.

Despite the encouraging results obtained by Kaneko and colleagues [58], the electrochemical performance of PBAs has not been widely studied for HER, which left a gap in tailoring the PBA composition, morphology, and electrolyte for this reaction. Recently, Karadas and coworkers [59] took advantage of the encouraging catalytic activities of numerous cobalt-based systems and stability of PBAs in harsh catalytic processes to evaluate the performance of transition metal hexacyanocobaltates in HER. Zinc hexacyanocobaltate does not exhibit a catalytic behavior towards HER, pointing that cobalt hexacyanocobaltate activity is originated by the presence of Co^{II} sites surrounded with nitrogen atoms, and that Co^{III} centers in coordination compound do not contribute in the catalytic hydrogen generation process. Actually, post-catalysis characterization showed a partial reduction of external cobalt and an increase of potassium ions in PBA, without losing its structural features. A Tafel slope of 80.2 $mVdec^{-1}$ was estimated, associating the process to a Volmer-Heyrosky reaction mechanism. This result suggests the formation of metal-hydride intermediates followed by reaction of hydrides and protons, resulting in H_2 evolution. Furthermore, the cobalt hexacyanocobaltate exhibited a turnover frequency of 0.090 s^{-1} at an overpotential of 250 mV, which is larger than that reported for cobalt base compounds.

Zhang et al. [44] synthesized nanocubes of $KCo[Fe(CN)_6]$ with a shell of PANI and evaluated its performance on HER in a 1 M KOH electrolyte. The shell of PANI improved the performance of the CoPBA, changing the Tafel slope from 203 $mVdec^{-1}$ for bare PBA to 173 $mV\ dec^{-1}$ for PBA@PANI, and the overpotential at a current of 10 $mAcm^{-2}$, from 200 mV to 170 mV, respectively. However, Pt/C catalyst outperformed the synthesized composite. Shrestha and co-workers [60] have reported a paramount performance of a binder-free Ni hexacyanoferrate film by potentiostatic anodization of a Ni plate in a bath containing $K_3[Fe(CN)_6]$ towards HER. The on-set for hydrogen generation was of just -15 mV vs RHE, close to that for Pt/C, with an overpotential of -200 mV at a current of 10 $mAcm^{-2}$. The film exhibited great stability with almost no change in its electrochemical performance after 4.5 h at -200 mV vs RHE. However, in these two studies, electrochemical characterization of the film was carried out in a 1 M KOH electrolyte, in which the PBAs are reported to transform into hydroxides[42,47]. Experimental evidence of the chemical and structural stability of formed PBA film is not provided. Then, it is highly probable that PBA served as a precursor for in-situ formed Ni-Fe hydroxides, which are responsible for such paramount performance. In fact, different groups have employed PBAs as precursors to obtain FeP [61], FeNi-P [62] and $(Ni,Co)Se_2$[63] with remarkable performance in HER.

6.3   Use as co-catalyst in photoelectrochemical cells

The efficiency of photoelectrocatalytic (PEC) fuels generation by water splitting is restricted by large recombination of photogenerated charge carriers, large overpotentials of conventional semiconductors to perform OER or HER, and slow charge transfer process. The last two inconveniences can be surpassed by modifying the surface of the semiconductor with properly designed co-catalysts that diminishes the activation energy of the charge transfer process, as illustrated in Fig. 10. In this way, the overpotential needed to perform OER or HER will be diminished, and the kinetics of the faradaic reactions is boosted. This usually means a current vs potential curve with a high fill factor, which will help to assemble dual-photoelectrode systems [64]. In fact, incorporating electrocatalysts as co-catalysts has been broadly used to diminish the activation energy of conventional semiconductors to perform photocatalytic water reduction reaction [65,66]. Earlier, the use of cyanometallates as robust electrocatalyst for OER and HER was established. In this section, the recent progress on the employment of PBAs and related solids as co-catalysts for photoelectrocatalytic water oxidation process are summarized.

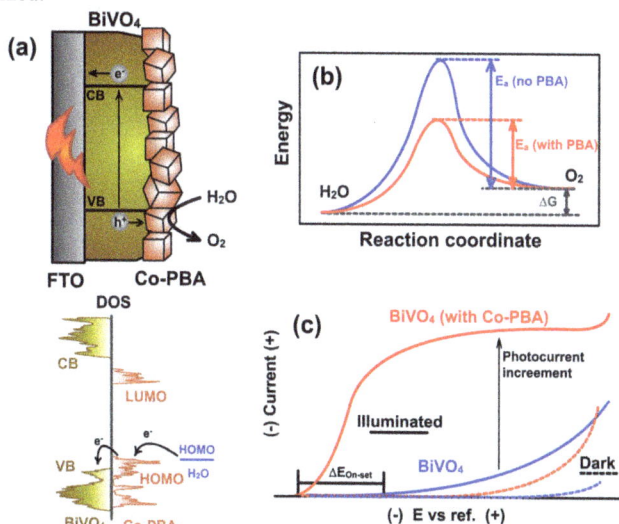

Fig. 10: Schematic representation of (a) photoanode modified with PBA; (b) the change in the activation energy for OER caused by the use of PBA as co-catalyst, and (c) current versus potential curves obtained under illumination and in the dark. BiVO$_4$ and a PBA with Co^{II} as external metal are illustrated as an example, since they are the most employed in literature [74,76].

Electrochemical Water Splitting: Materials and Applications   Materials Research Forum LLC
Materials Research Foundations **59** (2019) 179-214   doi: https://doi.org/10.21741/9781644900451-8

Several previous studies have shown the charge transfer process between semiconductors and PBAs under illumination [67–71]. In this sense, Tennakone *et al.* [72] showed that PBA film might trap holes photogenerated in CdS, and transfers them to I^- in solution, pointing the behavior of this material as a co-catalyst to improve a charge-transfer reaction on the surface of a semiconductor. However, the first use of a PBA as co-catalyst in the photoelectrochemical water oxidation reaction was reported in 2016, when Siuzdak *et al.* [73] synthesized TiO_2 nanotubes modified with $Fe_x^{III/II}[Fe^{III/II}(CN)_6]_y$ and pEDOT on top. The modification of TiO_2 nanotubes was performed by electropolimerization of EDOT in the presence of $[Fe(CN)6]^{3-/4-}$, followed by an anodic treatment in $FeCl_3$ to form an iron hexacyanoferrate. A change in the absorption edge of TiO_2 nanotubes was observed due to the π-π* transition in the polymer, as well as the metal-ligand charge transfer in the iron hexacyanoferrate. Authors observed a considerable increment in the photocurrent of TiO_2 nanotubes film when this semiconductor was modified with the polymer and the iron hexacyanoferrate in up to 4.7 folds. This behavior was ascribed to a combined sensitization and catalytic behavior of pEDOT-iron hexacyanoferrate on top of TiO_2, where electrons photo-excited to LUMO of the polymer are transferred to the conduction band of TiO_2 and holes in TiO_2 move towards the iron hexacyanoferrate, to perform the OER. Authors highlighted the importance of strong electronic coupling between the TiO_2 and the iron hexacyanoferrate through Ti-C≡N-Fe.

Hegner *et al.* [74] modified a $BiVO_4$ photoanode, which possesses sluggish kinetics for water oxidation reaction, with a film of $Co_x[Fe(CN)_6]_y$. CoFePBA film was formed by successive immersions (four to eight times) of $BiVO_4$ in $K_3Fe[(CN)_6]$ and $CoCl_2$ solutions, successively. A considerable increase for the measured photocurrent, and a displacement of the on-set potential were observed for $BiVO_4$ film modified with CoFePBA, compared to bare $BiVO_4$. This behavior was mainly attributed to the catalytic activity of CoFePBA towards water oxidation reaction, since displacement in the absorption edge of modified semiconductor was not observed from IPCE measurements, indicating that photons are mainly absorbed by $BiVO_4$ and interfacial characteristics improvement is provided by the presence of CoFePBA. Electrochemical impedance spectroscopy measurements evidenced a drastic diminution of charge transfer resistance when $BiVO_4$ was modified with CoFePBA, supporting the catalytic behavior of PBA on top of $BiVO_4$ film.

Furthermore, the modified film exhibited excellent photocurrent stability under illumination and a faradaic efficiency towards oxygen generation of around 95 %. The paramount behavior of modified films was also supported by the alignment of energetic states obtained from DFT calculation. It was found that holes in the valence band of $BiVO_4$ (O2s band) can be transferred to HOMO level in CoFePBA mainly composed by

Electrochemical Water Splitting: Materials and Applications   Materials Research Forum LLC
Materials Research Foundations 59 (2019) 179-214   doi: https://doi.org/10.21741/9781644900451-8

Co t2g orbitals. But authors warn about the 0.3 eV mismatch between the CoFePBA HOMO level and water HOMO level, indicating that it is necessary to design catalysts with active sites that lie closer to the HOMO level of water. The same authors modified hematite photo-anode with CoFe-PB catalyst following a similar procedure[75]. However, just a small improvement of photo-generated currents was observed. This behavior was ascribed to an inadequate level alignment between the hematite valence band and the CoFePBA HOMO. Since the later resides slightly below the first one, an overpotential is needed to favor hole transfer towards co-catalyst.

More recently, Trzciński et al. [76] reported the synthesis of $BiVO_4$ photoanodes modified with $Co_3[Co(CN)_6]_2$. The formation of the co-catalyst layer over photoanode was carried out by deposition of a metallic film of cobalt on top of the $BiVO_4$, followed by electrochemical oxidation of cobalt in a solution containing $[Co(CN)_6]^{3-}$ to obtain cobalt hexacyanocobaltate inform of regular cubic structures. Optical properties of $BiVO_4$ were not modified by the presence of $Co_3[Co(CN)_6]_2$ film, which was attributed to the small amount of the co-catalyst compared to that of the $BiVO_4$. A considerable increase for photogenerated currents and on-set potential for a current generation were observed when $BiVO_4$ was modified with $Co_3[Co(CN)_6]_2$ film, due to the catalytic behavior of the PBA. A charge extraction of around 60% was estimated for $BiVO_4/Co_3[Co(CN)_6]_2$, showing the potential of increasing the performance of the heterojunction (photoanode/co-catalyst) by optimizing the electrocatalytic film growth.

These results are triggering the development of highly efficient photoanodes modified with PBA, for photoelectrocatalytic OER in neutral electrolytes. Efforts need to be driven to the engineering of semiconductor/co-catalyst interface, by optimizing the deposition method, focusing in minimizing defects in the interface that might affect the charge carrier transport between both materials under illumination. Additionally, as it was already observed in the previous text, the processing of the PBA and its composition need to be tuned in order to obtain films with better electrocatalytic performance. Furthermore, it has been proven the stability of PBA in harshly oxidation conditions generated under OER, but there is scarce information regarding the stability of these materials during the photo-electrochemical water oxidation process since light might induce a charge transfer process in the considered metals.

The use of PBA in photoelectrocatalytic water oxidation is not only limited as co-catalyst. Rahut et al. [15], recently reported the synthesis and characterization of $Ag[Fe(CN)_5NO]$, which showed the possibility to generate a photocurrent of up to 1 $mAcm^{-2}$ when illuminated with a Xe lamp (300 W). However, the stability of the film after the illumination process has not been discussed. Actually, the measurement was performed in (0.1-2 M) KOH solution. In this solution, PBA decomposition into oxi-hydroxides is expected [42,47]. Nevertheless, the low band gap of this compound, of around 1.6 eV, shows this nitroprusside as a promising material for applications in photoelectrochemistry. Further research regarding the processing of the film and its evaluation are needed.

Concluding remarks

The design of an optimal electrocatalysts for OER or HER using PBAs and related solids is far from fully realized. The impact of several strategies to modulate the characteristics of these solids (morphology, structure, surface, number, and types of defects, among others), over the water oxidation process, need to be understood. Certainly, computational tools would help the *in silico* design of the catalyst with the most desired properties for this reaction, but there is no research in the literature devoted to this purpose. From the above discussed results, it is evident that cobalt(II) hexacyanoferrate(III) has been the most studied PBAs for the OER and HER reactions. This composition has two redox pairs with a low energy barrier height for the charge transfer between them. It would seem a coincidence, but this PBA is also the most studied composition as a prototype of a photo-induced molecular magnet [25]. Many other PBAs show an analogue magnetic behavior when they are irradiated with visible light and could be candidates to be evaluated for the water splitting reaction. Likewise, other transition metal nitroprussides could show a behavior similar to the one reported for silver nitroprusside, but such a possibility has not been explored. Transition metal tetracyanometallates have a layered structure, with layers of atomic thickness, which could be a favorable feature for their deposition as thin films on electrodes. This is another area of opportunity to be explored.

Acknowledgments

The preparation of this chapter was partially supported by the CONACyT (Mexico) Projects 2013-05-231461 and CB-2014-01-235840. The authors thank to LNCAE (Laboratorio Nacional de Conversión y Almacenamiento de Energía) the access to structural and electrocatalytic information on Prussian blue analogues and related transition metal cyanides.

References

[1]   U.N. Deparment of Economics and Social Affaris, 2018 Energy statistics pocketbook, United Nations Publication, New York, 2018.

[2]   J. Barber, P.D. Tran, J. Barber, From natural to artificial photosynthesis, J. R. Soc. Interface. 10 (2013). https://doi.org/10.1098/rsif.2012.0984

[3]   A.F. Collings, C. Critchley, Artificial photosynthesis, 1st ed., Wiley-VCH, 2005. https://doi.org/10.1002/3527606742

[4]   J. Kern, G. Renger, Photosystem II : Structure and mechanism of the water : plastoquinone oxidoreductase, Photosynth. Res. 94 (2007) 183–202. https://doi.org/10.1007/s11120-007-9201-1

[5]   S.J.A. Moniz, S.A. Shevlin, D.J. Martin, Z.X. Guo, J. Tang, Visible-light driven heterojunction photocatalysts for water splitting - A critical review, Energy Enviromen. Sci. 8 (2015) 731–759. https://doi.org/10.1039/C4EE03271C

[6]   D.G. Nocera, The artificial leaf, Acc. Chem. Res. 45 (2012) 767–776. https://doi.org/10.1021/ar2003013

[7]   D.R. Whang, D. Hazar, Artificial photosynthesis : Learning from nature, Chem. Photo. Chem. 2 (2018) 148–160. https://doi.org/10.1002/cptc.201700163

[8]   N.T. Suen, S.F. Hung, Q. Quan, N. Zhang, Y.J. Xu, H.M. Chen, Electrocatalysis for the oxygen evolution reaction: recent development and future perspectives, Chem. Soc. Rev. 46 (2017) 337–365. https://doi.org/10.1039/C6CS00328A

[9]   A. Fujishima, K. Honda, Electrochemical photolysis of water at a semiconductor electrode, Nature. 238 (1972) 37–38. https://doi.org/10.1038/238037a0

[10]   K. Maeda, K. Domen, New non-oxide photocatalysts designed for overall water splitting under visible light, J. Phys. Chem. B. 111 (2007) 7851–7861. https://doi.org/10.1021/jp070911w

[11]   A. Kudo, Y. Miseki, Heterogeneous photocatalyst materials for water splitting, Chem. Soc. Rev. 38 (2009) 253–278. https://doi.org/10.1039/B800489G

[12]   T. Kato, Y. Hakari, S. Ikeda, Q. Jia, A. Iwase, A. Kudo, Utilization of metal sulfide material of (cuga)1-xzn2xs2 solid solution with visible light response in photocatalytic and photoelectrochemical solar water splitting systems, J. Phys. Chem. Lett. 6 (2015) 1042–1047. https://doi.org/10.1021/acs.jpclett.5b00137

[13]   F. Le Formal, K. Sivula, M. Gra, The transient photocurrent and photovoltage behavior of a hematite photoanode under working conditions and the in fl uence of surface treatments, J. Phys. Chem. C. 116 (2012) 26707–26720. https://doi.org/10.1021/jp308591k

[14]   L. Reguera, N.L. López, J. Rodríguez-hernández, N.H. De Leeuw, E. Reguera, Synthesis , crystal structures , and properties of zeolite-like T3(H3O)2[M(CN)6]2·uH2O(T=Co, Zn ; M = Ru , Os), Eur. J. Inorg. Chem. 3 (2017) 2980–2989. https://doi.org/10.1002/ejic.201700278

[15]   S. Rahut, A. Bharti, J.K. Basu, Optical and electronic configuration of a novel semiconductor-silver nitroprusside for enhanced electrocatalytic and photocatalytic performance, Catal. Sci. Technol. 7 (2017) 6092–6100. https://doi.org/10.1039/C7CY01940H

[16]   C.J. Ballhausen, Introduction to ligand field theory, McGraw Hill, New-York, 1962.

[17]   A.M. Chippindale, S.J. Hibble, E.J. Bilbe, E. Marelli, A.C. Hannon, Mixed copper, silver, and gold cyanides, (MxM'1-x)CN: Tailoring chain structures to influence physical properties, J. Am. Chem. Soc. 134 (2012) 16387–16400. https://doi.org/10.1021/ja307087d

[18]   R. Martínez-Garcia, M. Knobel, E. Reguera, Thermal-induced changes in molecular magnets based on prussian blue analogues, J. Phys. Chem. B. 110 (2006) 7296–7303. https://doi.org/10.1021/jp0555551

[19]   K. Nakamoto, Infrared spectra of inorganic and coordination compounds, 4th ed., Wiley, 1986.

[20]   A. Cano, L. Lartundo-rojas, A. Shchukarev, E. Reguera, Contribution to the coordination chemistry of transition metal nitroprussides : A cryo-XPS study, New J. Chem. 43 (2019) 4835-4848. https://doi.org/10.1039/C9NJ00141G

[21]   J. Rodríguez-Hernández, E. Reguera, E. Lima, J. Balmaseda, R. Martínez-García, H. Yee-Madeira, An atypical coordination in hexacyanometallates : Structure and properties of hexagonal zinc phases, J. Phys. Chem. Solids. 68 (2007) 1630–1642. https://doi.org/10.1016/j.jpcs.2007.03.054

[22]   J. Rodríguez-Hernández, A.A. Lemus-Santana, C.N. Vargas, E. Reguera, Three structural modifications in the series of layered solids T(H2O)2[Ni(CN)4]·xH2O with T = Mn, Co, Ni: Their nature and crystal structures, Comptes Rendus Chim. 15 (2012) 350–355. https://doi.org/10.1016/j.crci.2011.11.004

[23]   E. Reguera, A. Dago, A. Gómez, J.F. Bertrán, Structural changes in insoluble metal nitroprussides on ageing, polyhedron, 15 (1996) 3139–3145. https://doi.org/10.1016/0277-5387(95)00582-X

[24] E. Reguera, Unique coordination in metal nitroprussides : The structure of Cu[Fe(CN)5NO]·2H2O and Cu[Fe(CN)5NO], J. Chem. Crystallogr. 34 (2004). https://doi.org/10.1007/s10870-004-7724-2

[25] J. Rodríguez-hernández, L. Reguera, A.A. Lemus-santana, E. Reguera, Silver nitroprusside : Atypical coordination within the metal nitroprussides series, Inorganica Chim. Acta. 428 (2015) 51–56. https://doi.org/10.1016/j.ica.2014.12.023

[26] O. Sato, Y. Einaga, A. Fujishima, K. Hashimoto, Photoinduced long-range magnetic ordering of a cobalt - iron cyanide, Inorg. Chem. 38 (1999) 4405–4412. https://doi.org/10.1021/ic980741p

[27] N. Ozaki, H. Tokoro, Y. Hamada, A. Namai, T. Matsuda, Photoinduced magnetization with a high curie temperature and a large coercive field in a Co-W bimetallic assembly, Adv. Funct. Mater. 22 (2012) 2089–2093. https://doi.org/10.1002/adfm.201102727

[28] Z. Gu, O. Sato, T. Iyoda, K. Hashimoto, A. Fujishima, Spin switching effect in nickel nitroprusside : design of a molecular spin device based on spin exchange interaction, Chem. Mater. 4756 (1997) 1092–1097. https://doi.org/10.1021/cm9606383

[29] V.D. Neff, Electrochemical oxidation and reduction of thin films of prussian blue, J. Electrochem. Soc. 125 (1978) 886–887. https://doi.org/10.1149/1.2131575

[30] D. Ellis, M. Eckhoff, V.D. Neff, Electrochromism in the mixed-valence hexacyanides. 1. Voltammetric and spectral studies of the oxidation and reduction of thin films of Prussian blue, J. Phys. Chem. 85 (1981) 1225–1231. https://doi.org/10.1021/j150609a026

[31] K. Itaya, I. Uchida, V.D. Neff, Electrochemistry of polynuclear transition metal cyanides: prussian blue and its analogues a p a n structure and properties of the transition metal hexacyanides, Acc. Chem. Res. 19 (1986) 162–168. https://doi.org/10.1021/ar00126a001

[32] C.A. Lundgren, R.W. Murray, Observations on the composition of prussian blue films and their electrochemistry, Inorg. Chem. 27 (1988) 933–939. https://doi.org/10.1021/ic00278a036

[33] P.J. Kulesza, M. a Malik, M. Berrettoni, M. Giorgetti, S. Zamponi, R. Schmidt, R. Marassi, Electrochemical charging, countercation accommodation , and spectrochemical identity of microcrystalline solid cobalt hexacyanoferrate, J. Phys. Chem. B. 5647 (1998) 1870–1876. https://doi.org/10.1021/jp9726495

[34]   A. Roig, R. Navarro, R. Tamarit, F. Vicente, Stability of Prussian Blue films on ito electrodes: Effect of different anions, J. Electroanal. Chem. 360 (1993) 55–69. https://doi.org/10.1016/0022-0728(93)87004-F

[35]   T. Abe, G. Toda, A. Tajiri, M. Kaneko, Electrochemistry of ferric ruthenocyanide (ruthenium purple), and its electrocatalysis for proton reduction, J. Electroanal. Chem. 510 (2001) 35–42. https://doi.org/10.1016/S0022-0728(01)00539-3

[36]   B.J. Feldman, O.R. Merloy, Ion Flux During Electrochemical Charging of Prussian Blue Films, J. Electroanal. Chem. 234 (1987) 213–227. https://doi.org/10.1016/0022-0728(87)80173-0

[37]   C.C.L. McCrory, S. Jung, J.C. Peters, T.F. Jaramillo, Benchmarking heterogeneous electrocatalysts for the oxygen evolution reaction, J. Am. Chem. Soc. 135 (2013) 16977–16987. https://doi.org/10.1021/ja407115p

[38]   J.R. Galán-Mascarós, Water oxidation at electrodes modified with earth-abundant transition-metal catalysts, Chem. Electro. Chem. 2 (2015) 37–50. https://doi.org/10.1002/celc.201402268

[39]   L. Catala, T. Mallah, Nanoparticles of Prussian blue analogs and related coordination polymers: From information storage to biomedical applications, Coord. Chem. Rev. 346 (2017) 32–61. https://doi.org/10.1016/j.ccr.2017.04.005

[40]   S. Pintado, S. Goberna-Ferrón, E.C. Escudero-Adán, J.R. Galán-Mascarós, Fast and persistent electrocatalytic water oxidation by Co-Fe Prussian blue coordination polymers, J. Am. Chem. Soc. 135 (2013) 13270–13273. https://doi.org/10.1021/ja406242y

[41]   H.T. Bui, D.Y. Ahn, N.K. Shrestha, M.M. Sung, J.K. Lee, S. Han, Self-assembly of cobalt hexacyanoferrate crystals in 1-D array using ion exchange transformation alkaline and neutral water, J. Mater. Chem. (2016) 9781–9788. https://doi.org/10.1039/C6TA03436E

[42]   L. Han, P. Tang, A. Reyes-carmona, J.R. Morante, J. Arbiol, J.R. Galan-mascaros, Enhanced Activity and Acid pH Stability of prussian blue-type oxygen evolution electrocatalysts processed by chemical etching, J. Am. Chem. Soc. 138 (2016) 16037–16045. https://doi.org/10.1021/jacs.6b09778

[43]   M. Aksoy, S.V.K. Nune, F. Karadas, A novel synthetic route for the preparation of an amorphous co/fe prussian blue coordination compound with high electrocatalytic water oxidation activity, Inorg. Chem. 55 (2016) 4301–4307. https://doi.org/10.1021/acs.inorgchem.6b00032

[44]   L. Zhang, T. Meng, B. Mao, D. Guo, J. Qin, M. Cao, Multifunctional Prussian blue analogous@polyaniline core–shell nanocubes for lithium storage and overall water splitting, RSC Adv. 7 (2017) 50812–50821. https://doi.org/10.1039/C7RA10292E

[45]   B. Rodríguez-García, Á. Reyes-Carmona, I. Jiménez-Morales, M. Blasco-Ahicart, S. Cavaliere, M. Dupont, D. Jones, J. Rozière, J.R. Galán-Mascarós, F. Jaouen, Cobalt hexacyanoferrate supported on Sb-doped SnO_2 as a non-noble catalyst for oxygen evolution in acidic medium, Sustain. Energy Fuels. (2018). https://doi.org/10.1039/C7SE00512A

[46]   E.P. Alsaç, E. Ülker, S.V.K. Nune, Y. Dede, F. Karadas, Tuning electronic properties of prussian blue analogues for efficient water oxidation electrocatalysis: experimental and computational studies, Chem. A Eur. J. (2017) 1–22.

[47]   A. Indra, U. Paik, T. Song, Boosting electrochemical water oxidation with metal hydroxide carbonate templated prussian blue analogues, Angew. Chemie - Int. Ed. 57 (2018) 1241–1245. https://doi.org/10.1002/anie.201710809

[48]   J. Su, G. Xia, R. Li, Y. Yang, J. Chen, R. Shi, P. Jiang, Q. Chen, Co3ZnC/Co nano heterojunctions encapsulated in N-doped graphene layers derived from PBAs as highly efficient bi-functional OER and ORR electrocatalysts, J. Mater. Chem. A. 4 (2016) 9204–9212. https://doi.org/10.1039/C6TA00945J

[49]   B.K. Kang, M.H. Woo, J. Lee, Y.H. Song, W. Zhongli, Y. Guo, Y. Yamauchi, J.H. Kim, B. Lim, D.H. Yoon, Mesoporous Ni–Fe oxide multi-composite hollow nanocages for efficient electrocatalytic water oxidation reactions, J. Mater. Chem. A. 5 (2017) 4320–4324. https://doi.org/10.1039/C6TA10094E

[50]   J. Nai, Y. Lu, L. Yu, X. Wang, X. Wen, D. Lou, Formation of Ni–Fe mixed diselenide nanocages as a superior oxygen evolution electrocatalyst, Adv. Mater. 29 (2017) 1703870. https://doi.org/10.1002/adma.201703870

[51]   P. Cai, J. Huang, J. Chen, Z. Wen, Oxygen-incorporated amorphous cobalt sulfide porous nanocubes as high-activity electrocatalysts for the oxygen evolution reaction in an alkaline / neutral medium zuschriften angewandte, Angewante Chemie. 129 (2017) 4936–4939. https://doi.org/10.1002/ange.201701280

[52]   S.V.K. Nune, A.T. Basaran, E. Ülker, R. Mishra, F. Karadas, Metal dicyanamides as efficient and robust water-oxidation catalysts, Chem. Cat. Chem. 9 (2017) 300–307. https://doi.org/10.1002/cctc.201600976

[53]   S. Rahut, S.K. Patra, J.K. Basu, Surfactant assisted self assembly of novel ultrathin Cu[Fe(CN)5NO] nanosheets for enhanced electrocatalytic oxygen evolution: Effect of nanosheet thickness, Electrochim. Acta. 265 (2018) 202–208. https://doi.org/10.1016/j.electacta.2018.01.152

[54]   R.L. Doyle, M.E.G. Lyons, The oxygen evolution reaction : Mechanistic concepts and catalyst design, in: S. Giménez, J. Bisquert (Eds.), Photoelectrichemical Sol. Fuel Prod. from Basic Princ. to Adv. Devices, Springer International Publishing, 2016: pp. 41–104. https://doi.org/10.1007/978-3-319-29641-8_2

[55]   S. Goberna-Ferrón, W.Y. Hernández, B. Rodríguez-García, J.R. Galán-Mascarós, Light-driven water oxidation with metal hexacyanometallate heterogeneous catalysts, ACS Catal. 4 (2014) 1637–1641. https://doi.org/10.1021/cs500298e

[56]   Y. Yamada, K. Oyama, R. Gates, S. Fukuzumi, High catalytic activity of heteropolynuclear cyanide complexes containing cobalt and platinum ions: Visible-light driven water oxidation, Angew. Chemie - Int. Ed. 54 (2015) 5613–5617. https://doi.org/10.1002/anie.201501116

[57]   Y. Yamada, K. Oyama, T. Suenobu, S. Fukuzumi, Photocatalytic water oxidation by persulphate with a Ca2+ ion-incorporated polymeric cobalt cyanide complex affording O2 with 200% quantum efficiency, Chem. Commun. 53 (2017) 3418–3421. https://doi.org/10.1039/C7CC00199A

[58]   T. Abe, F. Taguchi, S. Tokita, M. Kaneko, Prussian White as a highly active molecular catalyst for proton reduction, J. Mol. Catal. A Chem. 126 (1997) 89–92. https://doi.org/10.1016/S1381-1169(97)00156-8

[59]   E.P. Alsaç, E. Ulker, S.V.K. Nune, F. Karadas, A cyanide-based coordination polymer for hydrogen evolution electrocatalysis, Catal. Letters. 148 (2018) 531–538. https://doi.org/10.1007/s10562-017-2271-6

[60]   H.T. Bui, N.K. Shrestha, S. Khadtare, C.D. Bathula, L. Giebeler, Y.Y. Noh, S.H. Han, Anodically grown binder-free nickel hexacyanoferrate film: toward efficient water reduction and hexacyanoferrate film based full device for overall water splitting, ACS Appl. Mater. Interfaces. 9 (2017) 18015–18021. https://doi.org/10.1021/acsami.7b05588

[61]   N.K.A. Venugopal, S. Yin, Y. Li, H. Xue, Y. Xu, X. Li, H. Wang, L. Wang, Prussian Blue-derived iron phosphide nanoparticles in a porous graphene aerogel as efficient electrocatalyst for hydrogen evolution reaction, Chem. - An Asian J. 13 (2018) 679–685. https://doi.org/10.1002/asia.201701616

[62]   A.V. Narendra Kumar, Y. Li, H. Yu, S. Yin, H. Xue, Y. Xu, X. Li, H. Wang, L. Wang, 3D graphene aerogel supported FeNi-P derived from electroactive nickel hexacyanoferrate as efficient oxygen evolution catalyst, Electrochim. Acta. 292 (2018) 107–114. https://doi.org/10.1016/j.electacta.2018.08.103

[63]   X. Xu, H. Liang, F. Ming, Z. Qi, Y. Xie, Z. Wang, Prussian blue analogues derived penroseite (ni,co)se2 nanocages anchored on 3d graphene aerogel for efficient water splitting, ACS Catal. 7 (2017) 6394–6399. https://doi.org/10.1021/acscatal.7b02079

[64]   C. Ding, J. Shi, Z. Wang, C. Li, Photoelectrocatalytic water splitting: Significance of cocatalysts, electrolyte, and interfaces, ACS Catal. 7 (2017) 675–688. https://doi.org/10.1021/acscatal.6b03107

[65]   D. Guerrero-Araque, P. Acevedo-Peña, D. Ramírez-Ortega, H. Calderón, R. Gómez, Charge transfer processes involved in photocatalytic hydrogen production over CuO/ZrO2-TiO2 materials, Int. Jounal Hydrog. Energy. 42 (2017) 9744–9753. https://doi.org/10.1016/j.ijhydene.2017.03.050

[66]   D. Guerrero-araque, P. Acevedo-Peña, D. Ramírez-Ortega, L. Lartundo-Rojas, R. Gómez, SnO 2 -TiO 2 Structures and the effect of CuO, CoO metal oxide in the photocatalytic hydrogen production, J. Chem. Technol. Biotechnol. 92 (2017) 1531–1539. https://doi.org/10.1002/jctb.5273

[67]   J.P. Ziegler, Spectroscopic and electrochemical characterization of the photochromic behavior of prussian blue films on n-SrTiO3, J. Electrochem. Soc. 134 (1987) 358. https://doi.org/10.1149/1.2100460

[68]   N.R. De Tacconi, K. Rajeshwar, R.O. Lezna, Preparation, photoelectrochemical characterization, and photoelectrochromic behavior of metal hexacyanoferrate-titanium dioxide composite films, Electrochim. Acta. 45 (2000) 3403–3411. https://doi.org/10.1016/S0013-4686(00)00421-7

[69]   N.R. De Tacconi, K. Rajeshwar, R.O. Lezna, Photoelectrochemistry of indium hexacyanoferrate-titania composite films, J. Electroanal. Chem. 500 (2001) 270–278. https://doi.org/10.1016/S0022-0728(00)00315-6

[70]   K. Szaciłowski, W. Macyk, M. Hebda, G. Stochel, Redox-controlled photosensitization of nanocrystalline titanium dioxide, Chem. Phys. Chem. 7 (2006) 2384–2391. https://doi.org/10.1002/cphc.200600407

[71]   K. Szaciłowski, W. Macyk, G. Stochel, Synthesis, structure and photoelectrochemical properties of the TiO 2-Prussian blue nanocomposite, J. Mater. Chem. 16 (2006) 4603–4611. https://doi.org/10.1039/B606402G

[72]   K. Tennakone, A.R. Kumarasinghe, P.M. Sirimanne, Photocurrent enhancement in a cadmium sulphide anode coated with prussian blue, Thin Solid Films. 238 (1994) 101–103. https://doi.org/10.1016/0040-6090(94)90656-4

[73]   K. Siuzdak, M. Szkoda, J. Karczewski, J. Ryl, A. Lisowska-Oleksiak, Titania nanotubes infiltrated with the conducting polymer PEDOT modified by Prussian blue-a novel type of organic-inorganic heterojunction characterised with enhanced photoactivity, RSC Adv. 6 (2016) 76246–76250. https://doi.org/10.1039/C6RA15113B

[74]   F.S. Hegner, I. Herraiz-Cardona, D. Cardenas-Morcoso, N. López, J.R. Galán-Mascarós, S. Gimenez, Cobalt hexacyanoferrate on bivo4 photoanodes for robust water splitting, ACS Appl. Mater. Interfaces. 9 (2017) 37671–37681. https://doi.org/10.1021/acsami.7b09449

[75]   F.S. Hegner, D. Cardenas-Morcoso, S. Giménez, N. López, J.R. Galan-Mascaros, Level alignment as descriptor for semiconductor/catalyst systems in water splitting: The case of hematite/cobalt hexacyanoferrate photoanodes, Chem. Sus. Chem. 10 (2017) 4552–4560. https://doi.org/10.1002/cssc.201701538

[76]   K. Trzciński, M. Szkoda, K. Szulc, M. Sawczak, A. Lisowska-Oleksiak, The bismuth vanadate thin layers modified by cobalt hexacyanocobaltate as visible-light active photoanodes for photoelectrochemical water oxidation, Electrochim. Acta. 295 (2019) 410–417. https://doi.org/10.1016/j.electacta.2018.10.167

Electrochemical Water Splitting: Materials and Applications   Materials Research Forum LLC
Materials Research Foundations **59** (2019) 215-242   doi: https://doi.org/10.21741/9781644900451-9

Chapter 9

Ni-Based Electrocatalysts for Oxygen Evolution Reaction

K. Karthick[1] and Subrata Kundu[1,*]

[1] CSIR-Central Electrochemical Research Institute (CECRI), Karaikudi-630003, Tamil Nadu, India

kundu.subrata@gmail.com; skundu@cecri.res.in; karthickkchem@gmail.com

Abstract

In order to replace highly active noble metals, research on earth abundant based catalysts has been triggered in recent years. In this aspect, from iron group (Fe, Co, Ni) elements, mainly Ni based catalysts with the highly probable d-electronic configuration for hydroxide ion interaction and oxygen molecule cleavage assures the enhanced efficiency for alkaline water oxidation. Ni has been studied elaborately as oxides, hydroxides, sulphides, and selenides that resulted in unprecedented enhancements in OER and hence the overall applied cell voltage can be decreased in alkaline water electrolysis.

Keywords

Nickel, Oxygen Evolution Reaction, Overpotential, Oxide, Sulphide, Selenide, Tafel Slopes, Linear Sweep Voltametry

Contents

1. Introduction

The rapid increase in human population in recent decades increased the energy demand and hence the utilization of fossil fuels increases day by day to cover the growing energy requirements. Fossil fuels which are derived from the carbon derivatives that are formed from decayed animals and plants when used for the energy needs emit CO_2 to the atmosphere. As per the current situation, the atmospheric Co_2 level is around 400 ppm, which was around 300 ppm in the late 1950s [1]. This substantial growth is related to the industrial revolution, which corroborated the usage of fossil fuels. Because of the continuous supply of energy from these finite fossil fuels, they get depleted more than the expected rate [2]. Hence, in the next century, there will fewer or no more fossil fuel to extract and to deliver energy. These kinds of non-renewable energy sources cannot give sustainable energy supply and are also associated with environmental pollution [2]. Therefore, it is indeed necessary to move towards reliable and affordable 'renewable energy supplies to meet future energy demands. The pitfall with renewable energy systems like solar and wind is that they are intermittent and thus contiguous between demand and supply of energy becomes a question. Considering the current energy portfolio, the fossil fuels contribute to 87% of the energy demands followed by 6% of nuclear energy sources and the remaining from the renewable energy sources. 'Hydrogen' with the high volumetric energy density of 142 MJ/Kg compared to other energy sources like fossil fuels (below 50 MJ/kg) can act as a future energy carrier or fuel [3]. Considering hydrogen, it is the lightest element in the universe and not available as free hydrogen and mostly present as H_2O and also with carbons as hydrocarbons.

Amidst natural gas reforming, photo and electrochemical water splitting, 'Water electrolysis' offers pure hydrogen rapidly and utilizes electricity to dissociate water into constituents 'hydrogen' and 'oxygen'[4]. However, the process associated with the production of hydrogen is a bottleneck for its commercial scale usage. In water electrolysis, water oxidation (Oxygen evolution reaction (OER)) takes place at anodic counterpart, and water reduction (Hydrogen evolution reaction (HER)) takes place at cathodic counterpart [5]. So far, best electrocatalysts for water oxidation are IrO_2 and RuO_2 in acidic medium.

Similarly, in HER, the noble catalyst Pt with no alternates is being best at acidic conditions [6]. Since these are high cost and rare elements and available as trace amounts in the earth's crust, it is indeed necessary to switch over to reliable, cost-effective, less hazardous, earth abundant based catalysts to make water electrolysis process an economically viable one. Considering the mechanisms behind the HER and OER, the kinetic steps are more in OER which drags more anodic overpotentials. Moreover, the dissolution of IrO_2 and RuO_2 in the acidic medium as IrO_3 and RuO_4 that dissolve in solution during prolonged exposure results in a decrease in efficiency of the system [7,8].

The relative activity trends of iron group elements like Fe, Co, and Ni assured the commercialization for alkaline water electrolyzer with the increased OER activity with fewer overpotentials [9]. Among these three, Ni shows predominant activity in OER with high charge transfer kinetics. Hence, Ni based catalysts have been studied with different combinations as oxides, hydroxides, layered double hydroxides, selenides, and sulphides for the enriched OER activity [10-19]. This synergistic enhancement in activity of Ni compared to Fe and Co is related to its d-electronic configuration, which assured maximum intercalation of hydroxyl ions and cleavage of oxygen molecules. With the d^8 electronic configuration, the OH^- interaction is facile and also, in turn, facilitates the O_2 cleavage because of the 3d-2p repulsion (M-O) and as a result, O_2 molecule evolves swiftly [5]. Therefore, by tuning the changes in morphology, engineering the surface active sites and also doping with other elements at nano regime will result in an unprecedented increase in the rate of electron transfer at the electrode surface with fast oxygen evolution. This is advantageous in determining cell efficiency with low cost and vastly available catalysts to commercialize hydrogen production.

2.   The mechanism involved in oxygen evolution reaction and judging parameters

Even though the method of electrocatalytic water splitting was developed by the 19th century, the mechanistic steps involved in electrocatalytic oxygen evolution reaction are not still clear. The OER involves four protons and electron coupled reaction that drags more overpotentials. The main parameter behind the successful reactions is based on the Sabatier's principle, according to which the bonding between the oxygen and metal active site should not be too much stronger or weaker. The optimum levels of reaction will tend to give lesser overpotentials. In general, for water electrolysis, the equilibrium potential E_{eq} = 1.23 V. Moreover, because of the four individual protons combined charge transfer during OER, the kinetics is too sluggish to occur at lower overpotentials and ends up in decreasing the efficacy H_2 production [20]. Generalized reaction for water splitting by electricity is given by,

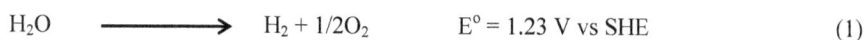

$$H_2O \longrightarrow H_2 + 1/2O_2 \qquad E° = 1.23 \text{ V vs SHE} \qquad (1)$$

Electrochemical Water Splitting: Materials and Applications   Materials Research Forum LLC
Materials Research Foundations **59** (2019) 215-242   doi: https://doi.org/10.21741/9781644900451-9

Because of the poor conductivity of pure H_2O, to enhance the efficiency of water splitting, it is performed generally in alkaline and acidic media. Plenty of H^+ ions from the acidic medium ensures the high HER rate, whereas the abundant amount of OH^- ions from the alkaline medium assures high OER rate respectively. However, the counter reactions like OER in acidic medium and HER in alkaline medium is a matter of debate which drags high overpotentials and results in an increase in overall cell voltage. The sequence of reactions occurs in acidic and alkaline media for HER and OER have been given below [5].

In acidic condition,

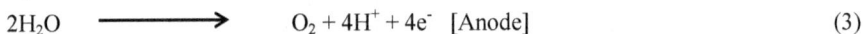

$$4H^+ + 4e^- \longrightarrow 2H_2 \qquad \text{[cathode]} \qquad\qquad (2)$$

$$2H_2O \longrightarrow O_2 + 4H^+ + 4e^- \quad \text{[Anode]} \qquad\qquad (3)$$

In alkaline condition,

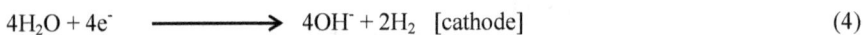

$$4H_2O + 4e^- \longrightarrow 4OH^- + 2H_2 \quad \text{[cathode]} \qquad\qquad (4)$$

$$4OH^- \longrightarrow O_2 + 2H_2O + 4e^- \quad \text{[Anode]} \qquad\qquad (5)$$

The corrosive dissolution of electrocatalysts in the acidic medium during long term usage restricts their commercial application. In alkaline conditions, even though many catalysts have been found to show encouraging activity for OER, still the overpotential is high because of the kinetic steps involved during anodization. The mechanistic steps involved in OER in alkaline condition are still under debate. However, there are many under discussions like oxide route, electrochemical metal peroxide route, electrochemical oxide route, and DFT peroxide route [5]. The mechanistic steps involved for OER in commonly used electrochemical metal peroxide path is given below.

For OER in alkaline condition,

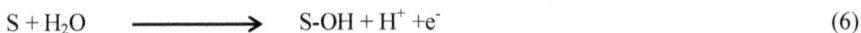

$$S + H_2O \longrightarrow S\text{-}OH + H^+ + e^- \qquad\qquad (6)$$

$$2S\text{-}OH \longrightarrow S\text{-}O + M + H_2O \qquad\qquad (7)$$

$$S\text{-}O + H_2O \longrightarrow S\text{-}OOH + H^+ + e^- \tag{8}$$

$$2S\text{-}OOH \longrightarrow S\text{-}O + O_2 + S + H_2O \tag{9}$$

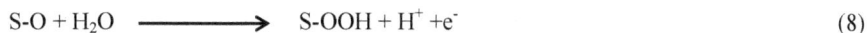

As given above from the equations (6-9) (S-active metal site), it is clear that the association of proton cum electron transfer kinetic steps tends to increase the activation energy for the cleavage of O_2 molecule from the electrode surface and hence simultaneously the overpotential is enhanced. Therefore, a search of metal based catalysts which are abundant, intrinsically active, safe, and easy to prepare is triggered by researchers across the world. From the 3d transition elements, nickel, cobalt, and iron with optimal free energy value for the oxygen to evolve has got prime attention [21]. In addition to the mechanistic steps involved in OER, the judging parameters mainly overpotential, Tafel slope, and Potentiostatic or Galvanostatic stability studies are to be analyzed with proper investigation [22]. Additional anodic potential from the equilibrium potential for driving the oxygen evolution is termed as 'overpotential'. As an activity descriptor, current densities of benchmarking levels like 10 mA cm^{-2} has been generally considered for comparison purposes of different catalysts for relating overpotentials of the catalysts and their corresponding efficiency. Tafel slope is extracted by the logarithmic current densities with respect to the overpotentials, and the corresponding linear fit gives the Tafel slope values. In general, the high approximation of the Butler-Volmer equation gives the Tafel equation at anodic and cathodic potentials. This one can be shown from Butler-Volmer equation as,

$$I \longrightarrow Io\,[\exp(\alpha_A nF/RT^*\eta) - \exp(\alpha_c nF/RT^*\eta)] \tag{10}$$

From the high overpotential regions, the Tafel equations are,

$$\ln I \longrightarrow \ln Io + (-\alpha_c nF/RT^*\eta)] \tag{11}$$

$$\ln I \longrightarrow \ln Io + (\alpha_A nF/RT^*\eta)] \tag{12}$$

From the equations (11 to 12), the observed slopes were extracted and give information about the electron transfer rate at the interface of electrode and electrolyte. Also, extrapolating linear slopes give exchange current density at equilibrium potentials for the catalysts taken [5, 21]. This exchange current density varies with respect to each catalyst

Electrochemical Water Splitting: Materials and Applications   Materials Research Forum LLC
Materials Research Foundations **59** (2019) 215-242   doi: https://doi.org/10.21741/9781644900451-9

based on the intrinsic activity of the catalyst studied. These two, overpotential and the Tafel slope value give an overview about the efficiency of the catalytic systems. Moreover, the sturdiness of the catalyst system studied is a foremost parameter to ensure the long-term usage for large scale utilization. In general, for a catalyst, stability was analyzed by chronoamperometry (PSTAT) or the chronopotentiometry (GSTAT) study. In PSTAT analysis, the potential is fixed at particular benchmarking current densities, and in general, 10 and 50 mA cm^{-2} is used whilst in GSTAT analysis, it is vice versa [22].

3.   Nickel based OER catalysts

In a corrosive condition of acidic medium, the OER is only facile in case of noble elements like IrO_2 and RuO_2, but still, the dissolution of these as IrO_3 and RuO_4 to the electrolytes respectively affects the efficiency [7,23]. For OER in alkaline conditions, there were considerable developments achieved with different metals based catalysts. A lot of reports emphasized the utilization of transition metals of 3d series based catalysts for electrocatalytic OER [21,24]. Among them, Ni, Co, Fe, Mn, and Cu have been frequently studied mostly as oxides, hydroxides, layered double hydroxides, sulphides, and selenides [24–28]. Considering the sluggish OER kinetics, the Ni based catalysts have been found to show tremendous enhancements in activities compared to others [29]. Ni has been studied with different stoichiometries with other earth abundant based metals to further develop the activities in alkaline medium. The dissolution of Ni derived electrocatalysts in acidic medium (OER) restricts its usage to alkaline conditions where it showed considerable activity and stability [30]. In this aspect, hydroxides of Ni has been developed as nickel hydroxide, Pt-doped Ni-Fe layered double hydroxides and hydroxy carbonate hydrates of nickel iron for electrochemical OER studies in alkaline conditions.

3.1   Ni-hydroxide based OER catalysts

The electrocatalytic OER results of Ni based catalysts, particularly hydroxides tend to deliver better OER activity. This is ascribed to the M^{2+}-OH bond strength. From the Ni based catalyst, Ni^{2+} with high electrons in 3d orbitals lower the strength of M-OH bond formation and also assists the cleavage of O_2 molecule [21]. Literature findings revealed that the better activity of $Ni(OH)_2$ is due to the phases of α and β and the activities were varied with respect to the morphology and also from the nanoparticles prepared. The recent findings enabled the researchers to understand that it is the β-NiOOH phase formed first and after this, it is converted to α-NiOOH during charging and discharging in a potential sweeping acted as a catalyst and delivered better activity [10, 29]. Based on this, β-$Ni(OH)_2$ has been developed by our research group by a solvothermal method and formed as ultra-thin nanosheets [14]. The surface faceting of (001) plane to (101) plane

Electrochemical Water Splitting: Materials and Applications   Materials Research Forum LLC
Materials Research Foundations **59** (2019) 215-242   doi: https://doi.org/10.21741/9781644900451-9

after the cycling study for 200 cycles delivered unprecedented results, and the overpotential was very low which was around 300 mV at current densities of 10 mA cm^{-2} and also the Tafel slope value was less (43 mV/dec). Similar kinds of reports have also been reported earlier. This work highlighted the surface faceting property during anodization over long potential sweeping of 200 cycles, which ensured the enhanced OER activity trend comparing commercial RuO_2 catalyst in alkaline condition.

From Fig. 1 (a-e), the nano-sheets of β-$Ni(OH)_2$ were found to have thin layers with the lattice planes of (002) before cycling. HAADF image also clearly showed the thin-sheets of β-$Ni(OH)_2$ with polycrystalline nature prepared by a solvothermal method. As mentioned above, this plane orientation is changed during cycling study and (101) phase has formed that directed the enhanced OER activity with the very low overpotential compared to the commercial IrO_2 catalyst.

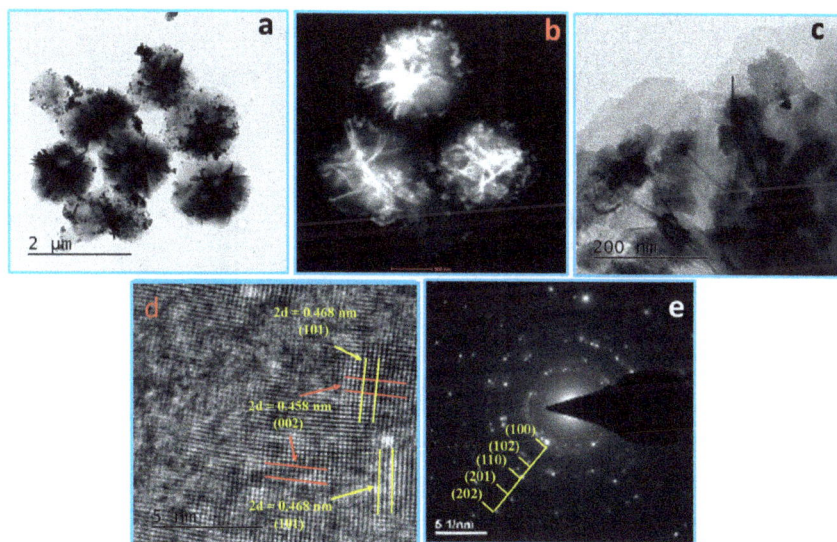

Fig. 1: a) Micrographs of β-$Ni(OH)_2$ from HR-TEM b) is the corresponding HAADF image, c) HR-TEM with nano-sheets and nano-hurls, d) lattice fringes with (101) planes and e) SAED pattern. Revised from Ref. 14.

From Fig. 2 (a-b), the activity of a catalyst before cycling was very low which after accelerated degradation study had shown tremendous enhancement with lesser overpotential at current densities of 10 mA cm^{-2} which was even better than the commercial RuO_2 electrocatalyst. This is because of the β-NiOOH phase formed with the plane (101) after the potential sweeping of a catalyst. Also, the intrinsic charge transfer kinetics was studied with Tafel slope values and the β-Ni(OH)$_2$ before cycling showed 98 mV/dec and the same after cycling had shown 43 mV/dec with improved kinetics (Fig. 2c). The stability of these kinds of Ni based hydroxides have been analyzed with GSTAT analysis and showed much stable nature (Fig. 2d).

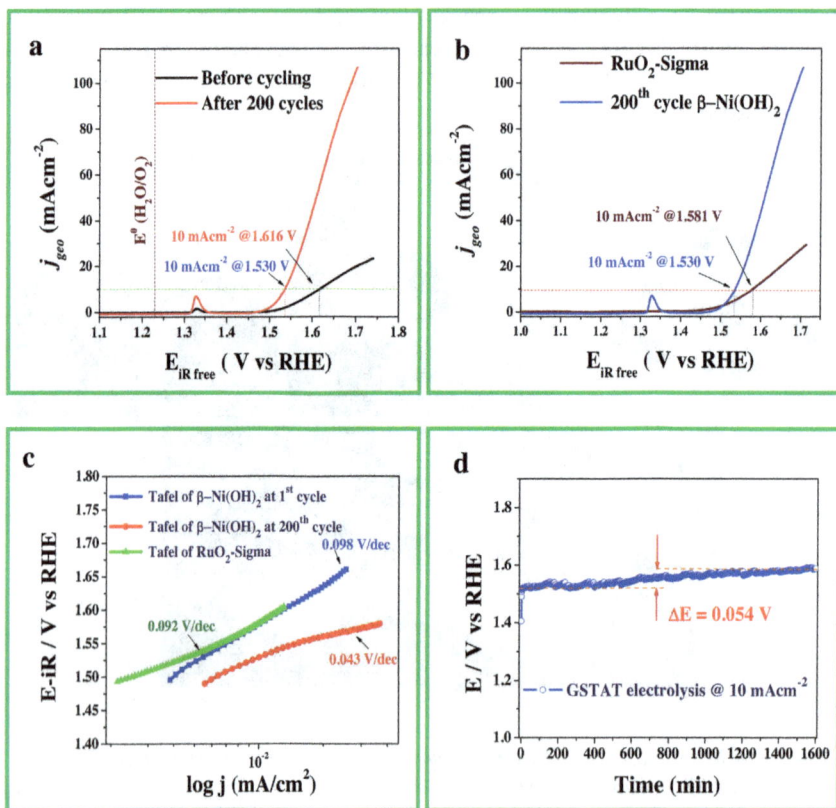

Fig. 2: a) LSV before cycling b) LSV after cycling c) Tafel slope and d) GSTAT analysis. Revised from Ref. 14.

Electrochemical Water Splitting: Materials and Applications
Materials Research Forum LLC
Materials Research Foundations **59** (2019) 215-242
doi: https://doi.org/10.21741/9781644900451-9

The activity of monometallic hydroxides can be enhanced by successful incorporation of other earth abundant based catalysts. Layered double hydroxides (LDH) with layered structure showed much attention as the ease of interaction of hydroxyl ions was too facile in LDH systems [24]. There were a lot of reports that highlighted the LDH systems with different combinations, and among them, Ni-Fe LDH was highly studied for OER with enriched activity. We recently developed Pt-incorporated NiFe-LDH for total water splitting. The successful incorporation of Pt into the LDH structure assured good HER activity also [31]. This resulted in enhanced activity with low cell voltage for alkaline water electrolysis. The method of preparation of NiFe-LDH was carried out by two methods, namely hydrothermal and co-precipitation method.

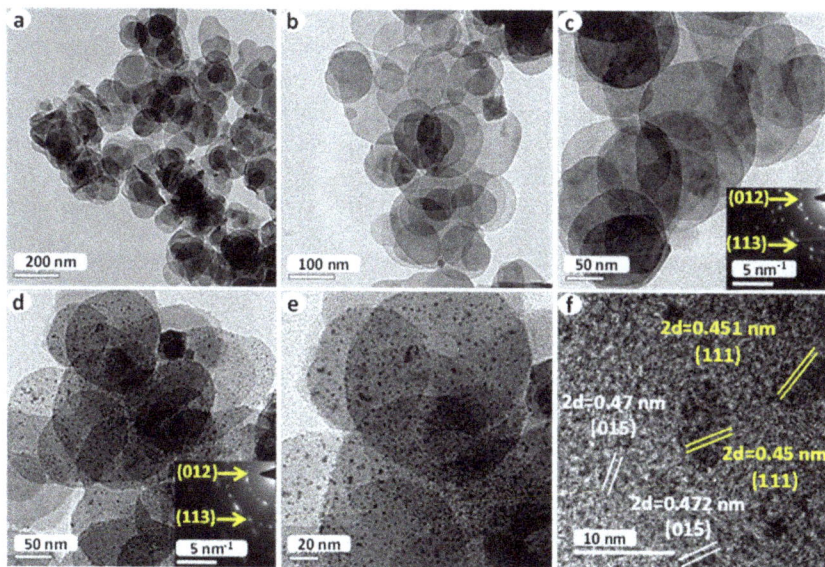

Fig. 3: (a-c) HR-TEM image of hydrothermally prepared Ni-Fe LDH, (d-e) is Pt incorporated Ni-Fe LDH and (f) SAED pattern. Revised from Ref. 31.

Electrochemical Water Splitting: Materials and Applications   Materials Research Forum LLC
Materials Research Foundations **59** (2019) 215-242   doi: https://doi.org/10.21741/9781644900451-9

The morphology as seen from Fig. 3 (a-f), showed sheet like structure with circular morphology for NiFe-LDH whereas Pt NPs after incorporation still showed circular sheets of NiFe-LDH systems with Pt NPs occupied here and there.

From the resulted activities from electrochemical studies, in case of LDH systems, it is enhanced much compared to monometallic hydroxides like $Ni(OH)_2$ as seen before. This is attributed to the LDH structure that assisted ease of electrolyte interaction, enriched metal lattices from bivalent and trivalent ions and the layered structure of sheets.

The observed overpotentials were 230 and 280 mV for NiFe-LDH at respective current densities of 10 and 50 mA cm^{-2} (Fig. 4a). The elecrochemical active surface area was also higher (Fig. 4c). The corresponding Tafel slope value was 33 mV/dec (Fig. 4d) that showed near four electron transfer with improvised kinetics at the interface.

Fig. 4b showed the LSV curves, and from the corresponding current density versus overpotential plot, the improvised activity was observed with the NiFe-LDH (hydrothermal method) compared to NiFe-LDH (co-precipitation method). This huge deviation in activity is ascribed to different morphological features of the catalyst. Moreover, the incorporation of Pt into the LDH system also ensured HER activity and thus gave a better result for total water splitting [32]. The water oxidation in neutral conditions is difficult, and it is rare to find materials to catalyze OER in neutral conditions [25].

Moreover, the conductivity of NiFe LDH systems is low and to make it active for OER in neutral conditions; we developed NiFeHCH (Nickel Iron Hydroxy Carbonate Hydrate) as a catalyst that gave unprecedented activity at different pH values like 14 and 8.5 [33]. Incorporating carbonates and water molecules in between the lattices of Ni and Fe ensured OER activity in neutral conditions also. Moreover, the ratio dependent study was carried out with respect to Fe concentration. The studies showed that NiFe (1:0.2) has shown tremendous activity in alkaline conditions, whereas in near neutral conditions, NiFe (1:0.4) showed better activity.

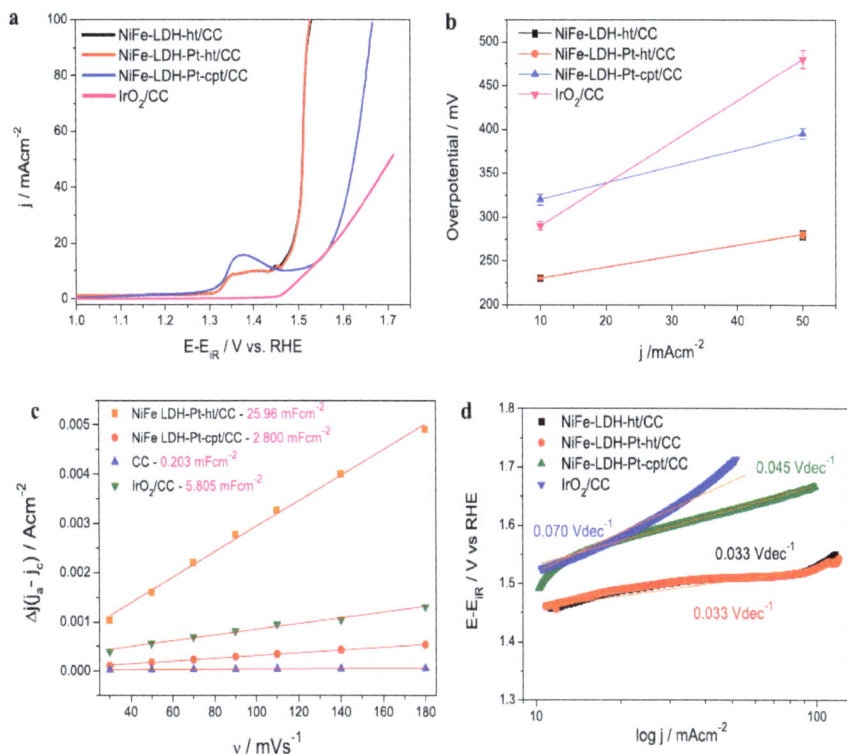

Fig. 4: a) LSV b) Overpotential versus current density c) ECSA and d) Tafel slope. Revised from Ref. 31.

The results of LSV and the linear Tafel slopes from Fig. 5 (a-b) showed that there is variance in activities that is relative to the added percentage of Fe in NiFeHCH. There was no clear cut information on why the concentration of Fe affects the activity. A study by Trotochaud *et al.* [15] gave information that at low concentrations around 5 % and 25% of Fe, the OER activity was getting enhanced that is related to the increased conductivity of the catalyst studied.

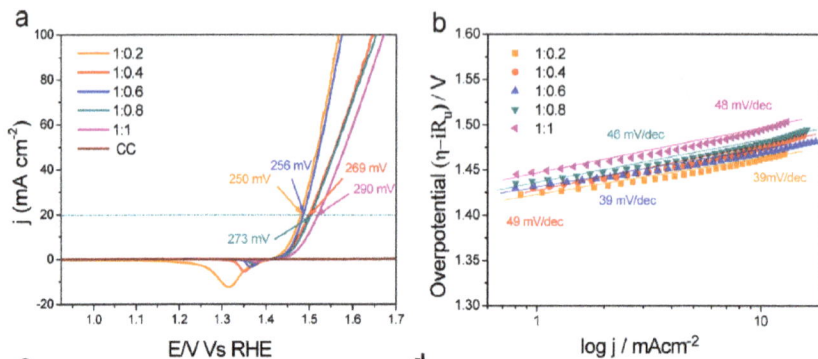

Fig. 5: (a) iR free LSV curves and (b) is the linear Tafel slopes. Revised from Ref. 33

The studied OER efficiency of catalysts in near neutral conditions showed tremendous enhancement because of the presence of carbonate anions and the hydrate molecules. The LSV showed overpotential of 389 mV for Ni:Fe (1:0.4) with lesser Tafel slope, which was very less compared to other stochiometries of Fe as seen in Fig. 6 (a-b). This gives way for other transition metals based catalysts along with Nickel for the OER activity in near neutral conditions with exceptional activities. Moreover, doping of metals as NPs along with these carbonate hydrates of nickel can also result in enhancement of cell efficiency.

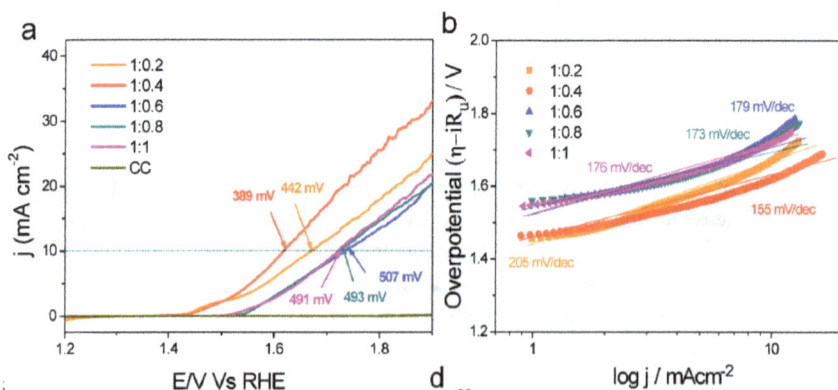

Fig. 6: (a) iR free LSV curves and (b) is the linear Tafel slope from the extracted LSV region. Revised from Ref. 33.

3.2 Ni-oxide based OER catalysts

The oxides of Ni have been studied elaborately with different stoichiometries and also in combination with different metal ions [25,31,34–36]. In general, high oxidation states of Ni have shown enriched activity [12]. Therefore, to prepare Ni oxides with high valence states, a lot of works has been carried out to make it highly active for OER. So far, Ni has been studied in combination with Co and Fe with different stoichiometries with better enhancements [37–40]. To make Ni based catalysts even cheaper and cost effective, we have chosen stainless steel as a substrate in which Ni in combination with other metal ions like Fe, Cr and Mn could give better OER performance in alkaline medium. First, stainless steel-304, NiO incorporated Fe_2O_3 was studied for electrocatalytic OER studies [41].

From the Fig. 7(a-d), FE-SEM and microstructural studies revealed the surface exposed growth of NiO with Fe_2O_3. In addition, color mapping studies revealed the successful presence of both Ni and Fe along with Cr and Mn as seen in Fig.7(e-j). In this particular study, Cr was leached from the SS-304 substrate to further enhance the activity trend. After the leaching of Cr by hypochlorite and KOH, the activity was enhanced abnormally.

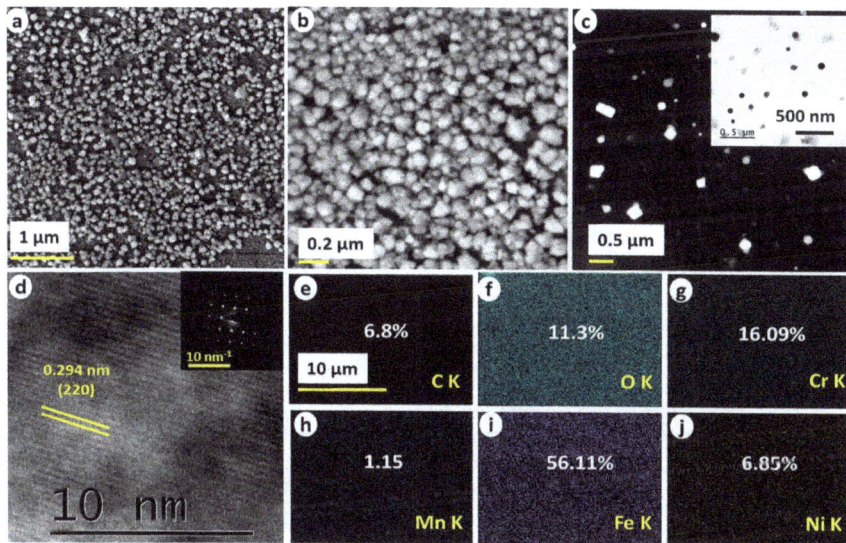

Fig. 7: (a and b) FE-SEM images, (c) is the corresponding HAADF image, (d) HR-TEM at higher magnification, (e-j) FE-SEM color mapping results of NiO incorporated Fe_2O_3. Revised from Ref. 41.

The observed activity trends were superior and the corresponding overpotentials were 260, 302 and 340 mV for the respective current densities of 10, 100 and 500 mA cm^{-2} as seen Fig. 8 (a-b). With this study, it is clear that the presence of other metals and the leaching of surface inactive metals resulted in tremendous enhancements in the activity observed. Moreover, a time-dependent study on leaching of Cr was analyzed and among which SS-304 over a reaction time of 12 h resulted in the enriched activity. These kind of related studies of Ni based oxides in combination with other elements are a fruitful advantage in increasing charge transfer kinetics and low R_{ct} (Fig. 8c-d) at low overpotentials for commercialization [37–40].

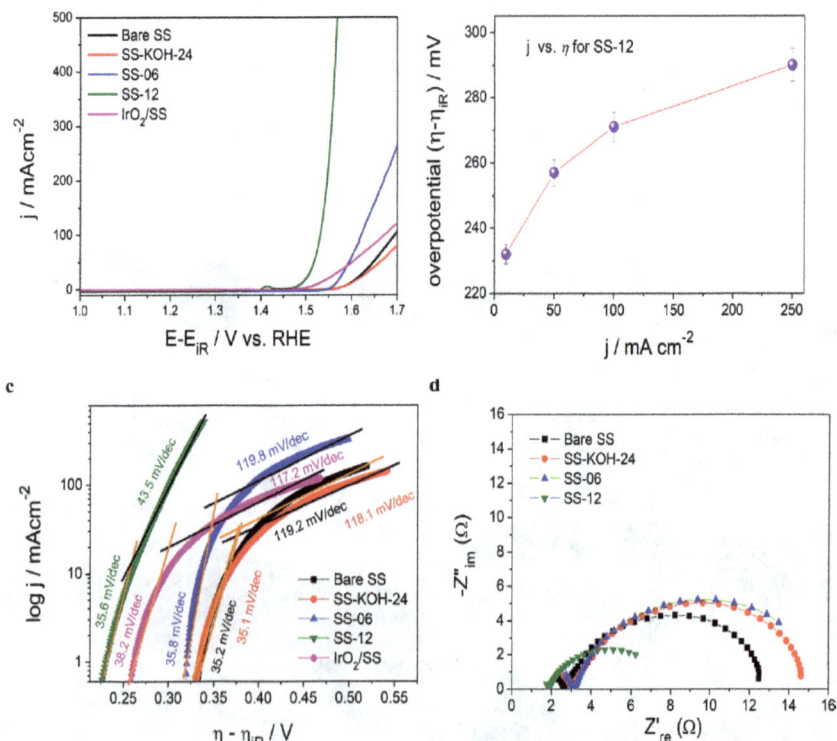

Fig. 8: a) LSV b) Overpotential versus current density c) Tafel slope and d) EIS anaysis. Revised from Ref.41.

Electrochemical Water Splitting: Materials and Applications   Materials Research Forum LLC
Materials Research Foundations **59** (2019) 215-242   doi: https://doi.org/10.21741/9781644900451-9

The activity was very high, particularly at high overpotentials, which is related to the high intrinsic activity of NiO when present along with Fe_2O_3. As observed from Tafel and electrochemical impedance spectrum, the charge transfer kinetics was too facile at both lower and higher overpotentials whereas the charge transfer resistance was very less ensuring the higher OER activity of SS-12 which is in accordance with the polarization studies. This study gives a clue that the less active phase NiO can become highly active with the assistance of other metal based oxides. In addition to this, low cost stain less scrubber has also been utilized as a catalyst in oxygen evolution studies at alkaline condition [42]. This report on stainless steel has resembled the type of SS-434 L and as expected, delivered better activity with very low cost.

From EDS, the stainless steel had elements of Fe, Mn, and Cr along with Ni as oxides to improve the activity as observed before (Fig. 9).

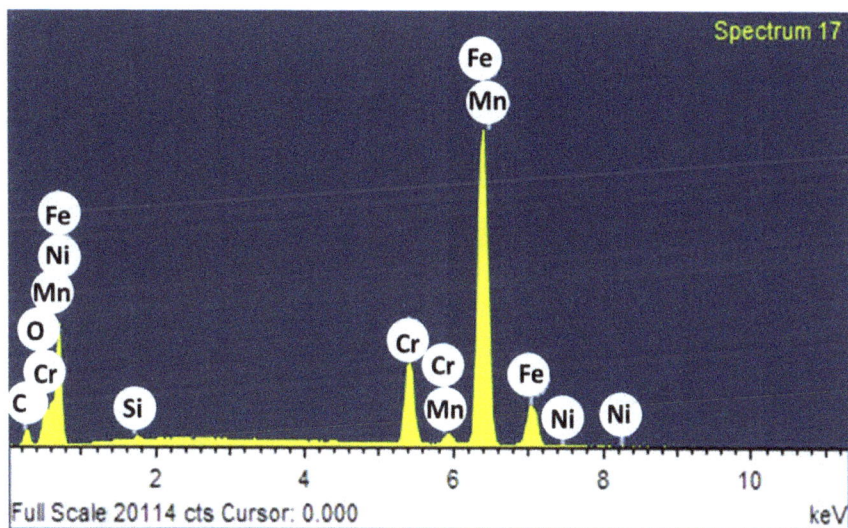

Fig. 9: EDS analysis of stainless steel scrubber. Revised from Ref. 42.

Here also, the activity was better with a very cheap scrubber as electrode material. From Fig. 10, the iR uncompensated LSV curves showed the OER activity with very less loading of SS scrubber. The overpotentials at current densities from 10 to 50 mA cm^{-2} were comparatively better comparing the cost efficiency of the catalyst system (Fig. 10a-c). The iR compensated LSV showed activity comparable to the Ni(OH)$_2$ and RuO$_2$

catalysts. The extracted linear Tafel slopes from LSV suggested that electron transfer was better in SS-scrubber and nearly as same as for Ni(OH)$_2$ catalyst (Figure 10d). This Ni based oxides with other earth abundant based catalysts is interesting in the development of cost-effective Ni based oxides as electrocatalysts for OER.

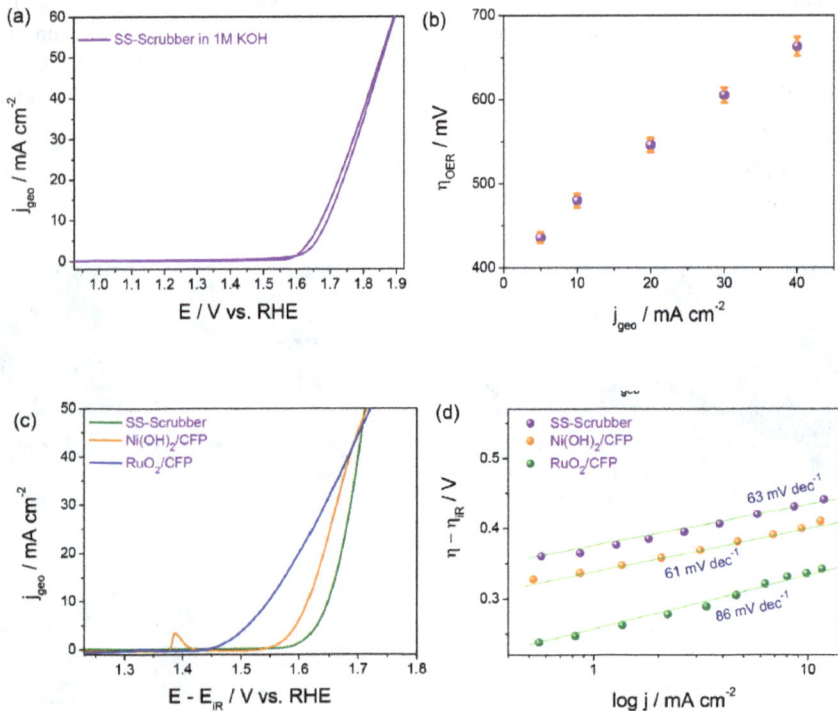

Fig. 10: a) LSV b) Overpotential versus current density c) Tafel slope and d) EIS anaysis. Revised from Ref. 42.

These Ni oxide based catalysts can give enhanced activity, which is in accordance with the Sabatier principle. Moreover, Ni oxides with different metal oxides as composites will have influence in the activity [37–40]. Like observed in previous work, the NiO was with the combination of Fe$_2$O$_3$, and in this work, in stainless steel scrubber, the active catalyst observed after cathodization was α-Ni(OH)$_2$ which converted into MOOH formation during anodization [41,42]. This is advantageous for large scale studies as the active catalyst varies with respect to the combination of metals present along with Ni [37–40]. Similarly, in another work, we utilized microwave heating to form $Ni(BO_3)_2$, which was applied for OER in alkaline medium (Fig. 11) [43].

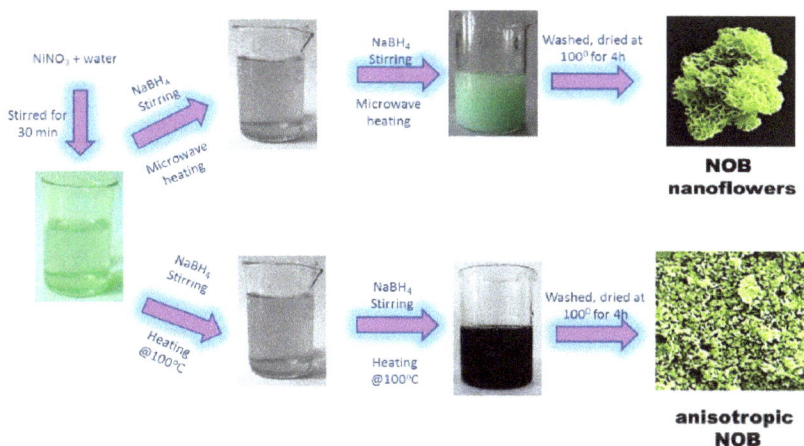

Fig. 11: Schematic representation of formation of NOB nanoflowers. Revised from Ref. 43.

The resultant overpotential was 317 mV to reach 10 mA cm^{-2} current density and the extracted linear Tafel slope value was found to be 74 mV/dec (Fig. 12a-b).

Depending upon the environment, Ni as an active catalyst gives different activity trend. This can be fine-tuned with other dopants to improvise the activities of a surface of the Ni based oxide catalysts [37–40]. Even considering other than oxide derived catalysts when undergoes OER conditions, the oxide phase only becomes an active catalyst [21].

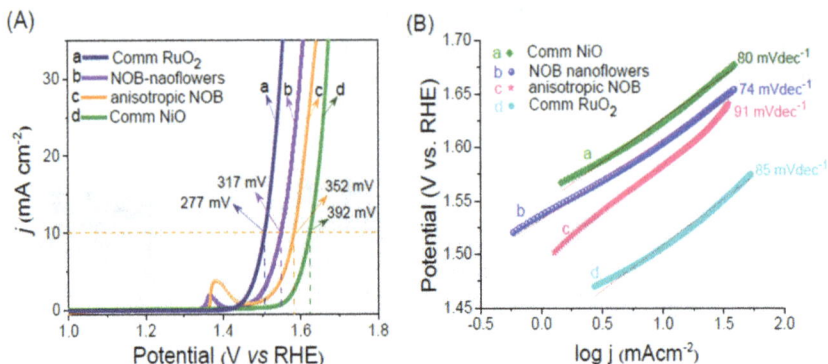

Fig. 12: (a) LSV curves and (b) Extracted linear Tafel slope. Revised from Ref. 43.

3.3   Ni-sulphides and selenides for OER

The sulphides of Ni have been studied for OER with different stoichiometries such as NiS, Ni_3S_2, Ni_xS_y, and also along with some other oxides to form as composites to give enhancements in OER [16,44,45]. These catalysts like sulphide and selenide, when undergone OER, at high anodic overpotentials get converted into oxides, and it is the oxide phase formed during anodization only that will act as a real catalyst in driving the OER [18]. Based on this, we have developed NiS-DNA assembly just by a wet-chemical route and formed a stable solution of NiS-DNA (Fig. 13) [27]. This NiS-DNA acted as a catalyst for driving the OER with impressive activity. The presence of DNA ensured that there is no need for the addition of the external binder. This also gave enhancement in the activity trend because of the presence of phosphate groups in DNA that could synergistically enhance the oxygen evolution at low overpotentials. Moreover, the utilization of DNA in this Ni based sulphide catalyst has also assured very less loading (minimal loading < 20 times) compared to normalized loading, which is in general 0.0123 mg cm^{-2}.

Materials Research Forum LLC
doi: https://doi.org/10.21741/9781644900451-9

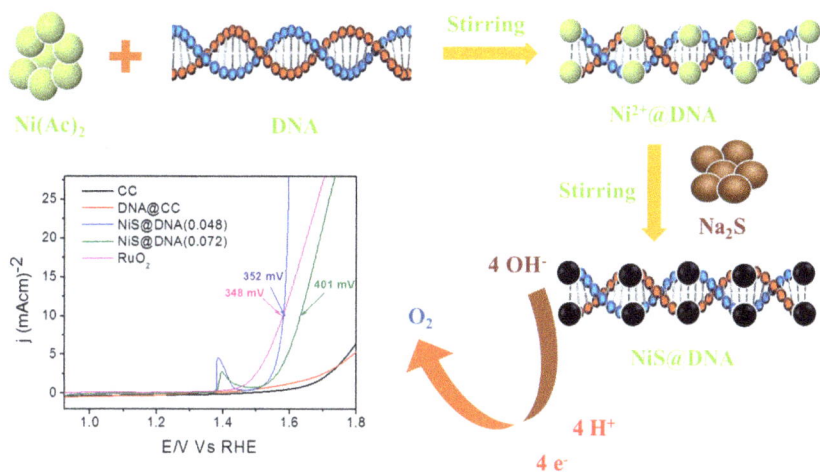

Fig. 13: Formation of NiS@DNA by a wet-chemical route for OER. Revised from Ref. 27.

The schematic representation, as showed in Fig. 13, at first the Ni^{2+} ions assembled over the DNA backbones where aromatic base pairs like adenine, guanine, cytosine, and thymine are present. In addition to this, there are phosphate groups and hydroxy groups from sugar moieties assists the ease of incorporation of metal ions like Ni over the self-assembly of DNA [46,47]. To this, when Na_2S is added, resulting in the subsequent formation of NiS over DNA and forms a stable solution. This NiS@DNA is directly for oxygen evolution studies at the alkaline condition. The prepared NiS NPs over DNA is an advantage as the loading is very less and gave impressive results during OER.

The LSV curves (Fig. 14a-b) of different concentrations of NiS@DNA highlighted the lesser overpotentials as 352 and 401 mV for attaining the current density of 10 mA cm^{-2} for NiS@DNA (0.048) and (0.072) respectively. Moreover, the mass activity was very high as the loading of the catalyst at stabilized DNA solutions is too small compared to the normalized loading (Fig. 14c). The charge transfer kinetics of NiS@DNA showed 58.6 mV/dec (Fig. 14d).

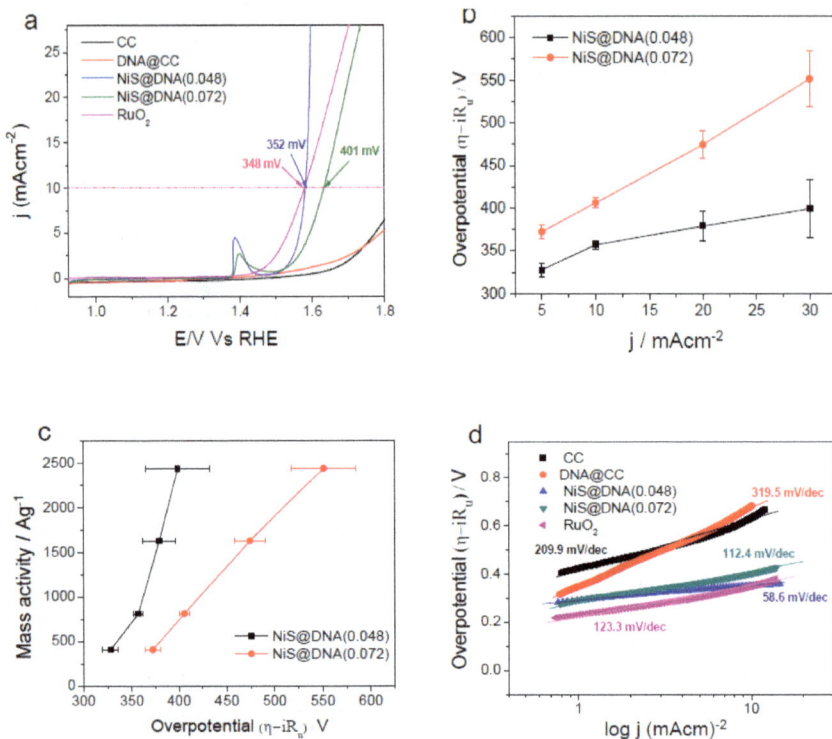

Fig.14: a) LSV b) η vs j (mA cm^{-2}) c) Observed Mass activity relationship and d) Extracted Tafel slope from LSV. Revised from Ref. 27.

These findings imply that NiS with the assistance of DNA can offer enhanced OER with very less loading and also with no such additional usage of external binders like Nafion which poisons the catalytically active sites at the electrode/electrolyte interface [32]. Therefore, these bio-molecule assisted Ni can be studied as selenides and tellurides also. Moreover, with the ease of preparation, using a greener scaffold like DNA, these kinds of Ni based catalysts can emerge as an important choice for OER in alkaline medium. Similarly, Nickel based selenides have also gained importance in recent years for OER [18,48].

Nickel selenides have been found to show exceptional OER activity in alkaline medium and also studied with different stoichiometries such as NiSe, NiSe$_2$, and Ni$_3$Se$_4$

[18,28,49]. Nickel selenide, in combination with other dopants and oxides, also resulted in increased OER activity. Moreover, this has also been added with HER active catalysts to drive overall water splitting in alkaline medium. Following these, we used a simple microwave irradiation to prepare Ni_3Se_4@Ni foam by simple addition of NaHSe into the solution (Fig.15) [28]. The as formed Ni_3Se_4@Ni foam was utilized for total water splitting in alkaline and in neutral conditions.

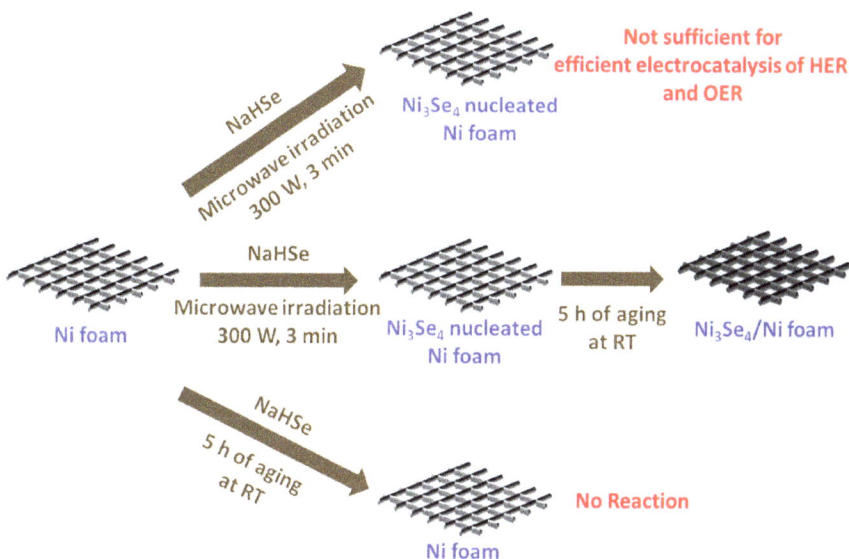

Fig. 15. Schematic representation of microwave assisted Ni_3Se_4 formation. Revised from Ref. 28.

The potentiality of Ni_3Se_4@Ni foam was testified in OER at three different pH conditions to evaluate the efficiency. Comparing bare Ni foam, the selenized one showed exceptional activity in all four pH values of 7, 13, 13.5, and 14.5. For reaching 10 mA cm^{-2} current density, the observed overpotentials for Ni_3Se_4@Ni foam were 480, 321, 244, and 232mV at pH values of 7, 13, 13.5 and 14.5 (Fig. 16a-b).

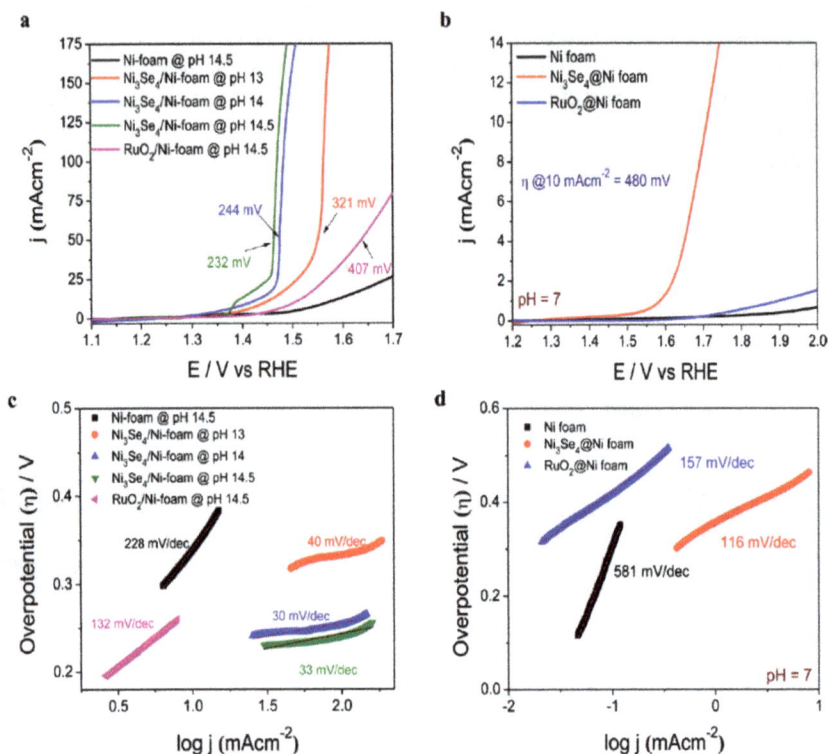

Fig. 16: a) LSV at KOH b) LSV at phosphate buffer c)Tafel slopes from LSV in alkaline medium d) Tafel slopes from LSV in neutral condition. Revised from Ref. 28.

The Ni_3Se_4 on Ni foam could deliver overwhelming activity in a neutral condition also. Moreover, the observed Tafel slope values were too low and showed superior electron transferring rate with the potential applied (Fig. 16c). For Ni_3Se_4@Ni foam, the corresponded value 116 mV/dec as Tafel slope at neutral state (Fig. 16d) had shown resultant better activity along with the more stable nature. The activities and the discussions of Ni_3Se_4@Ni foam made clear that it can also be explored by engineering it with surface modification and also by the incorporation of other active metal catalysts [28].

Conclusion

From this particular chapter, the readers can become well familiar with Ni based catalysts and also with the optimum bond energy between the metal active site and the hydroxyl ions for incorporation and optimum bond energy between formed oxygen molecules for cleavage, for enhanced OER. In addition, this chapter focused on Ni based systems which have particularly been studied mainly as oxides, hydroxides, sulphides, and selenides, resulted in enriched OER activity, particularly in alkaline medium. Moreover, some Ni based catalysts could also be able to provide reasonable OER activity in neutral conditions. However, the poor stable nature of Ni derived catalysts in acidic medium retards their application in acidic OER. Therefore, for total water splitting, we have to move on alkaline water electrolysis. It can also be understood that to have sustainable production of hydrogen through electrocatalytic water splitting in alkaline medium; the sluggish OER part is to be focused to attain increased activity and for that Ni is one of the best catalysts that drive OER compared to others. The counter reaction can be nullified with respect to the added HER active dopants to the Ni based catalysts. If the alkaline water electrolysis is commercialized with the assistance of Ni based catalysts with a very low applied voltage, then this will be the top notch one for deriving energy particularly from renewable energy sources. This chapter may pop new individual ideas in researchers to develop Ni based catalysts with different combinations for excellent activity in OER.

Acknowledgements

This chapter has been written by K. Karthick and Subrata Kundu. Authors thank the constant support from CSIR-CECRI, Karaikudi, India. K. Karthick wishes to acknowledge UGC, New Delhi, India for UGC-SRF award. We also wish to acknowledge the faculties from CIF-CECRI for the support.

References

[1]   T. Dietz, E.A. Rosa, Effects of population and affluence on CO_2 emissions, Proc. Natl. Acad. Sci. 94 (2002) 175–179. https://doi.org/10.1073/pnas.94.1.175

[2]   S. Shafiee, E. Topal, When will fossil fuel reserves be diminished? Energy Policy 37 (2009) 181–189. https://doi.org/10.1016/j.enpol.2008.08.016

[3]   N.L. Garland, D.C. Papageorgopoulos, J.M. Stanford, Hydrogen and fuel cell technology: Progress, challenges, and future directions, Energy Procedia 28 (2012) 2–11. https://doi.org/10.1016/j.egypro.2012.08.034

[4]   H. Wendt, G. Imarisio, Nine years of research and development on advanced water electrolysis. A review of the research programme of the Commission of the European Communities, J. Appl. Electrochem. 18 (1988) 1–14. https://doi.org/10.1007/BF01016198

[5]   H. Dau, C. Limberg, T. Reier, M. Risch, S. Roggan, P. Strasser, The mechanism of water oxidation: from electrolysis via homogeneous to biological catalysis, Chem. Cat. Chem. 2 (2010) 724–761. https://doi.org/10.1002/cctc.201000126

[6]   J.B. Raoof, R. Ojani, A. Kiani, S. Rashid-Nadimi, Fabrication of highly porous Pt coated nanostructured Cu-foam modified copper electrode and its enhanced catalytic ability for hydrogen evolution reaction, Int. J. Hydrogen Energy 35 (2010) 452–458. https://doi.org/10.1016/j.ijhydene.2009.10.069

[7]   M. Yagi, E. Tomita, S. Sakita, T. Kuwabara, K. Nagai, Self-assembly of active IrO_2 colloid catalyst on an ITO electrode for efficient electrochemical water oxidation, J. Phys. Chem. B Lett. 109 (2005) 21489. https://doi.org/10.1021/jp0550208

[8]   K.A. Stoerzinger, O. Diaz-Morales, M. Kolb, R.R. Rao, R. Frydendal, L. Qiao, X.R. Wang, N.B. Halck, J. Rossmeisl, H.A. Hansen, T. Vegge, I.E.L. Stephens, M.T.M. Koper, Y. Shao-Horn, Orientation-dependent oxygen evolution on RuO_2 without lattice exchange, ACS Energy Lett. 2 (2017) 876–881. https://doi.org/10.1021/acsenergylett.7b00135

[9]   C.C.L. McCrory, S. Jung, I.M. Ferrer, S.M. Chatman, J.C. Peters, T.F. Jaramillo, Benchmarking hydrogen evolving reaction and oxygen evolving reaction electrocatalysts for solar water splitting devices, J. Am. Chem. Soc. 137 (2015) 4347–4357. https://doi.org/10.1021/ja510442p

[10]   L.A. Stern, X. Hu, Enhanced oxygen evolution activity by NiO_x and $Ni(OH)_2$ nanoparticles, Faraday Discuss. 176 (2014) 363–379. https://doi.org/10.1039/C4FD00120F

[11]   X. Wang, H. Luo, H. Yang, P.J. Sebastian, S.A. Gamboa, Oxygen catalytic evolution reaction on nickel hydroxide electrode modified by electroless cobalt coating, Int. J. Hydrogen Energy 29 (2004) 967–972. https://doi.org/10.1016/j.ijhydene.2003.05.001

[12]   J. Landon, E. Demeter, N. Inoğlu, C. Keturakis, I.E. Wachs, R. Vasić, A.I. Frenkel, J.R. Kitchin, Spectroscopic characterization of mixed Fe–Ni oxide electrocatalysts for the oxygen evolution reaction in alkaline electrolytes, ACS Catal. 2 (2012) 1793–1801. https://doi.org/10.1021/cs3002644

[13]   K. Fominykh, P. Chernev, I. Zaharieva, J. Sicklinger, G. Stefanic, M. Doblinger, A. Muller, A. Pokharel, C. Bocklein, C. Scheu, T. Bein, D. Fattakhova-Rohlfing, Iron-doped nickel oxide nanocrystals as highly efficient electrocatalysts for alkaline water splitting, ACS Nano 9 (2015) 5180–5188. https://doi.org/10.1021/acsnano.5b00520

[14]   S. Anantharaj, P.E. Karthik, K. Subrata, Petal-like hierarchical array of ultrathin $Ni(OH)_2$ nanosheets decorated with $Ni(OH)_2$ nanoburls: A highly efficient OER electrocatalyst, Catal. Sci. Technol. 7 (2017) 882–893. https://doi.org/10.1039/C6CY02282K

[15]   L. Trotochaud, S.L. Young, J.K. Ranney, S.W. Boettcher, Nickel–iron oxyhydroxide oxygen-evolution electrocatalysts: The role of intentional and incidental iron incorporation, J. Am. Chem. Soc. 136 (2014) 6744–6753. https://doi.org/10.1021/ja502379c

[16]   P. Luo, H. Zhang, L. Liu, Y. Zhang, J. Deng, C. Xu, N. Hu, Y. Wang, Targeted synthesis of unique nickel sulfide (NiS, NiS_2) microarchitectures and the applications for the enhanced water splitting system, ACS Appl. Mater. Interfaces 9 (2017) 2500–2508. https://doi.org/10.1021/acsami.6b13984

[17]   J.S. Chen, J. Ren, M. Shalom, T. Fellinger, M. Antonietti, Stainless steel mesh-supported nis nanosheet array as highly efficient catalyst for oxygen evolution reaction, ACS Appl. Mater. Interfaces 8 (2016) 5509–5516. https://doi.org/10.1021/acsami.5b10099

[18]   A.T. Swesi, J. Masud, W.P.R. Liyanage, S. Umapathi, E. Bohannan, J. Medvedeva, M. Nath, Textured NiSe2 film: bifunctional electrocatalyst for full water splitting at remarkably low overpotential with high energy efficiency, Sci. Rep. 7 (2017) 1–11. https://doi.org/10.1038/s41598-017-02285-z

[19]   C. Tang, N. Cheng, Z. Pu, W. Xing, X. Sun, NiSe nanowire film supported on nickel foam: an efficient and stable 3D bifunctional electrode for full water splitting, Angew. Chemie - Int. Ed. 54 (2015) 9351–9355. https://doi.org/10.1002/anie.201503407

[20]   T. Reier, M. Oezaslan, P. Strasser, Electrocatalytic oxygen evolution reaction (OER) on Ru, Ir, and Pt catalysts: a comparative study of nanoparticles and bulk materials, ACS Catal. 2 (2012) 1765–1772. https://doi.org/10.1021/cs3003098

[21]   S. Anantharaj, S.R. Ede, K. Sakthikumar, K. Karthick, S. Mishra, S. Kundu, Recent trends and perspectives in electrochemical water splitting with an emphasis on sulfide, selenide, and phosphide catalysts of Fe, Co, and Ni: A review, ACS Catal. 6 (2016) 8069–8097. https://doi.org/10.1021/acscatal.6b02479

[22] S. Anantharaj, S.R. Ede, K. Karthick, S. Sam Sankar, K. Sangeetha, P.E. Karthik, S. Kundu, Precision and correctness in the evaluation of electrocatalytic water splitting: revisiting activity parameters with a critical assessment, Energy Environ. Sci. 11 (2018) 744–771. https://doi.org/10.1039/C7EE03457A

[23] S. Anantharaj, S. Kundu, Enhanced water oxidation with improved stability by aggregated RuO_2-$NaPO_3$ Core-shell nanostructures in acidic medium, Curr. Nanosci. 13 (2017) 333–341. https://doi.org/10.2174/1573413713666170126155504

[24] S. Anantharaj, K. Karthick, S. Kundu, Evolution of layered double hydroxides (LDH) as high performance water oxidation electrocatalysts: A review with insights on structure, activity and mechanism, Mater. Today Energy 6 (2017) 1–26. https://doi.org/10.1016/j.mtener.2017.07.016

[25] J. Liu, D. Zhu, T. Ling, A. Vasileff, S.Z. Qiao, S-$NiFe_2O_4$ ultra-small nanoparticle built nanosheets for efficient water splitting in alkaline and neutral pH, Nano Energy 40 (2017) 264–273. https://doi.org/10.1016/j.nanoen.2017.08.031

[26] L. Zhou, X. Huang, H. Chen, P. Jin, G. Li, X. Zou, A high surface area flower-like Ni–Fe layered double hydroxide for electrocatalytic water oxidation reaction, Dalt. Trans. 44 (2015) 11592–11600. https://doi.org/10.1039/C5DT01474C

[27] K. Karthick, S. Anantharaj, S. Kundu, ACS Sustain, Chem. Eng. 6 (2018) 6802-6810. https://doi.org/10.1021/acssuschemeng.8b00633

[28] S. Anantharaj, J. Kennedy, S. Kundu, Microwave-initiated facile formation of Ni_3Se_4 nanoassemblies for enhanced and stable water splitting in neutral and alkaline media, ACS Appl. Mater. Interfaces 9 (2017) 8714–8728. https://doi.org/10.1021/acsami.6b15980

[29] W. Sun, Y. Chen, K. Rui, J. Zhu, S.X. Dou, Recent progressonnickel-based oxide/(Oxy)hydroxide electrocatalysts for the oxygen evolution reaction, Chem. - A Eur. J. 25 (2018) 703–713. https://doi.org/10.1002/chem.201802068

[30] R. Schlogl, P. Strasser, T. Reier, H.N. Nong, D. Teschner, Electrocatalytic oxygen evolution reaction in acidic environments - reaction mechanisms and catalysts, Adv. Energy Mater. 7 (2016) 160127. https://doi.org/10.1002/aenm.201601275

[31] S. Anantharaj, K. Karthick, M. Venkatesh, T.V.S.V. Simha, A.S. Salunke, L. Ma, H. Liang, S. Kundu, Enhancing electrocatalytic total water splitting at few layer Pt-NiFe layered double hydroxide interfaces, Nano Energy 39 (2017) 30–43. https://doi.org/10.1016/j.nanoen.2017.06.027

[32]   S. Anantharaj, P.E. Karthik, B. Subramanian, S. Kundu, Pt nanoparticle anchored molecular self-assemblies of DNA: An extremely stable and efficient HER electrocatalyst with ultralow Pt content, ACS Catal. 6 (2016) 4660–4672. https://doi.org/10.1021/acscatal.6b00965

[33]   K. Karthick, S. Anantharaj, S.R. Ede, S. Kundu, Nanosheets of nickel iron hydroxy carbonate hydrate with pronounced oer activity under alkaline and near-neutral conditions, Inorg. Chem. 58 (2019) 1895–1904. https://doi.org/10.1021/acs.inorgchem.8b02680

[34]   U.Y. Qazi, C.Z. Yuan, N. Ullah, Y.F. Jiang, M. Imran, A. Zeb, S.J. Zhao, R. Javaid, A.W. Xu, One-step growth of iron–nickel bimetallic nanoparticles on feni alloy foils: highly efficient advanced electrodes for the oxygen evolution reaction, ACS Appl. Mater. Interfaces 9 (2017) 28627–28634. https://doi.org/10.1021/acsami.7b08922

[35]   J. Chi, H. Yu, G. Li, L. Fu, J. Jia, X. Gao, B. Yi, Z. Shao,Nickel/cobalt oxide as a highly efficient OER electrocatalyst in an alkaline polymer electrolyte water electrolyzer, RSC Adv. 6 (2016) 90397–90400. https://doi.org/10.1039/C6RA19615B

[36]   M. Gong, H. Dai, A mini review of NiFe-based materials as highly active oxygen evolution reaction electrocatalysts, Nano Res. 8 (2014) 23–39. https://doi.org/10.1007/s12274-014-0591-z

[37]   A. Wang, Z. Zhao, D. Hu, J. Niu, M. Zhang, K. Yan, G. Lu, Tuning the oxygen evolution reaction on a nickel–iron alloy via active straining, Nanoscale 11 (2019) 426–430. https://doi.org/10.1039/C8NR08879A

[38]   K. Li, T. Tian, Y. Ding, H. Gao, J. Wu, L. Zheng, X. Zhou, Study of the active sites in porous nickel oxide nanosheets by manganese modulation for enhanced oxygen evolution catalysis, ACS Energy Lett. 3 (2018) 2150–2158. https://doi.org/10.1021/acsenergylett.8b01206

[39]   C. Zhu, D. Wen, S. Leubner, M. Oschatz, W. Liu, M. Holzschuh, F. Simon, S. Kaskel, A. Eychmüller, Nickel cobalt oxide hollow nanosponges as advanced electrocatalysts for the oxygen evolution reaction, Chem. Commun. 51 (2015) 7851–7854. https://doi.org/10.1039/C5CC01558H

[40]   H.Z. Cao, J. Xia, L.K. Wu, G.Y. Hou, Y.P. Tang, G.Q. Zheng, W.Y. Wu, A nanostructured nickel–cobalt alloy with an oxide layer for an efficient oxygen evolution reaction, J. Mater. Chem. A 5 (2017) 10669–10677. https://doi.org/10.1039/C7TA02754K

[41]   S. Anantharaj, M. Venkatesh, A.S. Salunke, T.V.S.V. Simha, V. Prabu, S. Kundu, High-performance oxygen evolution anode from stainless steel via controlled surface oxidation and Cr removal, ACS Sustain. Chem. Eng. 5 (2017) 10072–10083. https://doi.org/10.1021/acssuschemeng.7b02090

[42]   S. Anantharaj, S. Chatterjee, K.C. Swaathini, T.S. Amarnath, E. Subhashini, D.K. Pattanayak, S. Kundu, Stainless steel scrubber: A cost efficient catalytic electrode for full water splitting in alkaline medium. ACS Sustain. Chem. Eng. 6 (2018) 2498–2509. https://doi.org/10.1021/acssuschemeng.7b03964

[43]   S.R. Ede, S. Anantharaj, B. Subramanian, A. Rathishkumar, S. Kundu, Microwave-assisted template-free synthesis of $Ni_3(BO_3)_2$(NOB) hierarchical nanoflowers for electrocatalytic oxygen evolution, Energy and Fuels 32 (2018) 6224–6233. https://doi.org/10.1021/acs.energyfuels.8b00804

[44]   X. Shang, X. Li, W.H. Hu, B. Dong, Y.R. Liu, G.Q. Han, Y.M. Chai, Y.Q. Liu, C.G. Liu, In situ growth of NixSy controlled by surface treatment of nickel foam as efficient electrocatalyst for oxygen evolution reaction, Appl. Surf. Sci. 378 (2016) 15–21. https://doi.org/10.1016/j.apsusc.2016.03.197

[45]   J. Jian, L. Yuan, H. Qi, X. Sun, L. Zhang, H. Li, H. Yuan, S. Feng, Sn–Ni_3S_2 Ultrathin nanosheets as efficient bifunctional water-splitting catalysts with a large current density and low overpotential, ACS Appl. Mater. Interfaces 10 (2018) 40568-40576. https://doi.org/10.1021/acsami.8b14603

[46]   L.Y.T. Chou, K. Zagorovsky, W.C.W. Chan, DNA assembly of nanoparticle superstructures for controlled biological delivery and elimination, Nat. Nanotechnol. 9 (2014) 148–155. https://doi.org/10.1038/nnano.2013.309

[47]   A. Kuzuya, Y. Ohya, DNA nanostructures as scaffolds for metal nanoparticles, Polym. J. 44 (2012) 452–460. https://doi.org/10.1038/pj.2012.38

[48]   A.T. Swesi, J. Masud, M. Nath,Nickel selenide as a high-efficiency catalyst for oxygen evolution reaction, Energy Environ. Sci. 9 (2016) 1771–1782. https://doi.org/10.1039/C5EE02463C

Keyword Index

About the Editors

Dr. Inamuddin is currently working as Assistant Professor in the Chemistry Department, Faculty of Science, King Abdulaziz University, Jeddah, Saudi Arabia. He is a permanent faculty member (Assistant Professor) at the Department of Applied Chemistry, Aligarh Muslim University, Aligarh, India. He obtained Master of Science degree in Organic Chemistry from Chaudhary Charan Singh (CCS) University, Meerut, India, in 2002. He received his Master of Philosophy and Doctor of Philosophy degrees in Applied Chemistry from Aligarh Muslim University (AMU), India, in 2004 and 2007, respectively. He has extensive research experience in multidisciplinary fields of Analytical Chemistry, Materials Chemistry, and Electrochemistry and, more specifically, Renewable Energy and Environment. He has worked on different research projects as project fellow and senior research fellow funded by University Grants Commission (UGC), Government of India, and Council of Scientific and Industrial Research (CSIR), Government of India. He has received Fast Track Young Scientist Award from the Department of Science and Technology, India, to work in the area of bending actuators and artificial muscles. He has completed four major research projects sanctioned by University Grant Commission, Department of Science and Technology, Council of Scientific and Industrial Research, and Council of Science and Technology, India. He has published 138 research articles in international journals of repute and eighteen book chapters in knowledge-based book editions published by renowned international publishers. He has published forty-two edited books with Springer, United Kingdom, Elsevier, Nova Science Publishers, Inc. U.S.A., CRC Press Taylor & Francis Asia Pacific, Trans Tech Publications Ltd., Switzerland and Materials Research Forum LLC, U.S.A. He is the member of various editorial boards of the journals and serving as associate editor for journals such as Environmental Chemistry Letter, Applied Water Science, Euro-Mediterranean Journal for Environmental Integration, Springer-Nature, Frontiers Section Editor of Current Analytical Chemistry, published by Bentham Science Publishers, editorial board member for Scientific Reports-Nature and editor for Eurasian Journal of Analytical Chemistry. He has attended as well as chaired sessions in various international and national conferences. He has worked as a Postdoctoral Fellow, leading a research team at the Creative Research Initiative Center for Bio-Artificial Muscle, Hanyang University, South Korea, in the field of renewable energy, especially biofuel cells. He has also worked as a Postdoctoral Fellow at the Center of Research Excellence in Renewable Energy, King Fahd University of Petroleum and Minerals, Saudi Arabia, in the field of polymer electrolyte membrane fuel cells and computational fluid dynamics of polymer electrolyte membrane fuel cells. He is a life member of the Journal of the Indian

Chemical Society. His research interest includes ion exchange materials, a sensor for heavy metal ions, biofuel cells, supercapacitors and bending actuators.

Dr. Rajender Boddula is currently working as Chinese Academy of Sciences-President's International Fellowship Initiative (CAS-PIFI) at National Center for Nanoscience and Technology (NCNST, Beijing). His academic honors include University Grants Commission National Fellowship and many merit scholarships, study-abroad fellowships from Australian Endeavour Research fellowship and CAS-PIFI. He has published many scientific articles in international peer-reviewed journals and has authored six book chapters, and also serving as editorial board member and referee for reputed international peer-reviewed journals. He has published edited books with Materials Research Forum LLC, U.S.A. His specialized areas of expertise is energy conversion and storage, which include nanomaterials, graphene, polymer composites, heterogeneous catalysis, photoelectrocatalytic water splitting, biofuel cell, and supercapacitors.

Dr. Rizwana Mobin is working as Assistant Professor in the Department of Industrial Chemistry, Govt. College for Women, Cluster University, Srinagar, India. She received her B.Sc. Hons., Masters and Ph.D (Applied Chemistry) from Aligarh Muslim University, Aligarh, India on the topic "Studies on Thin-Layer Chromatographic Analysis of Surfactants". She has been the recipient of the Gold medal at Masters level. She has published several research articles in international journals of repute and six book chapters in knowledge-based book editions published by renowned international publishers. She has edited books with Materials Research Forum LLC, U.S.A. Her research expertise includes thin-layer chromatography, development of new methodologies involving green solvent system for the analysis of surfactants and food dyes.

Prof. Abdullah M. Asiri is the Head of the Chemistry Department at King Abdulaziz University since October 2009 and he is the founder and the Director of the Center of Excellence for Advanced Materials Research (CEAMR) since 2010 till date. He is the Professor of Organic Photochemistry. He graduated from King Abdulaziz University (KAU) with B.Sc. in Chemistry in 1990 and a Ph.D. from University of Wales, College of Cardiff, U.K. in 1995. His research interest covers color chemistry, synthesis of novel photochromic and thermochromic systems, synthesis of novel coloring matters and dyeing of textiles, materials chemistry, nanochemistry and nanotechnology, polymers and plastics. Prof. Asiri is the principal supervisors of more than 20 M.Sc. and six Ph.D.

theses. He is the main author of ten books of different chemistry disciplines. Prof. Asiri is the Editor-in-Chief of King Abdulaziz University Journal of Science. A major achievement of Prof. Asiri is the discovery of tribochromic compounds, a new class of compounds which change from slightly or colorless to deep colored when subjected to small pressure or when grind. This discovery was introduced to the scientific community as a new terminology published by IUPAC in 2000. This discovery was awarded a patent from European Patent office and from UK patent. Prof. Asiri involved in many committees at the KAU level and on the national level. He took a major role in the advanced materials committee working for KACST to identify the national plan for science and technology in 2007. Prof. Asiri played a major role in advancing the chemistry education and research in KAU. He has been awarded the best researchers from KAU for the past five years. He also awarded the Young Scientist Award from the Saudi Chemical Society in 2009 and also the first prize for the distinction in science from the Saudi Chemical Society in 2012. He also received a recognition certificate from the American Chemical Society (Gulf region Chapter) for the advancement of chemical science in the Kingdome. He received a Scopus certificate for the most publishing scientist in Saudi Arabia in chemistry in 2008. He is also a member of the editorial board of various journals of international repute. He is the Vice- President of Saudi Chemical Society (Western Province Branch). He holds four USA patents, more than one thousand publications in international journals, several book chapters and edited books.

www.ingramcontent.com/pod-product-compliance
Lightning Source LLC
Chambersburg PA
CBHW071157210326
41597CB00016B/1575